END TIMES

END TIMES

A BRIEF GUIDE TO
THE END OF THE WORLD

ASTEROIDS, SUPERVOLCANOES,
ROGUE ROBOTS, AND MORE

BRYAN WALSH

SEVEN DIALS

First published in the United States in 2019 by Hachette Books
An imprint of Perseus Books, a division of Hachette Book Group, Inc.
1290 Avenue of the Americas, New York, NY 10104

This edition published in Great Britain in 2019 by Seven Dials
an imprint of The Orion Publishing Group Ltd
Carmelite House, 50 Victoria Embankment
London EC4Y 0DZ

An Hachette UK Company

1 3 5 7 9 10 8 6 4 2

A CIP catalogue record for this book is
available from the British Library.

ISBN (Trade Paperback) 9781841884035
ISBN (eBook) 9781841884059

MIX
Paper from
responsible sources
FSC   FSC® C104740
www.fsc.org

www.orionbooks.co.uk

To Siobhan and to Ronan

CONTENTS

INTRODUCTION

The average human can expect to live more than 2 billion seconds,[1] but there are only a few moments when everything can change at once. It might be the second after you receive the worst news of your life, or the moment when the person you had always waited for says yes. For me that moment is captured in a photograph. I'm in the hospital room on the day my son is born, standing to the left of my wife, Siobhan. A smile is surfacing through the fatigue bunched around my eyes. My father is standing on the right, beside my mother, as she looks into the lens with an expression of pure joy. She is holding our first child. His name is Ronan. He's just a few hours old, his fine, thin skull dusted with reddish-blond hair, his fingers curled tightly in fists, his eyes shut against the light. Ronan is here, one of the newest inhabitants on planet Earth, and for us nothing will be the same again.

What I see when I look at that photograph today is the future coming into being. My father, who loomed throughout my child-hood, is not just my father any longer, but a grandfather. My mother, the first person I remember being conscious of, is not just my mother any longer, but a grandmother. And I, a son for thirty-nine years, am no longer just a son, but a father, as my wife is now a mother. We're part of a chain that turns toward the future, one human link at a time. And those links are as fragile as a newborn baby.

Until that moment I'd never really thought about the future, which is ironic, because for a decade and a half as a professional journalist the future was my subject. The first years were spent as a foreign correspondent for *Time* magazine in East Asia, where I witnessed the greatest victory over poverty the world has ever experienced, an economic and political earthquake that will reverberate for decades. I reported from ground zero on SARS, the first emerging global disease of the twenty-first century, a virus that came out of nowhere and exposed just how vulnerable our interconnected world was to the peril of sickness. I worked for a year in Japan as *Time*'s Tokyo bureau chief, reporting from a country that lives on the very edge of the future.

After six years in Asia I moved to *Time*'s headquarters in New York to cover climate change, a force that will do more than any other to reset the boundaries of our future. I attended historic conferences like the 2009 United Nations climate change summit in Copenhagen, and ventured to the vanishing ice sheets of the Arctic. I trekked to the dwindling rain forests of South America and the drought-stricken mountains of northern India. Everywhere I went, I witnessed the diminishing of humanity's future, melting away like the glaciers I once watched calving off Greenland.

When people found out that I covered climate change—and if they believed that climate change was real—they would usually ask me if I found the beat depressing. Weren't we all doomed? I'd tell them something about how climate change was vitally important because it represented the intersection of business and politics and science, all while allowing me to earn plenty of exotic stamps in my passport. Which was true enough. Climate change was important, and I did feel lucky to cover it. I could read the studies, and I could write articles—so many articles—warning that our species was headed for doom if we didn't make radical changes in the way we lived. But I never really felt it. I didn't feel the future—its weight, its uncertainty, its importance, and, like my newborn son, its fragility.

But I would.

———————

In a 2012 poll by Reuters covering more than twenty countries, 15 percent of respondents predicted that the world would end in their lifetimes.[2] A 2015 survey of Americans, British, Canadians, and Australians found that a majority rated the risk of our way of life ending within the next one hundred years at 50 percent or greater, while a quarter believed humanity had a better than even chance of being wiped out altogether over that time frame.[3] More Americans believe that life was better fifty years ago—when a nuclear holocaust was an everyday possibility—than it is today.[4] In 2018, a UN scientific panel reported that the world had just twelve years to sharply reduce carbon emissions or risk a global catastrophe.[5] Meanwhile, the tone of the news in the era of President Donald Trump has become nothing short of apocalyptic on both sides of the political divide. And when we're not reading about the real-life end of the world, we're watching a fictionalized version: *The Walking Dead*, *The Hunger Games*, *Avengers: Endgame* and half the new shows on Netflix. The bloodier and more dystopic, it seems, the more we love it—as long as we're watching, and not participating. If we fear the end times, part of us seems to crave them—and perhaps believes we deserve them.

What's ironic is that this existential panic unfolds against the backdrop of a world that—for most of humanity—is better than it has ever been. In 2018, for the first time in history, more than half of the world's population qualified as "middle class" or "rich."[6] Infant mortality has fallen by more than half over the past twenty-five years.[7] Even as weapons have grown far more lethal, the global death rate from conflict is less than it was six hundred years ago.[8] And if numbers like that seem too dry, ask yourself this question: Would I have preferred to be born fifty years ago, not long after a global war killed more than 60 million people? One hundred years ago, before the age of antibiotics, when a simple infection could end your life? One thousand years ago, when human life expectancy was about thirty years?[9] I doubt it.

If we don't appreciate the present, it's in part because we don't fully understand the past—even as we make the mistake of assuming the future will be like the present. Psychologists have a name for this trait: the *availability heuristic*, the human tendency to be overly influenced by what feels most visible and salient in our experience. The availability heuristic can cause us to overreact, as when we hear about reports of a suicide bombing and become fixated on the danger from terrorists, ignoring the longer-term data that shows such incidents are on the decline.[10] Risks that are most available to the mind are the ones that we care about, which is why so much of our regulation is driven by crisis, rather than by reason.[11] As a longtime journalist, I plead guilty here—the standard definition of the news is the recent and the memorable, so the media plays a role in our overemphasis of now at the expense of the historical perspective. No newspaper has ever led its front page with the story that 100,000 people rose out of extreme poverty yesterday[12]—yet for years, that is exactly what has been happening almost daily. The myopia of the availability heuristic leaves us fixated on everything that seems to be going wrong today, and blind to how far we've come.

But that same psychological bias can also lead us to underreact to far greater dangers and threats that we've never experienced. The internet may remember everything but human memory is short and spotty. Few of us have experienced in our own lives catastrophes truly worthy of the name, and no human has seen an asteroid on a collision course with our planet, or witnessed a disease rise and threaten our very existence. These threats have no availability to us, so we treat them as unreal—even if science and statistics tell us otherwise. Our failure to understand that the future could be radically different than the past is above all else a failure of human psychology. And that failure could prove fatal for our species.

In life, as in the stock market, past performance is no guarantee of future results. It's not just the rising tide of climate change, or the creeping instability at home and abroad, or the deadly natural disasters that seem to be piling up with each passing year. It's not just the nauseating sensation that our world is spinning out of control, one presidential tweet at a time. Our very future is in danger, as it has never been before, both from an array of cosmic and earthbound threats and from the very technologies that have helped make us so prosperous.

We think we know how bad it can get, but the worst catastrophes that have ever befallen the human race—two world wars; the Black Death, which killed as many as 200 million people in the fourteenth century; the biggest hurricanes and most devastating earthquakes—are mere speed bumps compared to the risks this book will cover, the risks we now face. These risks are darker than the darkest days humanity has ever known. They're called existential risks, risks capable of putting an end to the existence of humankind, for all time. They are the mistakes we can't recover from, the disasters that could end the human story in midsentence.

Our species has always lived under the shadow of existential risk—we just didn't know it. At least five times over the course of our planet's 4.5-billion-year history, life has been virtually wiped out in great extinction waves, often punctuated by a natural catastrophe that struck on a planetary scale. Asteroid impacts, supervolcanic eruptions, even gamma rays from space—the universe is not a safe space.

The death of the dinosaurs some 66 million years ago, thanks largely to the impact of a six-mile-wide asteroid, was a mass extinction event. Ninety-nine point nine percent of the species that have ever lived on Earth have gone extinct. Some evolved into new species, but most, including every other *Homo* species we've ever shared the planet with, simply died out. And the same fate could befall us.

But if the universe has always wanted to kill us, at least a little bit, what's new is the possibility that we might destroy ourselves, whether by error or intention. What are called man-made or anthropogenic existential risks were born with the successful test of the first nuclear weapon at Trinity Site in New Mexico on July 16, 1945. The bomb gave us the power to do to ourselves what natural selection had done to most other species before us.

Nuclear war, though, is just the first man-made existential risk, one that has grown no less lethal even as it has receded from our attention. With every passing year, billions upon billions of tons of man-made greenhouse gas emissions are added to the atmosphere, increasing man-made climate change. Given enough time—along with some bad luck—global warming could begin to threaten our existence. Even more frightening—and far harder to predict or control—are the existential risks arising from new technologies like synthetic biology or artificial intelligence, technologies that could create threats we can hardly imagine, bombs that could explode before we even know they're armed.

How much danger are we in? The Canadian philosopher John Leslie, who helped invent the field of existential risk studies with his 1996 book, *The End of the World*, gave a 30 percent chance that humans would go extinct over the next five centuries.[13] In his final published remarks, the late Stephen Hawking put our species on an extinction clock, writing: "One way or another, I regard it as almost inevitable that either a nuclear confrontation or environmental catastrophe will cripple the Earth at some point in the next 1,000 years."[14] At a 2008 symposium put on by Oxford University's Future of Humanity Institute (FHI)—one of a new array of academic groups formed to study existential risk—a group of experts collectively put the overall chances of human extinction before the year 2100 at 19 percent.[15] That may leave us with a better than four-in-five chance of making it to the twenty-second century, but as the existential risk expert Phil Torres points out, even a 19 percent chance of human extinction over the next century means that the average American would be *1,500 times* more likely to die in an end times catastrophe than they would in a plane crash.[16]

In a 2003 book, Martin Rees, Britain's Astronomer Royal and the cofounder of the Centre for the Study of Existential Risk (CSER) at the University of Cambridge, put the odds of humanity successfully making it through the century on par with a coin flip—fifty-fifty. "Just a few people or even an individual can, by error or by design, cause a catastrophe that cascades very widely, even globally," Rees told me when we spoke in 2018. "I like to say the global village will have its village idiots." And now those village idiots are armed and dangerous.

Rees is right to focus primarily on the existential risk that emerging technologies will super-empower individuals and small groups that may harbor apocalyptic intentions. We're as vulnerable to planetary disasters like asteroids and supervolcanoes that have wiped out life on Earth before as we ever were. But the very fact that *Homo sapiens* has survived and thrived for hundreds of thousands of years means that we can reasonably hope our run of luck will continue through the next century, and even longer. By one estimate the probability of human extinction from a natural catastrophe over the next century is almost certainly lower than 0.15 percent—tiny, though not zero.[17] And we have something the dinosaurs and other long extinct species lacked—scientists and engineers who can defend us from the dangers above and below, provided we give them the resources and the authority they need.

But the same brains that could protect us from natural existential risks have introduced entirely new ones into the world, technological risks far greater than anything this planet could throw at us. We're only beginning to understand how these technologies might be used, and how they might be abused. What sets them apart from existing man-made threats like nuclear weapons is that they come not just with risks, but with benefits. Synthetic biology offers us the potential to create immortal organs, powerful drugs, and crops that could keep a growing and warming planet fed. Artificial intelligence may be the most important invention in human history—and possibly the last one we'll ever need. These technologies are "dual use"—the same science can be used for good, including to counter other existential risks, and for ill. We may not be able to tell which is which until it's too late. There are no easy answers when it comes to the end of the world.

In 2017, I left *Time* to begin working on the book you're reading now, a book that would raise the alarm about the existential threats our world faces, and ask how we might counter them. But even as I began researching the subject and speaking to experts in the field, something about it remained unreal to me, distant and abstract. This is an occupational hazard of existential risk studies. The human mind reels at the numbers—hundreds of millions of deaths, billions of deaths, total extinction. There is a term for this, too: *scope neglect*, our psychological inability to scale up from the small numbers of a human-level story to the vast figures of mass death. The words may not have been said by Joseph Stalin, but that doesn't make them any less true: "One death is a tragedy; a million deaths are a statistic." I was treating the end of the world as a statistic, just as I had done for so long as a reporter with global warming.

That changed as I began to understand the most salient fact about existential risk: It's not about us. It's about our sons, daughters, nephews, and nieces, and all those unnamed billions, even trillions, who might come after them—but won't, if our human story ends now.

The Oxford moral philosopher Derek Parfit proposed a thought experiment. Imagine three possible futures. In the first, there is peace. In the second, there is a nuclear war that kills 99 percent of the world's population, leaving a sliver of survivors to carry on. In the third, a nuclear war kills 100 percent of the world's population—every man, woman, and child, resulting in the total eradication of the human race. It doesn't take an Oxford PhD to conclude that the first future is the best of the three, or that the third one—the extinction scenario—is the worst. And most people would instinctively conclude that the difference between peace and a nuclear holocaust that killed all but 1 percent of the world is much greater than the difference between the death of 99 percent of the global population and the death of 100 percent. Certainly that's what the raw numbers would say. But Parfit disagreed, writing the following in his 1984 book *Reasons and Persons*: "Civilization began only a few thousand years ago. If we do not destroy mankind, these few thousand years may be only a tiny fraction of the whole of civilized human history. The difference between [possibilities] 2 and 3 may thus be the difference between this tiny fraction and all of the rest of history."[18]

Extinction, as the environmental slogan goes, is forever.[19] The true horror of the end of the world is measured not by our own deaths, and the deaths of everyone we know and love, not just by the deaths of our children and grandchildren, but by the nullification of all who would come after them, all those who would live and love and carry this species forward. An existential risk realized is the death of the future.

Basic morality calls us to do what we can to save a single life. If the world were in immediate existential peril—if, for example, a very large asteroid were bearing down on Earth—we would do whatever we could, spend whatever we had, to try to save the billions of people who live here. By that same token, shouldn't we be even more motivated, even more desperate, to protect the future generations who would take their turn on Earth—provided we don't destroy it all now, or let it be destroyed by inaction? If we include the future—all of the future—the stakes of what we do or don't do in the present moment become unimaginably enormous. We could have thousands, tens of thousands, even millions of years of civilization ahead of us, but that future depends on those of us alive in the present moment.

That is what I feel now when I look at that photograph of my new family. It is the past and future made tangible in a single moment. I think of the endless chain of events that had to fall into place in the past to make my son a reality. And I think of the events that are unfolding even now to bring the entire human future into being. That chain could have ended at any time, cut short by disease or catastrophe or simple bad luck. And until very recently, if that end had come—if this species had gone extinct like so many others— there would have been little we could have done about it. Our ancestors couldn't deflect an asteroid, or invent a vaccine to cure a killer disease. But we can. This could be the end of our times—or just the beginning. The choice and the responsibility are ours.

In the pages to come, I'll offer a tour of existential risk, and plot a path to survival. I'll survey the threat of asteroids and comets from space, and hunt with astronomers for the near-Earth objects that could extinguish our future on this planet. I'll explore the underappreciated danger from the supervolcanoes that have disrupted life on this planet over and over again, including one that sits beneath America's first national park. I'll travel to the birthplace of man-made existential risk—Trinity Site in New Mexico, where a terrible beauty was born. I'll tell the inside story of the climate change conferences that have failed to stop the frightening pace of global warming, and ask just what the present owes to the future. I'll share what it was like to live through the first global disease outbreak of the twenty-first century, and why a simple virus can wreak havoc on an interconnected world. I'll stand looking over the shoulders of the scientists remaking life with synthetic biology, and I'll ask whether the rise of superintelligent artificial intelligence (AI) is something to be welcomed, feared, or merely disbelieved. I'll search for extraterrestrial civilizations, and scour the so far silent cosmos for any clues it might offer for our own fate. And I'll explain how our species can survive the unsurvivable, should existential catastrophe finally arrive.

Make no mistake—we are in mortal danger. But the existential risks that follow in these pages have called forth dedicated scientists and experts who are doing their part and more to defend our future from the end times. New organizations have been created to study existential threats across disciplines: the Future of Humanity Institute at the University of Oxford, the Centre for the Study of Existential Risk at the University of Cambridge, the Future of Life Institute in Boston. Combatting existential risk isn't just a matter of devising asteroid deflectors or ensuring our future robot overlords are peaceful. It demands a new kind of scientific method, a willingness to grapple with planetary uncertainties and cosmic numbers. Original thinkers like Nick Bostrom, Anders Sandberg, Milan Ćirković, Olle

Häggström, and others have shed fresh light on the ultimate fate of human beings, a subject that, despite its obvious importance, has been less studied than the life of the humble dung beetle.[20] Their work inspired this book, and I will return to it again and again through these pages.

To save ourselves we need to think about the unthinkable, and not merely understand the future but feel its gravity. Our greatest existential challenge isn't technical or political, but conceptual. We have to believe that the end of the world can happen, and at the same time we have to believe that we can do something about it. But our track record is poor.

The *Bulletin of the Atomic Scientists* is an academic journal founded after World War II by some of the same people who helped develop the atomic bomb. In 1947 the artist Martyl Langsdorf was tasked with creating a cover for its first issue, and she channeled her dread of the new weapon into what would become one of the most iconic symbols of the Cold War: the Doomsday Clock, its hands inching toward midnight.

From 1947 on, the people behind the *Bulletin* have shifted the hands of the Doomsday Clock to represent, crudely but effectively, just how close our end times might be. In 1949 it was moved to three minutes to midnight in response to the Soviet Union's first successful test of an atomic bomb, which kick-started the nuclear arms race. After Washington and Moscow signed the 1963 Partial Test Ban Treaty, putting an end to aboveground nuclear bomb tests, the hands were moved back to twelve minutes to midnight. And while the first decades of the Doomsday Clock focused exclusively on the threat of a nuclear holocaust, the clock continued to keep time after the Cold War ended, broadening out to include new dangers from climate change and emerging technologies. The Doomsday Clock is the closest thing we have to a thermometer of existential risk.

Each fall, the science and security board of the *Bulletin*—a group of top-level scientists and defense experts—meet and ask themselves two questions: Is humankind safer or at greater risk this year than the last? And is humankind safer or at greater risk this year relative to the entire history of the clock? It's an imprecise symbol—existential risk isn't divided up into neat sixty-minute intervals—but a grimly effective one nonetheless. It's why one of the first trips I took for my reporting for *End Times* was to Washington, D.C., on January 25, 2018, to witness the *Bulletin* reveal the new time for the Doomsday Clock at the National Press Club. They did not disappoint. "We have come to a grim assessment," Rachel Bronson, the president and CEO of the *Bulletin*, announced to the world. "As of today, it is two minutes to midnight."

Only in 1953, after a year in which both the United States and the Soviet Union exploded their first hydrogen bombs—weapons of mass destruction far more powerful than the atomic bombs dropped on Hiroshima and Nagasaki—had the clock been this close to striking midnight. The construction of the Berlin Wall, the Cuban Missile Crisis, the Vietnam War—none posed as great a threat to the further existence of the human race as the events of 2017, at least according to the hands of the Doomsday Clock. The *Bulletin* experts cited an array of factors: North Korea's atomic breakout, uncertainty over the future of the Iranian nuclear deal, a planned U.S. withdrawal from the Paris climate agreement, the spread of cyberhacking, and of course, an unpredictable President Donald Trump. But what mattered more than the reasons behind the clock's move—all debatable—was the time itself, and how little might remain for us. "These are dangerous, dangerous times," Bronson told me after the announcement. "This is not your father's Cold War."

In the movie version of this story, Bronson and her *Bulletin* colleagues would have been delivering their alarm to an overflowing crowd of journalists in the nation's capital. Cable news shows would have interrupted their regular programming to cover the announcement live, and print newspapers would have broken out their wartime headline font. The Doomsday Clock is about nothing less than the fate of the entire human race, all seven and a half billion of us, and all those who might come after. There should be nothing more important.

Yet only a handful of journalists were with me in the audience that January morning in the National Press Club's First Amendment Lounge, asking only a handful of desultory follow-up questions. The Doomsday Clock would not go unmentioned by the media—fear makes for good press—but most outlets would treat it as one more data point in a world going madder by the day, to be overtaken almost immediately by the next story, the next scandal.

As I walked out of the National Press Club that morning, I could witness the forgetting already unfolding on cable news playing in the building's lobby. The breaking stories were about how Trump would perform at the upcoming World Economic Forum summit in Davos, Switzerland, how global demand for iPhones was slowing down, how the midterm elections were shaping up. On one screen ninety-four-year-old Henry Kissinger, a ghost from an earlier doomsday, croaked his testimony to the Senate Armed Services Committee. The clock kept ticking—our clock, our time. And no one seemed to care.

We must care. We can't give in to apathy, just as we can't give in to panic or despair. We face enormous challenges, and so many of them are of our own making. But we can overcome them, for our sake and for the sake of generations to come. I know the future that I'm fighting for. I can look in his eyes. And we can begin that fight together, by casting our eyes to the skies above.

ASTEROID

The Universe Is Trying to Kill Us

This is a book about the threat of human extinction, but it's impossible to begin the subject without considering the eradication of a group of animals that also once seemed secure in its reign over the Earth: the dinosaurs. Dinosaurs existed for more than 180 million years,[1] hundreds of times longer than *Homo sapiens*'s so-far brief stay on the planet—and then they met a sudden and violent end. What matters for us is not merely the fact that the dinosaurs went extinct, however, but how it happened. We might be reading our own future in the chronicle of catastrophe that is the fossil record.

About 66 million years ago, at what we now know as the close of the Cretaceous period, the orbit of a single asteroid intersected that of Earth. Asteroids are hardly rare—hundreds of thousands of these chunks of rock and metal circle the sun, just as our planet does. At more than six miles wide, this leftover piece of the solar system's birth was unusually large, yet it was still tiny compared to the planet it would impact, the equivalent of a marble colliding with a beach ball. But as the asteroid met the Earth it plunged through the planet's atmosphere at a steep angle, like a high diver slashing toward the bottom of the pool, speeding toward what is now present-day Mexico at a velocity of 12.4 miles per second—20 times faster than a gunshot.[2] Friction ignited the air around the plummeting asteroid into blue-hot plasma.[3] It took just twenty seconds for the asteroid to hit the sea below.

The ocean at the impact point was shallow, and as the asteroid reached the water it punched through the surface, the seafloor, and the crust below, burrowing into red-hot streams of underground magma. The energy released by the resulting explosion had the force of 100 million tons of TNT, greater than 6,500 times more powerful than the nuclear bomb dropped on Hiroshima. A mega-tsunami flooded the surrounding coasts. More than a thousand cubic miles of vaporized rock were blown into the sky, punching a hole in the planet's atmosphere. Some of the debris condensed into sand-sized, solid particles called spherules, and as they fell back to Earth they were so numerous that 10,000 of them were found on every square inch of the planet's surface. Thermal radiation from the hot air started fires globally and burned much or all of the land biomass. Some four inches of the Earth's oceans boiled off. "The asteroid hit and everything on the surface burned. It was like being inside an oven with the broiler on," said Brian Toon, an atmospheric researcher at the University of Colorado, Boulder, I spoke to who has studied the aftermath of the impact. "It would have baked the dinosaurs alive."

Those dinosaurs were, above all else, victims of catastrophically terrible luck. As we'll see, objects from space strike the Earth all the time, but the chance of being hit by something the size of the Chicxulub asteroid—named for the modern Mexican town near the underwater impact crater the asteroid left, one of the largest in the planet's history—is minuscule. That was the first turn of bad fortune. Even an asteroid as large as Chicxulub may not have wiped out the dino-

saurs had it not landed precisely where it did. The shallow seabed at
the impact point was unusually rich in sulfur, so the debris cloud
caused by the asteroid was spiked with sulfur droplets. Sulfur reflects
sunlight, and that cloud suffused with sulfur—and massive amounts
of soot—blocked much of the sun's heat from reaching the Earth's
surface. Global temperatures plummeted, perhaps by as much as 50
degrees Fahrenheit over land and 36 degrees over the oceans. For as
long as two years following the asteroid strike, the darkness was so
total that photosynthesis all but stopped, leaving only whatever food
might have been left on the land and below the surface. "Light levels
would have fallen below one percent for a couple of years," said Toon.
"When you have something that extreme, animals and plants can't
compensate."

That was the second turn of bad fortune—and the last one for
the dinosaurs. As traumatic as the actual asteroid strike would have
been, especially near the impact site, it was the global loss of light and
food that ultimately killed the dinosaurs and some three-quarters of
the animals and plants living on the Earth at the time. Wrenching
environmental change unfolded faster than anything living could
adapt, deadly to anything that couldn't hide in a hole. It's what would
likely kill us if a comparably sized asteroid collided with the Earth
today—unless we're smart enough to save ourselves.

The story of the asteroid that wiped out the dinosaurs has become
so ingrained in our imagination as a stand-in for a natural apocalypse
that it's easy to forget how recently the theory was born. The first identification of an asteroid crossing Earth's orbit didn't occur until 1932,
and even as Neil Armstrong and Buzz Aldrin walked on the moon
in 1969, scientists were only beginning to understand that the craters
pocking the lunar surface were the result of ancient asteroid impacts.

Here on Earth, geologists were still in the grip of a theory called
uniformitarianism. It may sound like a little-known Protestant sect,

but uniformitarianism was the idea that the geological features we see today—from oceans to craters to mountains—were all formed by gradual and uniform processes like erosion. That includes extinctions—scientists at the time largely believed that species went extinct gradually, rather than in sudden mass waves. It was only in 1980 that the father-and-son team of physicist Luis Alvarez and geologist Walter Alvarez published research showing that clay from the layer of sedimentary rocks marking the end of the Cretaceous contained an unusual concentration of iridium, an element found in far greater amounts in asteroids than in the Earth's crust. The iridium was like the fingerprint of a suspect on a murder weapon, and the Alvarezes theorized that the dinosaurs' extinction had been caused by the impact of a gigantic asteroid or comet.[4]

While there are still some dissenters who believe that massive volcanic eruptions were the chief cause of the dinosaurs' demise—a possibility we'll examine in the next chapter—the asteroid theory was accepted by most scientists, especially after the underwater site of the Chicxulub crater was discovered in 2001.[5] The Alvarezes' research also helped dispel the theory of uniformitarianism as it became clear to scientists that along with work of slow, everyday geological processes, the Earth had been utterly reshaped by instantaneous catastrophes—hence the name of the new theory, catastrophism. And if such a catastrophe had happened once, it could happen again.

Dinosaurs are the poster species for total extinction, and so their destroyer has come to symbolize a universe that, as the astrophysicist Neil deGrasse Tyson has said, "is trying to kill us."[6] Unlike some of the other natural existential risks we'll discuss in this book, we'd likely be able to see asteroids coming, which adds to their dreadful power, even as the cataclysmic effects of impact would be felt in an instant. No wonder Hollywood has made asteroids its favorite nonhuman existential villain. They may be chunks of rock and metal, but asteroids are inherently cinematic.

There's something else that sets asteroids apart. Unlike most other existential risks—especially natural ones—this is one world-ending threat we can potentially eliminate now. Astronomers can find and track asteroids that might intersect with the planet's orbital path— what are called near-Earth objects (NEOs)—years and even decades in advance of a possible impact. And if an NEO menaces, we have theoretical methods to target and deflect it before it endangers life on Earth. As Tom Jones, a veteran NASA astronaut who has an academic background in asteroid detection, told me, "We are learning how to change the way the solar system works to preserve our species."

When Chicxulub came for them sixty-six million years ago, there was nothing the dinosaurs could do but die. And if a similar-sized NEO struck again during humanity's watch on the planet, there would have been nothing we could have done, either—until today. Now we can alter the mechanics of the heavens to keep ourselves and our future safe. We can save the world—from this threat at least—but we'll need to act soon.

Humanity's first real inklings of the peril posed by NEOs came on March 24, 1993, when a team of three astronomers at the Palomar Observatory outside Los Angeles discovered a comet on a collision course. The weather at the Peninsular Ranges was windy and cold that night, with clouds obscuring the sky from an incoming storm. The geologist Eugene Shoemaker; his wife, Carolyn, an astronomer; and their colleague David Levy were carrying out a multiyear NEO survey, painstakingly searching for asteroids, comets, and anything else they might be able to find arcing through our solar system. But hopes that night were as dim as the sky. The best conditions for aster-oid hunting are clear and still. March 24 was neither.[7] Even today, with the aid of digital cameras and tracking software, the practice of discovering and cataloging asteroids is as much art as it is science, and one that favors the meticulous. In the early 1990s, though, asteroid hunters still used analog film. Aiming their cameras through telescopes, they took multiple minutes-long exposures of specific sections of the night sky, then compared those images in the hopes of locating a heavenly body moving against the known background of stars and planets. As if the bad weather that night weren't obstacle enough, the Shoemakers and Levy had discovered a couple of nights earlier that a stack of their highly sensitive film—which cost nearly four dollars per shot—had been accidentally exposed to the light, making it unusable except for a portion around the center of the slide. Was it even worth potentially wasting their precious remaining good film on such a cloudy night?

Eugene Shoemaker, known as Gene, was a legend in asteroid studies. He was one of the first scientists to conclude that the craters found throughout the Earth were physical evidence that the planet had been struck by asteroids and comets in the past.[8] But between the cost of the film and the bad weather, even the stalwart Gene Shoemaker was ready to pack it in for the night. Levy, though, argued that they might as well try to use the damaged slides—if the conditions remained poor, at least they wouldn't have squandered any of the unexposed film. Gene eventually agreed. Before the gathering clouds finally drew a curtain on the night, Levy managed to take exposures of three fields of sky they hoped could be compared to existing images they had taken previously.[9]

Once the exposures had been shot and developed, Carolyn Shoemaker began scanning them, two at a time, using a stereo-microscope that caused any moving images—like asteroids—to rise above the background field of stars. Carolyn's story is an inspirational one. She had spent decades as a wife and homemaker to Gene before making a later career change to become an astronomer, and at age fifty-one now spent her nights hunting for asteroids at Palomar and Northern Arizona University. As she scanned a pair of images the following day, Carolyn spotted something near the center of the field a few degrees away from the planet Jupiter. "I don't know what this is," she told Gene and Levy, "but it looks like a squashed comet."

Over the course of her second act, Carolyn Shoemaker discovered more than 800 asteroids and 32 comets.[10] Yet no find would prove more consequential than the object she spotted that evening. What Gene and Levy photographed and Carolyn picked out was indeed a comet, one that would eventually be designated Shoemaker-Levy 9, as it was the ninth such comet discovered by the team. It appeared squashed because it had fragmented, shattered by the force of Jupiter's gravitational field when the comet's orbit had taken it near the planet in 1992. But even a broken comet is dangerous, and now what remained of Shoemaker-Levy 9 was on a direct course for the largest planet in our solar system.

Beginning on July 16, 1994, with the telescopes of the world watching—along with the spacecraft *Galileo*, then just 150 million miles from Jupiter—the lead fragments of Shoemaker-Levy 9 slammed into the gas giant.[11] Shoemaker-Levy 9 marked the first time that astronomers were able to directly observe a collision between two extraterrestrial objects in the solar system, and the show did not disappoint. Most of the cometary fragments were at least as large as 1.25 miles across and were traveling at 125,000 miles per hour, fast enough to cross the width of the continental United States in little more than a minute. A single piece of Shoemaker-Levy 9 delivered the energy equivalent of 6 million megatons of TNT, a force hundreds of times greater than the explosive power of the world's combined nuclear arsenal at its peak.[12] The impacts created fireballs that reached heights of more than 1,800 miles—higher than three hundred Mount

Everests—and burned at temperatures as hot as 42,660 degrees.[13] The great dark scars from the collisions were as large as 7,400 miles across, and were visible even from a child's backyard telescope on Earth.[14]

For astronomers, the Shoemaker-Levy 9 collision with Jupiter was an unprecedented cosmic fireworks spectacle, one so powerful that the thin rings circling the planet were left tilted, like a football player whose helmet had been knocked askew by a blind-side hit.[15] When one 2.5-mile-wide fragment struck the planet, the flash generated by the impact was so bright that it temporarily blinded many of the infrared telescopes trained on the event. Yet if Carolyn Shoemaker hadn't picked out that "squashed comet"—and if David Levy hadn't urged his team members to keep working through that cold and cloudy night at the Palomar Observatory—scientists might never have witnessed the collision.

But Shoemaker-Levy 9 wasn't merely a scientific milestone. Jupiter is the biggest planet in the solar system, so large that you could fit 1,300 Earths inside it and still have room to spare. If any planet in the solar system would be capable of shrugging off a blow from a comet, it would be Jupiter. Before the impact event, astronomers wondered if Shoemaker-Levy 9 would simply disappear into the bowels of the gas giant. Yet months after the collision the scars from Shoemaker-Levy 9 were even more pronounced than Jupiter's Great Red Spot, that eye-shaped spinning storm in the planet's atmosphere that is twice the size of the entire Earth. It was as if a new inmate had picked the biggest, baddest guy in prison—and beat him to a pulp. If a cosmic collision could do that much damage to Jupiter, what would it do to our little blue planet? Suddenly the solar system seemed like a much more dangerous place.

Expert concern about the threat posed by NEOs had begun to mount in the years before Shoemaker-Levy 9, spurred by the Alvarezes' discovery and by a close call with a half-mile-wide asteroid that passed within 500,000 miles of the Earth in 1989.[16] In 1990 Congress included language in NASA's budget authorization requiring the agency to prepare a report on NEOs—how to track them and how to stop them. But when then vice president Dan Quayle endorsed an idea that same year for the federal government to buy telescopes to track potentially hazardous asteroids—and use modified Strategic Defense Initiative antimissile weapons in orbit to destroy them—it became a punch line, not policy.[17] Planetary defense was plagued by the "giggle factor"[18]—a real term for when a scientific subject appears too ridiculous on its face to be taken seriously. The highly visual lesson of Shoemaker-Levy 9 helped change that. "We woke up to the fact that asteroid impacts had an effect on the evolution of life on the earth," said Steve Larson, a coinvestigator at the Catalina Sky Survey at the University of Arizona, one of the first programs dedicated to searching for NEOs.

Hollywood even got in on the act, eager in those immediate post–Cold War years for disaster scenarios on an apocalyptic scale. Two separate films about efforts to prevent an extinction-level collision event, *Deep Impact* and *Armageddon*, were released within two months of each other in the summer of 1998. *Deep Impact*, in the likely event that you've blocked it from your mind, is the comet one, and Gene and Carolyn Shoemaker were even credited as "comet advisors" on the film. *Armageddon* is the one with Ben Affleck and Bruce Willis as oil rig workers drafted by NASA to drill into an incoming asteroid and blow it up with a nuclear bomb.[19] (In the DVD commentary, Affleck made the astute point that it would probably have been easier to train astronauts to become oil drillers than vice versa.) Despite being critically loathed Armageddon still made $553 million to *Deep Impact*'s $349 million. I saw them both in the theaters.

The same year those films hit screens, NASA established its NEO Program and, under a congressional directive, dramatically increased its participation in the Spaceguard Survey, which was tasked with discovering and tracking at least 90 percent of potentially hazardous NEOs larger than 1 kilometer (0.62 miles) within the following ten years. Impactors that big could potentially cause continental or even planetary damage. In a field where cost overruns and blown deadlines are common, NASA almost met that target on time, and achieved its goal in 2010. Because the movement of celestial bodies is predictable with sufficient data, scientists could forecast the orbits of identified asteroids decades into the future, and determine whether any have a chance of colliding with the Earth in the decades to come.

The results were a relief. Thanks to the work of Spaceguard and the astronomers around the world who contribute to it, we can be confident that our civilization is unlikely to be ended by any *Armageddon*-class asteroids for the foreseeable future. But humanity shouldn't become complacent. While NASA has pinpointed nearly all of the largest NEOs, of which there are some 2,000,[20] a new congressional mandate in 2005 to find at least 90 percent of potentially hazardous NEOs larger than 140 meters (460 feet) by 2020 has proven far harder to fulfill. While an impact from a smaller asteroid may not threaten civilization, it could easily wipe out a city, and potentially much more. And there's a good chance that we might not even see it coming before a collision was imminent, far too late to deflect the asteroid or even evacuate those in harm's way. "This is the only natural hazard that is predictable and preventable," said Clemens Rumpf, a research fellow at the University of Southampton who studies asteroid strikes. "But we have to know that the asteroid is coming."

To do that—to fully "retire the risk," as asteroid hunters say— we'll need to spend more on planetary defense. That includes executing a mission that would rehearse tracking and deflecting an asteroid in space. Such an effort would ensure that should the time come to save the world, we'll at least have a practice run under our belts. It would cost more than the roughly $60 million NASA currently dedicates to planetary defense each year,[21] but the math of existential risk proves that it's worth spending far more to reduce even a tiny risk of planetary catastrophe. The biggest challenge isn't what we can budget for, but what we can imagine. "There's no technical obstacle to protecting ourselves," said Franck Marchis, a senior planetary astronomer and asteroid expert at the SETI Institute in Silicon Valley. "We know how to protect ourselves. We've just never tried."

———————————

To understand the risk from NEOs, you need to understand what they are and where they come from. Our solar system was formed some 4.6 billion years ago, as the sun emerged from a cloud of gas and dust known as a solar nebula. Over a hundred million years the matter that remained following the sun's formation began to clump together. The planets, including our own, grew through a process of accretion, like sand castles built one grain at a time.

Our planet's youth was a wild one, as the Earth endured a rain of asteroid and comet strikes. Not long after the planet was formed, a celestial object the size of Mars collided with our planet, blasting billions of fragments of rock and molten magma into space. Some of that rock eventually cooled and formed a sphere—our moon.[22] That was just one massive impact—between 3.9 and 4.5 billion years ago, the Earth, moon, and many of the other planets of the solar system were struck again and again during what scientists call the Late Heavy Bombardment. Extrapolating from the impact evidence on the pockmarked moon—where there is no air or water to erode the silent craters—the Earth may have been hit by more than 20,000 asteroids or comets capable of leaving craters from 6.2 miles to 620 miles in diameter.[23] The Late Heavy Bombardment was, as the author Craig Childs writes in his book *Apocalyptic Planet*, "what everything else is measured against, an apocalypse unmatched in Earth's history."[24] Though there was still creation amid the destruction. Some of those comets may have borne water to a young and barren Earth, along with the carbon-based molecules that form the foundation of life. Whatever asteroids and comets may do to us in the future, we might not even be here without them.

Even after the solar system finally settled down into a more sedate middle age, there were still millions of pieces of rock and metal wandering unattached through space like leftover pieces of a jigsaw puzzle. These were asteroids and comets. Comets are composed of rock, frozen gases, and ice—the characteristic tail in a comet is a result of the sun vaporizing some of its material, releasing dust particles that trail behind it. They originate in the Kuiper Belt and the Oort Cloud, on the outermost fringes of the solar system. Asteroids tend to be rocky or metallic, and true to their name are usually found within what's known as the asteroid belt, between Mars and Jupiter. While films like *The Empire Strikes Back* make it seem as if an asteroid belt is so dense with rock that the odds of successfully navigating it are—to quote C-3PO—approximately 3,720 to 1, our belt is actually so sparse, as the planetary scientist Carrie Nugent points out in her delightful book *Asteroid Hunters*, that the actual odds are closer to 1:1.[25] If you took all the asteroids in the asteroid belt and squashed them together, you'd have an object less than 4 percent of the mass of our moon.[26]

Every once in a while, gravity perturbations from Jupiter or Mars can kick asteroids out of the main belt, like a big child nudging a smaller one off a merry-go-round. Those loose asteroids might spin out of the solar system altogether or be swallowed up by the sun, but some can end up in shallower orbits that bring them close enough to our planet to be classified as an NEO. An NEO's orbit must bring it within about 30 million miles of the Earth's orbit, around 125 times the distance between the Earth and the moon. A potentially hazardous object, or PHO—the ones to watch out for—is any NEO big enough to potentially survive the plunge through our atmosphere, meaning larger than 100 or so feet wide, and with an orbital path that can bring it within about 5 million miles of the Earth.[27]

That we rarely notice asteroid impacts is in part a sign of just how empty space is—by one estimate, only 0.0000000000000000000042 percent of the universe actually contains matter.[28] But in truth we're under near-constant bombardment, though mostly by the cosmic equivalent of BBs. Each day the Earth is battered by about 100 tons of dust and sand-sized particles;[29] keep a close enough watch on the night sky and you'll see the larger pieces flaring briefly as they burn up in the atmosphere as what we call, inaccurately, shooting stars. The size and number of asteroids in the solar system follows a proportional relationship called a power law, so that smaller asteroids are far more numerous than large ones. That's why it has been easier for NASA to locate asteroids above 1 kilometer than it has been to locate those above 140 meters—the big ones are bigger, so they were easier to spot, and there are fewer of them to find.

Though the English astronomer Edmond Halley—he of the eponymous comet—theorized as early as the seventeenth century that impacts could create craters on the Earth's surface,[30] the first NEO wasn't discovered until 1898.[31] For years astronomers found asteroids—which showed up as smudges on the pictures they took from telescopes, just as any moving object is blurred in a photo—more annoying than intriguing. They were searching for the interesting stuff—planets, moons, and stars, all of which are far rarer than what appeared to be the dregs of the solar system. In her book, Nugent writes that the astronomer Edmund Weiss found asteroids so irritatingly common that he began referring to them as "those vermin in the sky."[32]

One scientist's vermin can be another's life work, however, and by the 1960s interest in hunting asteroids surged. The Palomar Observatory, where the Shoemakers and David Levy discovered their comet; the Leiden Observatory in the Netherlands, led by another husband-and-wife team, Cornelis and Ingrid van Houten; the LONEOS survey in the mountains near Flagstaff, Arizona—all were staffed by astronomers dedicated to finding NEOs.[33] But no search program has been more successful than the Catalina Sky Survey at the University of Arizona, which as of 2018 had found nearly half of the roughly 18,000 NEOs that have been discovered to that point.[34] As Eric Christensen, Catalina's primary investigator, told me when I visited him at the University of Arizona: "Finding near-Earth objects is our only mission, our only goal, and we're free to optimize everything we can towards it."

If a civilization-threatening asteroid or comet does zero in on Earth, there's a better than even chance that the NEO hunters of Catalina will be the ones to spot it. They are the first line of defense this planet has against existential threats from space—which is why I decided to travel to Arizona to see their work firsthand.

———————

Greg Leonard was right—I should have brought a jacket. The Mount Lemmon Observatory, where the Sky Survey does its work, is perched high in the Santa Catalina Mountains north of Tucson, Arizona. Tucson is a desert town and warm even in March, but the observatory is more than 9,000 feet above sea level. By the time I'd navigated the switchback mountain roads, the sun had vanished and temperatures were plunging, so much so that I was shivering as I walked the last few steps to the telescope. I was warned to point my car away from the observatory when I parked—light is the enemy of asteroid hunting, and even a brief flash of headlights could be enough to spoil an observation. The sky was mostly clear as the evening began, save for a few spindly clouds framed against the materializing stars. I was lucky—had the cloud cover been fuller, I would have spent my time on Mount Lemmon watching Leonard, the observer on duty that night, filing old data.

A few things to know about modern asteroid hunting. You don't directly look through the telescope, which is fortunate, because the room that houses it is open to the night sky, and therefore the high-altitude chill. Instead, Leonard and his apprentice Brian sat in a cramped—though blessedly heated—control room, surrounded by stacks of humming servers, walls of computer screens, and lots of coffee. The computers direct the telescope and display the images, the servers store the data from observations, and the coffee keeps the observers awake through their all-night duty rotation. Whereas the Shoemakers had to use limited and costly analog film to take glimpses of the sky—and then had to develop that film themselves and manually scan each image for moving objects— Catalina and other asteroid surveys now use what are called "charge-coupled devices," or CCDs, thin wafers of light-sensitive silicon on top of an array of pixels. When light is concentrated on the CCD, photons fall onto the pixels and are stored there, like raindrops collected in a bucket.[35] Far more images can be collected now than was ever possible during the days of analog photography, and those pictures are of higher quality and easily shareable. As a result, the pipeline of data has gone from a trickle to a torrent. And data is the lifeblood of asteroid hunting.

But while the tools have improved, the techniques of the search haven't changed much. After a tour of the observatory—which didn't take long, given its compact size—I balanced on a stool and watched as Leonard directed the telescope to a single patch of sky. He paused for thirty seconds to gather enough light to make an exposure. Another benefit of CCDs is that astronomers no longer have to keep their telescope fixed for minutes to make a single image, which means far more of the sky can be covered in a single night. He shifted the telescope slightly—or slewed it, the technical term for rotating a telescope—to observe a different region of the sky, and took another exposure. The only sound inside the observatory was the grinding of gears as the telescope, originally built in the late 1960s, slid by a few degrees. Two more exposures were taken, and then Leonard returned to the first target and took another exposure, each separated in time by about fifteen minutes, until he had four exposures of the same target.

Leonard motioned for me to come over to the computer screen. He clicked a button and the images he'd taken over the hour ran together, creating the illusion of animation, like a child's crude flipbook. The stars, so much brighter under the telescope's magnification than they were even through the cold and clear Arizona night, twinkled from frame to frame, the result of turbulence in the Earth's atmosphere refracting their distant shine.

But Leonard wasn't looking at the stars. He was searching for something, anything, moving against the background of space, reflecting the spare light of the distant sun. (Asteroids—like planets, moons, and virtually every other object in the heavens that isn't a star—are visible only because they reflect the light of the sun.) Catalina, like other modern sky surveys, uses software that automatically scans the exposures and highlights anything that might be an asteroid or a comet. But the program is prone to false positives, mistaking dust or bent light for a moving object. NEO hunting still requires the eyes of a trained observer, one who can parse the signal from the noise.

Once an observing team locates a possible NEO, they relay the data to the Minor Planet Center (MPC) in Cambridge, Massachusetts, the global catalog for all things asteroid and comet. The MPC is the final arbiter on asteroid discovery, processing some 50,000 observations every day[36] sent in by professional surveys like Catalina as well as amateur astronomers from around the world. Nearly all of those reports are false alarms—either outright mistakes, or NEOs that are authentic but have already been discovered by someone else, or objects that are still confined to the asteroid belt, and which therefore pose no threat to the Earth. Eric Christensen compared the process to factory trawlers that fish the sea: "We have a big net, and we're just trying to catch whatever's in the ocean, and most of what we catch are not NEOs."

Searching for asteroids is an almost monkish discipline, one that demands a fanatical attention to detail as well as the stamina to work on a mountaintop when the rest of the world is asleep. Observers spend several nights in a row at the observatory, sleeping through the days in nearby bunks, and then take several days off. "It tends to wreak havoc with your social life," Leonard told me during a break in the observation. "The ones who make it are the ones who can keep themselves busy when they're not on the mountain."

For Leonard, who has cropped graying hair and the build of a triathlete, that means exercise, in part to compensate for the eye-straining hours crouched over a computer screen on Mount Lemmon. Gene Shoemaker himself initially brought Leonard into the business. On the wall next to a bank of computer servers was a poster memorializing Shoemaker, featuring a star field above lines from Shakespeare's *Romeo and Juliet*: "When he shall die, / Take him and cut him out in little stars, / And he will make the face of heaven so fine / That all the world will be in love with night / And pay no worship to the garish sun."

As lonely as life in the observatory could seem, it soon became clear to me that NEO hunting was a team effort. Throughout the night, Leonard and his apprentice fielded a string of requests to try to locate possible NEOs that have been spotted by other observers, somewhere else on Earth. Because asteroids won't stay still, they need to be observed multiple times over multiple days before astronomers can know where they're going, how large they are, and whether they might pose a threat to Earth—and they need to make those observations around the sun, which blocks much of the sky from view. Carrie Nugent compares the process to being a teacher trying to count children running around a field at different speeds, the view obstructed by a large tree.[37] The faster children will appear soon enough, but the slower-moving kids may remain blocked behind the tree for some time, and the teacher—and the asteroid hunter—can only wait and watch.

Given those challenges, it's not surprising that mistakes do happen. On January 13, 2004, an automated telescope in New Mexico recorded an observation of a possible NEO. Staff at the MPC in Cambridge then posted the information on their website—as they always do—so that amateur astronomers could target the candidate for further observations. One spotter in Germany found it and calculated that it was on pace to grow in brightness by an astounding 4,000 percent over the next day. This was concerning, in much the same way that observing a pair of headlights getting rapidly brighter would be concerning if you were standing in the middle of the road. A NASA researcher did further work and calculated that there was a one-in-four chance that the 100-foot-wide asteroid—now named 2004 AS1—was bound to hit somewhere in the Northern Hemisphere, and that it could do so within a couple of days.

You might think that if there were a 25 percent chance that an asteroid was about to strike, NASA would wake up the president in the middle of the night, drag them into the Situation Room, and warn them that the Earth was in imminent danger. But in 2004 there was no clear protocol for responding to the possibility of a possible hit, even from a smallish asteroid. So the astronomers kept doing what they were doing—watching the sky in the hopes of gathering additional observations that would dispel the uncertainty. The heavy cloud cover over both Europe and the United States at the time made any further observations impossible, however, until the air cleared over Colorado and an amateur astronomer named Brian Warren was able to use his telescope to search the portion of the sky where 2004 AS1 would be if it were indeed on a collision course with Earth. But the asteroid was nowhere to be found. In fact, the 2004 AS1 was more than 7.4 million miles away when it passed by the Earth, some 32 times the distance between our planet and the Moon.

As it turned out, though, we were doubly lucky, and doubly wrong. 2004 AS1 missed the Earth, but it was closer to 1,000 feet wide, not 100, which made it about the height of the Eiffel Tower— large enough to potentially create devastation on a continental scale had it collided with the Earth.[38]

Despite all the challenges they face—bad weather, thin budgets, the fact that the sun stubbornly obstructs their view—asteroid hunters like Greg Leonard are very good at their jobs. They've located more than 8,000 NEOs in the 140-meter-plus category that NASA has been charged to track, and thousands more below that size. (Remember: there are proportionally more small asteroids out there to be found.) The Catalina Sky Survey alone discovers around three new NEOs per observation session on average.

Toward the end of my night on Mount Lemmon—around the time I began wondering if I would be able to stay awake on the twisty drive down the mountain—Leonard and his partner zeroed in on a small asteroid, one that hadn't been registered by the MPC yet.

It was small, and later calculations showed that it posed no threat to the Earth, but the act of finding an unknown asteroid provides the kind of instant gratification that is rare in the sciences, where years or even decades can pass between the first steps to a discovery and the final recognition of publication. "We have the ability to go to the telescope and know we've discovered something that night," Eric Christensen told me when I met him the next morning, once I'd come down from the mountain and had dosed myself with near-lethal amounts of caffeine.

Those discoveries have immediate practical value. Asteroids and comets differ from other existential risks because with sufficient data, astronomers can predict the future. Take enough observations, mix in some math, and scientists can determine with remarkable precision where any NEO is likely to be in five, ten, or even a hundred years time. That's what makes the act of asteroid hunting on lonely mountaintop observatories so necessary. It's only through standing watch and scanning the skies, night after night, that we're able to know what threats the cosmos may be sending our way.

This isn't mere science; asteroid hunting is about the preservation of the species. The jolt of excitement I felt when the Catalina team zeroed in on their new asteroid came from witnessing two human beings, in a remote observatory, playing their small roles in keeping the other seven and a half billion people on this planet safe from extinction. No other animal can do that, and neither could human beings until very recently. Asteroid hunters like Greg Leonard take that charge seriously. "I know that the chances of me dying in an asteroid impact is less than dying from a lightning strike," Leonard told me toward the end of my time on Mount Lemmon. "But I also know that if we do nothing, sooner or later, there's a one hundred percent chance that one will get us. So I feel privileged to be doing something."

Intelligence-gathering alone won't keep the Earth safe, though. As Leonard said, given enough time, a large NEO will end up on a collision course with our planet. It's happened before and it will happen again. So when the day comes, what will we do about it?

———————————

The Earth's first line of defense against incoming fire is actually the gas giant that sits fifth from the sun. Jupiter's gravity sweeps up some of the most dangerous Earth-threatening comets and asteroids. Our second line of defense is our atmosphere. Most objects that collide with the Earth never reach the surface. Asteroids travel through the frictionless vacuum of space at speeds that reach tens of thousands of miles per hour. But when a meteor—which is what an asteroid is called once it reaches Earth—breaches the atmosphere, it hits air, which quickly piles up. Friction from air resistance causes the meteor to glow brightly and heat up to temperatures as high as 3,000 degrees.[39] Up to 95 percent of the meteors that enter the Earth's atmosphere burn up completely, and most of the rest rarely leave behind more than tiny fistfuls of rock or metal known as meteorites.[40] As antimissile systems go, the atmosphere is superior to anything developed by the Pentagon.

But even NEOs that never make it to the ground can cause substantial damage. On June 30, 1908, an asteroid or comet, perhaps 130 to 200 feet wide, exploded in the skies above the Stony Tunguska River in the heart of Siberia. The airburst produced the same amount of destructive energy as 185 Hiroshima-scale nuclear bombs. The comparison is apt—nuclear warheads are detonated in the air, rather than at ground level, to distribute the destructive force over a wider area. Tunguska was nothing less than a natural nuclear strike—albeit without the radiation—and one more powerful than any bomb humans have ever employed in wartime. The explosion annihilated more than 770 square miles of forest, pulverizing an estimated 80 million trees. It is the largest known NEO impact in recorded human history.

Fortunately, then as now, Siberia is a place where trees vastly outnumber human beings, and no one is known to have been hurt by the strike. But Tunguska was still a close call—had the NEO arrived just four hours later, the rotation of the Earth would have brought the Russian city of St. Petersburg into the crosshairs. A 2018 White House report found that if a Tunguska-sized impactor were to hit New York, it would obliterate virtually the entire city and many nearby suburbs, taking out the world's financial nerve center and potentially killing millions.[41] "Something this size wouldn't take down civilization, but if it hit in the wrong place, plenty of people would be dead," said Ed Lu, a former astronaut and the executive director of the B612 Foundation, a Bay Area nonprofit dedicated to asteroid defense.

The Tunguska strike occurred only a decade after the first NEO had even been discovered. There were no sky surveys, no astronomers searching for incoming fire, no warning, and no defense. It was only luck that the victims of the Tunguska were trees, not people. But what could we do today if astronomers discovered another largeish asteroid—say, over 300 feet—set to impact a major population center like New York or London or Tokyo?

If the impact were predicted to occur within a few days—as briefly seemed possible in 2004—our best hope would be to move as many people as possible out of harm's way. Were the object to hit inside the United States, the Federal Emergency Management Agency (FEMA) would be charged with preparing for the disaster and its aftermath. An incoming asteroid's path and power can be tracked in advance, allowing astronomers to create a likely impact zone, just as meteorologists do with hurricanes. In fact, NEOs are far more predictable than weather here on Earth, let alone natural disasters like earthquakes that strike with no warning at all. And while a Tunguska-sized asteroid could obliterate a city with a direct hit, just 3 percent of the world's land surface is covered in urban areas,[42] meaning it's much more likely that an asteroid would either strike a largely unpopulated area like Siberia or land in the oceans that cover two-thirds of the planet. That's the good news.

The bad news is that if an asteroid the size of the one that killed off the dinosaurs were to hit our planet today, it would have global effects no matter where it landed, and according to one study could result in fatality rates of up to 100 percent—in other words, extinction.[43] In a 2013 congressional hearing, Representative Bill Posey of Florida asked then NASA administrator Charles Bolden what the strategy would be for dealing with an Earth-threatening asteroid that was discovered with three weeks' warning. Bolden was blunt. Our strategy, he replied, would be to "pray."[44]

If we're smart and forward-thinking, we won't have to depend on supernatural intervention. "With enough warning—let's say at least ten years—we could design a space mission to protect ourselves," said Ian Carnelli, a program manager at the European Space Agency who works on asteroid surveillance and defense.

The way to stop an NEO is to deflect it, though that word is deceptive. Rather than trying to knock an asteroid to the side, we would try to either slow down or accelerate the asteroid along its given orbital path. Remember that an impact occurs when an asteroid and the Earth intersect while traveling along their separate orbital paths. Asteroid experts compare the process to cars merging on a highway. To avoid a collision, one driver has to speed up or slow down. There's no speeding up or slowing down the Earth, so we have to alter the velocity of the asteroid, ensuring that it arrives either too late or too early for its appointment with our planet.

One option is to take advantage of a fundamental force of the universe: gravity. Here's a quick high school physics refresher: all objects with mass or energy—planets, asteroids, even light itself—are attracted toward each other through the force of gravity. If a large object—called a gravity tractor in this case—could be placed in space near an incoming asteroid, its gravitational attraction could be just enough to slightly tug the NEO's orbital path away from an inter- section with the Earth.

We could also take advantage of what is known as the Yarkovsky effect. Just like the Earth, asteroids rotate as they journey along their orbits, which means each half of the asteroid has a day and a night that alternate as the object spins. When the warmer daylight side of the asteroid rotates to face away from the Sun, it releases infrared photons that carry a bit of momentum from the asteroid, acting like a minuscule rocket thrust. That's the Yarkovsky effect. As anyone who has worn a black T-shirt on a hot, sunny day knows, dark colors absorb light, while paler colors reflect it. By painting one side of the asteroid—perhaps by using paintballs made of dry powder with an electrical charge, which could theoretically survive the vacuum of space—we could use the Yarkovsky effect to tweak the asteroid's speed. A similar method would involve employing a laser to burn away one surface of the asteroid; as it ejected the vaporized rock and metal, the asteroid would be pushed ever so slightly in the opposite direction.

Each of these methods would create only tiny changes in the orbital path of an NEO, but if we act decades before it is predicted to hit the Earth, those tiny adjustments could accumulate over the years to ensure that the asteroid would miss our planet. But we may not have decades of warning, and if the fate of the Earth is at stake, we'd have to opt for a more direct application of Newtonian physics. An object in motion—like an asteroid—stays in motion with the same speed and the same direction unless acted upon by an unbalanced force. We could bring that unbalanced force to bear on an asteroid by ramming an unmanned spacecraft called an impactor into it. Newton's second law—force equals mass times acceleration—would do the rest, slowing down or speeding up the NEO. If we know how large an asteroid is and how fast it is traveling, we should be able to figure out how large and how fast our impactor needs to be.

NASA knows this method can work, because they've tried it—though not exactly on purpose. In 2005, the Deep Impact spacecraft rendezvoused with the comet Tempel 1, some 266 million miles from Earth. Upon arrival, Deep Impact—which, for the record, was not named after the film—released an 820-pound impactor that rammed into the comet at about 23,000 mph, delivering a jolt of force equivalent to 4.8 tons of TNT.[45] Given that the comet was nearly four miles across while the impactor was the size of a washing machine, there was no measurable deflection to speak of—that's Newtonian physics for you—but the collision did leave a measurable crater, and gave NASA at least an outline of how a kinetic impactor could work on a smaller asteroid, or with a bigger impactor. Which brings us to nukes.

The more force we can deliver to an NEO, the more we can alter its orbit—and for better or for worse, there is nothing in the human arsenal more forceful than a nuclear weapon. If we needed to deflect a large asteroid, or one that was already close to Earth—so the change in the NEO's orbit would need to be more extreme—nukes would likely be our only alternative. Erika Nesvold, an astrophysicist formerly with the Carnegie Institution for Science, devised an algorithm called Deflector Selector that simulated 18 million attempts to prevent an asteroid impact. She concluded that a nuclear option was the right call for as many as half of the simulations. "It's not all that surprising," she told me. "This is a physics problem, and nukes have the most energy."

What we wouldn't do is simply fire a bunch of intercontinental ballistic missiles at the asteroid and hope to blow it to smithereens, as if we were playing a real-life game of Missile Command. It may sound counterintuitive, but you don't want to blow up an asteroid if you're trying to defend the Earth. There's no telling where the resulting debris might hit, and as Shoemaker-Levy 9 demonstrated, a broken impactor can still pack a serious punch. One 2019 computer model study found that if an impactor did break up an asteroid on a collision course, the space rock would eventually pull itself back together.[46] As with other deflection techniques, the aim is to speed up or slow down the asteroid along its orbital path. Nuclear weapons just provide extra oomph.

One method would be to explode a nuclear device several hundred feet away from the asteroid. Space is a vacuum, so there is no air to carry the destructive force of a shockwave as on Earth, but the high-energy gamma rays, X-rays, and neutrons released by the detonation would hit the nearby surface of the asteroid and vaporize part of it, creating plasma that ejects particles back into space and so thrusts the asteroid in the opposite direction. Hopefully nothing gets blown up—especially the Earth. "This isn't about sending Bruce Willis to the asteroid with a bomb," said Carnelli.

About that. You can't discuss asteroid defense—even among PhD-holding astrophysicists—without Bruce Willis and *Armageddon* coming up sooner or later. On the one hand *Armageddon*—and the somewhat more scientifically sound *Deep Impact*—introduced audiences to the existential threats posed by NEOs in visceral fashion, and proved that we weren't helpless to stop them. On the other hand, certain licenses were taken with the science. In *Armageddon* the killer asteroid is described as being "the size of Texas," reportedly because Michael Bay didn't think that audiences would believe than an NEO six or seven miles across would possibly be big enough to wipe out the human race. (It would be.) A group of scientists at the University of Leicester in Britain calculated that the bomb Willis and his crew planted after drilling into the asteroid would have needed at least 50 billion megatons of kinetic energy in order to blow apart a Texas-sized NEO. For the sake of comparison that's a billion times more powerful than the biggest nuclear bomb ever built, the Soviet Union's Tsar Bomba.[47]

NASA also wouldn't send astronauts—let alone a team of untrained oil rig drillers—to intercept an incoming asteroid. Any mission would be unmanned. But it is possible, as a last-ditch effort should a large NEO be discovered with little time to spare, that NASA might take a page from the Michael Bay playbook and try to plant a nuclear bomb inside an asteroid. The agency has studied using a Hypervelocity Asteroid Intercept Vehicle, a theoretical spacecraft that would crash into an oncoming asteroid at high speeds, burrowing several feet deep into the object, before setting off a nuclear device.[48] Detonating a nuclear warhead below the surface of an asteroid, rather than on it, could increase the explosive power by as much as twentyfold. It's worth noting that one of the reasons the U.S. National Nuclear Safety Administration gave for not dismantling America's largest atomic warheads after the Cold War was the possibility that they might be required for planetary defense.[49]

So we have theories for how we might protect our planet in the event of an NEO strike. But what we don't have—yet—is an organized and practiced strategy for that defense. Creating and executing that plan is largely the job of one man: Lindley Johnson, NASA's first planetary defense officer.

In 1994, around the time that the Shoemakers and David Levy were searching for NEOs at the Palomar Observatory, Lindley Johnson was a major in the U.S. Air Force, studying advanced space operations at the Air Command and Staff College in Montgomery, Alabama. Johnson was well versed in threats from above. He had already served at the North American Aerospace Defense Command (NORAD), the U.S.-Canadian military unit that provides surveillance against incoming ballistic missiles, as well as the Space and Missile Systems Center in Los Angeles. Unusually for the time, and especially for a military officer, Johnson was already convinced of the danger from NEOs. Shortly after the Shoemaker-Levy 9 comet was discovered in March 1994—but, crucially, before astronomers watched it collide with Jupiter, inspiring the first sustained efforts at planetary defense—Johnson published a paper describing early theories about how to deflect an NEO on a collision course with the Earth. It was met with a shrug—few people at the time, especially in the government, believed that an object from space could threaten our planet.

The Shoemaker-Levy 9 collision in July 1994 changed the course of Johnson's career. He was assigned to the Air Force's Space Command, where he pushed for collaboration with NASA, including on projects that employed space surveillance telescopes to search for asteroids and comets. After retiring from the military in 2003, Johnson went to work for NASA. For years he was the space agency's only senior manager working on NEOs, but he helped push for a steady increase in the asteroid-hunting budget. In 2015 NASA created the Planetary Defense Coordination Office and put Johnson in charge of, effectively, defending the Earth.

Planetary Defense Officer is a very grand title for someone whose work by his own admission mostly consists of interdepartmental coordination. After arriving at NASA headquarters in Washington each morning, Johnson checks what he calls "the night's catch"—the collection of NEOs that have been discovered overnight by observatories like the Catalina Sky Survey and cataloged by the Minor

Planets Center. He asks questions: Do any of these NEOs fall into the category of potentially hazardous, meaning they could pass within 5 million miles of the Earth? Have astronomers calculated any new asteroid approaches that could threaten the planet? Imagine it as the President's Daily Briefing, only the threats are space rocks, not terrorists and hostile governments.

If something does pose a potential impact risk to the Earth, it would be Johnson's responsibility to push the alert up through NASA, to the White House Office of Science and Technology and eventually to the president. That's only happened once so far during Johnson's time with NASA, in 2008, when astronomers correctly predicted that a small asteroid would reach the Earth and largely burn up before impact.[50] It marked the first time scientists had managed to success-fully forecast an impact event in advance of impact. Johnson made the prediction public, mostly so the sight of a sudden flash in the sky wouldn't be misinterpreted as something military. "We didn't want to get anyone too excited about a big explosion in the atmosphere," he told me.

More recently, Johnson and his colleagues took advantage of a near-miss by a house-sized asteroid on October 12, 2017, to test out the Earth's nascent NEO warning system. The asteroid never posed a direct threat, but it did come within 28,000 miles of the planet—as close to the Earth as man-made satellites—which made it perfect as a dry run. "We picked it up, tracked it, got a high-precision predic-tion of the close approach, and then we used that to exercise our notification system to the U.S. government and our international partners," Johnson said.

Johnson and his team have found that while asteroids and comets may be more predictable than other natural threats, they still have an element of ambiguity. Even if scientists suspect that a newly discovered NEO could be on a future impact course with

Earth, they won't be certain—at least not at first, especially if a potential collision is years or even decades away. As asteroids whiz close to the Earth, their orbit will be tweaked by the planet's strong gravitational field. Should an asteroid travel through a precise range of altitudes called a "keyhole" during a first pass by Earth, its path can be changed in such a way that it will be all but guaranteed to impact the planet during its next orbital go-round. (Remember that asteroids, like the Earth, orbit the sun, so one that passes by will come around again.)

Scientists, though, can only estimate the chances an asteroid will hit that precise keyhole—hence the slight uncertainty around a future impact. Shortly before Christmas 2004, a 1,200-foot-wide asteroid known as 99942 Apophis—named after the Egyptian snake god symbolizing chaos—was initially calculated to have a 2.7 percent probability of hitting the Earth on April 13, 2029.[51] That probability might seem minuscule, but ask yourself this: would you get on a plane if there were a 2.7 percent chance it would explode in midair? Fortunately scores of additional observations helped clarify Apophis's position, and astronomers concluded that there was zero chance of it hitting the Earth on its 2029 or 2036 go-round. In fact, the chance of Apophis hitting the Earth anytime over the next century is a microscopic 0.00089 percent[52] as of early 2019—though that is higher than your chance of dying in a plane crash.[53]

Should it strike the Earth, an asteroid the size of Apophis would hit with the power of more than a billion tons of TNT[54] and lay waste to an entire state, if not much more. But it would still take years of additional observation before astronomers could be more certain whether or not an impact was going to occur. Even once an impact is determined to be likely, astronomers would produce not a bull's-eye on the planet, but rather a "risk corridor"—a narrow band stretching across the Earth that marks the territory where an NEO is predicted to land. There's an inevitable trade-off in asteroid deflection—the more time astronomers are given to observe an asteroid, the more precise impact probability and any risk corridor will be, but that leaves less time for a deflection effort, which would almost certainly take a decade or longer to plan and complete.

And who, ultimately, would make the decision about whether to give the green light to such a mission, one that would cost billions of dollars and might not even work? Large NEOs may be the definition of a global threat, but NASA's budget is greater than that of all other space-faring countries combined,[55] and only the United States has the experience and the technology to lead a deflection mission. In the event of an incoming asteroid, the president would likely put together a Planetary Impact Emergency Response Team, which would include NASA, the Department of Defense, the Department of Energy (which oversees the U.S. nuclear arsenal), and FEMA, among others. Internationally, there are other space-capable nations like Russia or China that could convene under the auspices of the United Nations Committee on the Peaceful Uses of Outer Space, the closest thing the world has to a legal body for space law.[56]

One might assume that if a major asteroid were discovered well on its way toward our planet, we'd come together as a species and do whatever it takes to defend ourselves, up to and including mobilizing our most powerful nuclear weapons. But the 1967 Outer Space Treaty explicitly bans the use of atomic weapons in space, presenting a legal obstacle. And consider this—in the early 1990s, the astronomer Carl Sagan identified what he called the "deflection dilemma." Any technology that would be powerful enough to alter the course of an asteroid away from the planet could theoretically be used to direct an asteroid *toward* a target.[57] Like an enemy country, for example.

That might seem paranoid. What leader would play geopolitical games at a moment of existential peril? But in a world where international trust is evaporating and science itself is no longer believed, how much trust would the rest of the world have in the word of the U.S. president—especially the current one—that an asteroid deflection mission using untested technology would truly be peaceful, or successful? Even a deflection mission launched with the best of intentions could inadvertently nudge an incoming asteroid away from one country and toward another. Like physicians, planetary defenders should first do no harm, but the pressure to do something, anything, in the event of an incoming asteroid would surely be intense.

The quickest way to instill global confidence in our ability to deflect an asteroid when the pressure is on would be to deflect an asteroid when the pressure is off. Practice makes perfect. That's why NASA and its partners at the European Space Agency (ESA) have plans to launch the Double Asteroid Redirection Test (DART) spacecraft in 2022. Its target will be the binary asteroid Didymos, which came within 4.5 million miles of the Earth during an orbital pass in 2003. Didymos means "twin" in Greek—one larger asteroid, Didymos A, is orbited by a smaller rock called Didymos B.[58] The DART probe, which is about the size of a refrigerator, will impact Didymos B in an attempt to alter its trajectory. The mission will mark the first time that earthlings have purposefully tried to modify the orbit of a heavenly body, and it could serve as proof of concept for a future deflection mission.

NASA deserves some credit for the steps it has taken on planetary defense in recent years, including the creation of Johnson's job. Before the Planetary Defense Coordination Office was established in 2015, responsibility for asteroid hunting and defense was scattered around a number of departments—a lack of focus NASA's own inspector-general criticized the agency for in a damning 2014 report.[59] In September 2018, the White House announced a plan to nearly triple spending on planetary defense to $150 million in 2019, with much of the money earmarked for the DART mission. That additional funding needs to be approved by Congress,[60] although Americans seem to support it—a 2018 Pew poll found that protecting the planet should be the number two priority of the U.S. space program, after researching climate change, another existential threat.[61]

"In the order of things that people should be worried about, [NEOs] isn't highest on the list," Johnson told me. "But it does have the potential to be the most devastating natural disaster known to man. All the money that we will have spent on it would have been worthwhile if it prevents an event that could take hundreds of billions of dollars to recover from—if we are even able to recover. It's definitely worth governments spending a bit of their treasure to find these things ahead of time, because you can't do anything unless you find them."

But what if we can't find them?

On February 15, 2013, Lindley Johnson was in Vienna, preparing to share an asteroid defense plan with the United Nations Committee on the Peaceful Uses of Outer Space. The timing was perfect. That day an asteroid designated DA14 came within just 17,000 miles of the Earth. NASA and its partner observatories had been tracking the asteroid for a year, and knew with 100 percent certainty that it would pass near but not threaten the Earth. It was set to be the ideal demonstration of how smart NEO surveillance can keep a dangerous asteroid from taking us by surprise.

Except that isn't how we'll remember February 15, 2013. At 9:20 a.m. local time that same day, a different asteroid, some 65 feet across, penetrated the Earth's atmosphere over Russia. Speeding at more than 40,000 mph, the now meteor exploded in an airburst some four-teen miles above Chelyabinsk Oblast, a rugged, industrial region in the Ural Mountains known for its military and nuclear industries.

The light from the explosion—captured in dash cams and home videos—burst yellow, and then orange, and briefly glowed brighter than the sun, with a force estimated at 400 to 500 kilotons of TNT.[62] Some 1,500 people were injured in the blast, mostly from glass shattered by the shock wave, and more than 7,200 buildings across the region were damaged. People were blown off their feet, and in the aftermath the air smelled of gunpowder and sulfur; monitoring stations as far away as Antarctica could detect the detonation. "I opened the window from surprise—there was such heat coming in, as if it were summer in the yard," wrote one witness. "In several seconds there was an explosion of such force that the window flew in along with its frame, the monitor fell and everything that was on the desk."[63] Later analyses showed that the Chelyabinsk meteor was the largest natural object to enter the Earth's atmosphere since the Tunguska event, more than a century earlier. And scientists didn't have a clue the asteroid was coming until it was too late.

This is the dirty secret of asteroid hunting—while we've managed to locate more than 95 percent of the NEOs that are larger than 1 kilometer, it's proven much more difficult to find the far more numerous asteroids below that size. Recall that in 2005 Congress charged NASA with locating 90 percent of the NEOs larger than 140 meters by 2020. As of 2018, NASA and its partners had found about one-third of the estimated 24,000 or so NEOs in the target class. As for meeting the deadline, Johnson told me, "That ain't going to happen."

Chelyabinsk was a violent reminder for those in the NEO community that space can still throw things at us that we can't see coming. The asteroid approached from the direction of the sun, so it couldn't be spotted by any Earth-based telescope. (Point a powerful telescope in the direction of the sun and its expensive innards will melt.) Its size was well below even the 140-meter cutoff that NASA is tasked with finding. Although observatories like Catalina have found and cataloged about 500 asteroids of comparable size, astronomers estimate there are probably *millions* of Chelyabinsk-class NEOs out there, waiting to be discovered.

Chelyabinsk also showed that it doesn't take a large asteroid, or even one that makes it all the way to the ground, to cause panic and destruction—especially if it strikes without warning. Chelyabinsk Oblast is home to one of the biggest nuclear processing facilities in Russia. I visited the region in 2007 while reporting a story for *Time* magazine on Russia's environmental woes. (The Mayak processing facility in Chelyabinsk Oblast was the site in 1957 of one of the worst nuclear accidents on record, though the full extent of the damage was long kept secret by the Soviet Union.) Even in 2007, when relations between the United States and Russia were less strained, police pulled over my car to ask what a foreign reporter and photographer were doing nosing around a militarily sensitive area. The airburst from the 2013 meteor could have easily been mistaken for a nuclear strike by the United States, which was indeed the first reaction of many witnesses on the ground. Nuclear tensions between Washington and Moscow are even higher now than they were in 2013. It's not difficult to imagine—but horrifying to picture—what a knee-jerk Russian reaction to a seeming nuclear attack could have led to.

The Chelyabinsk meteor proved there are still holes in NEO surveillance, as did news that a meteor exploded above Russia's Kamchatka peninsula in December 2018 with the force of ten Hiroshima bombs. But there are new tools coming online soon, most notably the Large Synoptic Survey Telescope (LSST), a wide-field telescope under construction in the mountains of northern Chile. Set to begin full operations in 2022, the LSST will be able to photograph the entire available sky every few nights.[64] That should help multiply the rate of NEO discovery as much as tenfold.[65]

Asteroid hunting will also benefit from advances in artificial intelligence, as machines get better at picking out actual NEOs from the background visual noise. As I found on my visit to the Mount Lemmon Observatory, machines have a tendency to produce false positives, which is why human observers are still needed to screen the initial results. But NASA's Frontier Development Lab has brought together AI experts and astronomers to trawl the enormous data sets generated by observatories and develop algorithms that can rapidly pick out potentially hazardous asteroids.[66] In 2017, researchers from Google and the University of Texas fed data from NASA's Kepler space telescope into a machine-learning algorithm and found two new planets that had been previously missed.[67] The algorithm was correct 96 percent of the time—a batting average any asteroid hunter would love to have. And AI doesn't need coffee.

What could really turbocharge the NEO search would be a space-based, infrared telescope dedicated to planetary surveillance. Orbital telescopes can run day and night, and they aren't subject to bad weather. Because NEOs are warmer than other space bodies, they're easier to locate through an infrared telescope that hunts for heat rather than an optical telescope that uses visible light. "We've had several studies over the years, and they come back with the same answer," Johnson told me. "We really need to have a space-based capability that is able to get above the Earth's atmosphere so that we can hunt for them twenty-four seven."

The downside is cost. Space is not cheap. It's expensive to get objects successfully off the planet, and it's expensive to keep them running when every repair mission requires an orbital rendezvous. NASA has a project on the books for a space-based infrared telescope called NEOCam, but as of early 2019 it remains in what the agency calls "extended phase A study," hovering between budgetary life and death.[68] Without the aid of space-based telescopes like NEOCam, NASA says it could take thirty years or more to meet its congressional mandate to find 90 percent of the NEOs above 140 meters[69]—a mandate, it should be noted, that came with no extra funding, just expectations. That means decades when the Earth could be surprised by the next Chelyabinsk—or something much worse.

───────────────

When I began working on this book I knew I wouldn't be satisfied with merely investigating the ways our world might end. I wanted to determine what we can do to protect ourselves, which policies and which priorities need to be put into place to give human beings the best chance of making it to the next century and beyond. These aren't easy questions—the world has no shortage of needs, many of them far more immediate than the remote chance of a species-ending catastrophe. That's why existential risks tend to be overlooked and underfunded. Which is understandable—until the day that delay becomes fatal.

So how much should we spend and how hard should we work to try to avert a disaster that would be as catastrophic as it is unlikely? This is where asteroids provide a useful starting point for breaking down the tricky math of existential risk. The true probability of most other existential threats is unknown and unknowable. Scientists have no way of accurately forecasting the chances of an AI uprising, epidemiologists can't foresee a pandemic before the first cough, and no one knows if aliens even exist, let alone if they want to turn Earth into their latest colony. But if they can track all the asteroids out there, the very smart astronomers and physicists dedicated to planetary defense can predict with near-clockwork precision the probability of a too-close encounter with an NEO over the next century, or even longer. And that predictability means that we can make a case—not merely invent one—for spending much more to protect the planet.

Impacts from asteroids larger than about three miles across could plausibly lead to human extinction. A hit from an asteroid that big is predicted to occur once every 20 million years, which translates to a 0.000005 percent chance of it happening in any given year. This is lower than your chance of dying in a lightning strike, which amounts to 0.000009 percent.[70] Yet people do die from lightning strikes—sixteen Americans were that unlucky in 2017[71]—while there was not a single recorded human death from an asteroid impact last year, the year before, or, for that matter, all of recorded human history.[72] This raises the question: if we've made it this long on Earth without a single asteroid-related fatality, why spend a single dollar on NEO defense?

This is where we need to understand what makes existential risks different from ordinary risks. Decades of experience have told us that on average lightning will kill a couple dozens of Americans a year at most—but we also know that there's a zero percent chance that some kind of mega-lightning strike will zap everyone on Earth and bring an end to the human race. That's how it works for conventional risks. The number of people who die each year from airplane crashes or hurricanes or from accidents while taking selfies (dozens annually in that last category, according to scientists[73]) can rise or fall based on external factors or the preventative measures we might take, but there won't be one year when the entire population of the world dies in a plane crash. These risks—the risks we navigate every day—are finite.

Existential risks, by contrast, are highly improbable but carry the threat of extreme and even infinite consequences, which can skew the numbers in unexpected ways. The National Research Council has estimated that an average of 91 people a year will die in an asteroid strike, but of course those 91 people did not die last year and it's unlikely that 91 people will be killed by an asteroid next year.[74] Rather, it means that at some point—unless we develop an impregnable asteroid defense system—we can expect that millions or even billions of people will die in one major impact event. Average that global death toll over the very long period of time that's likely to pass before that asteroid strike, and you get 91 deaths a year. That's more than the number of people who died globally in airliner crashes in 2017[75]—and that same year the Federal Aviation Administration alone spent nearly $2 billion on aviation safety.[76]

This is the mathematics of end times. It may seem impossible to put a dollar figure on human extinction, if for no other reason than that if the world ended, there'd be no one left to pay the bill. That plays havoc on the conventional cost-benefit analyses economists use to determine whether a given policy is worth its price tag, but we can at least attempt to approximate the value of avoiding extinction. The authors of a report by the Global Challenges Foundation, a Swedish nonprofit that raises awareness of existential risks, estimated that if human beings could protect themselves from global catastrophic threats like asteroids, the species could expect to survive for 50 million years. That would be enough time for 3 *quadrillion* future humans, or three thousand trillion people. If each of those lives, and the lives of the 7.6 billion people on Earth today, were valued at the bargain-basement price of $50,000—far stingier than the price the U.S. government puts on a single American life when judging the impact of regulations—the cost of a total extinction event would be $150 quintillion, nearly two million times larger than the current value of all the money in the world.[77]

As the jurist and economist Richard Posner points out in his book *Catastrophe: Risk and Response*, one way to try to determine how much should be spent to avert a risk is to multiply its expected cost by its probability.[78] Remember that 0.000005 percent chance of an extinction-level asteroid strike happening in a given year? As minuscule as those odds are, the $150 quintillion cost of extinction is so high that the math suggests we should be spending $750 *trillion* a year to avert NEO strikes. That would be a thousand times more than the United States currently spends in total on defense. It's more than nine times the value of the global economy. It is a *lot* of money.

It would be insane, not to mention impossible, to spend that much money on NEO defense. There are other risks that demand attention and dollars, including existential risks we'll learn about later in the book. Even NASA, which spent nearly $150 billion in today's dollars to put men on the moon, would run out of ideas well before it could spend anything close to three-quarters of a quadrillion dollars on shooting down asteroids. But what this thought experiment does tell us is that spending $50 million a year, as NASA has been doing, or even $150 million, is almost certainly too little to keep the Earth safe from a threat that has ended life on this planet before, and which will do so again.

What's true of asteroids is true of existential risks more generally. The cost of losing the world to any cause, however unlikely, is so great that we should be spending far more of our money and effort to offset those risks however we can. But as I discovered again and again while working on this book, we're held back less by what we can budget than by what we can imagine. Human beings are terrible at evaluating risk—especially existential risk. We rely on feeling rather than fact, and privilege emotional memories over hard numbers. "We treat something as impossible unless there is an experiential aspect to it, and no one has experienced an asteroid strike," said Paul Slovic, a professor of psychology at the University of Oregon and an expert in risk perception. "The human tendency is to take that small probability and sweep it to zero." That's how we end up ignoring risks that could wipe us off the face of the planet. Not because we're making a reasoned decision to spend money on one need over another, but because we're not being reasonable at all.

That's an understandable tendency. It's also one that just may get us all killed, unless we're brave enough to come face-to-face with the end of the world. The universe may be trying to kill us, but that doesn't mean we have to let it.

2

VOLCANO

A Decade Without a Summer

Seventy-four thousand years ago, give or take a few millennia, *Homo sapiens* had a very bad day, perhaps the worst day that we've ever experienced. On what is now the Indonesian island of Sumatra, a mountain called Toba exploded—though the word doesn't do justice to the act of sheer geophysical violence perpetrated that day. What happened to Toba was so destructive that scientists would eventually coin a term for volcanic disasters of its scale: a supereruption.[1] But the eruption was only the beginning. Toba's aftereffects dimmed the sun and draped a volcanic winter around the world. It might have brought our species closer to the brink of extinction than we have ever been before or since. At a moment when *Homo sapiens* was far from the world-dominating force we are today, Toba was our ultimate trial. It was also a warning—the most dangerous natural existential risk we face comes not from the skies above, but from the ground beneath our feet.

As a journalist with *Time* in Hong Kong, I helped report on the aftermath of the 2004 Indian Ocean tsunami. More than 230,000 people, in fourteen different countries, died in the disaster, which was triggered by a 9.1-magnitude earthquake that ripped through the bed of the Indian Ocean off the coast of that same Indonesian island of Sumatra. The blunt power of the quake created a rupture 600 miles long; the scale of human death and injury was so great it extended across an ocean. I didn't think the Earth was capable of anything worse. But as I researched the Toba supereruption, I learned that catastrophe had categories beyond my imagination.

A volcano is a mechanism to transport what is under the ground into the air, and so eruptions are judged first on the sheer amount of rock they emit. As much as 9 million tons of sulfuric rock and dust *per second* erupted from the Toba volcano over what scientists estimate was about two weeks of devastation. By the time the eruption had finished, Toba had spewed the equivalent of as much as 700 cubic miles of volcanic ash and magma into the air.[2] Collect it all in one place and there would be enough to load the Grand Canyon nearly three-quarters full.

If that doesn't help you imagine the scale of destruction, picture Mount St. Helens in Washington State. Its eruption in 1980 was the largest and most destructive in U.S. history, killing fifty-seven people and causing a billion dollars in damages. It was captured on video and even made the cover of *Time* magazine. Yet on the basis of the amount of volcanic tephra—rock, ash, and tiny microscopic pieces of glass with a telltale hook shape—that was emitted in the eruption, Toba was the equivalent of *2,800* Mount St. Helens eruptions, or one a day for more than seven and a half years.[3]

Closer to Toba's eruption site on Sumatra, a dense mass of molten rock and gas—some of it as hot as 1,200 degrees, with boulders stampeding as fast as 60 mph—flowed for hundreds of miles. Anything that could burn caught fire, including plants or animals.[4] The eruption pulverized what had been a mountain and carved out a caldera—a volcanic crater—that today measures more than 60 miles by 18 miles, with a depth of a third of a mile.[5] That caldera would eventually fill with water, forming the lake that now bears Toba's name.

Eruptions are ranked through the Volcanic Explosivity Index (VEI), a scale that runs from 1 at the lowest to 8 at the highest. Toba was far and away an 8, a level technically described as "mega-colossal." The eruption cloud reached as high as nineteen miles above the surface, well into the stratosphere.[6] The noonday sun would have gone dark. "Toba was undoubtedly the biggest eruption the planet has seen over the last one hundred thousand years, if not far longer," Steve Sparks, a volcanologist at University of Bristol in the United Kingdom, told me.

Toba made its immediate surroundings hell on Earth, but its significance for the story of existential risk came in the effects felt thousands of miles away. Eruptions produce ash with a unique chemical signature, a volcanic fingerprint that allows scientists to trace Toba's legacy around the globe. Beginning in the 1990s, researchers found evidence of large Toba ash deposits scattered in marine sediments in the Indian Ocean. Further studies found ash in the South China Sea, the Arabian Sea, even in Lake Malawi in southeastern Africa, more than 4,000 miles away from the eruption site.[7] In India and Pakistan ash drifts accumulated as deep as 20 feet,[8] and today researchers estimate that nearly 1 percent of the planet's surface would have been covered by Toba ash.[9]

Volcanic ash is toxic in its own right—breathing ash can cause lung impairment and lasting respiratory damage, and it can poison pasture and cropland. But the climatic effects of the Toba ash cloud were far more devastating. The ash and dust blown into the stratosphere darkened the sky. While dust particles are usually washed out of the sky within weeks by rainfall, the eruption cloud also contained huge concentrations of sulfur dioxide—just as the debris cloud made by the Chicxulub asteroid impact did. The SO_2 combined with water in the stratosphere to generate sulfuric acid droplets, producing a lingering haze that could have reduced the amount of incoming sunlight by as much as 90 percent.[10]

We know from direct experience that major volcanic eruptions can trigger temporary global cooling by blunting sunlight. After Mount Pinatubo in the Philippines blew in 1991, the resulting clouds of sulfur aerosols reduced global temperatures by nearly one degree Fahrenheit over the next couple of years.[11] Pinatubo was one of the strongest eruptions in the twentieth century, but it only scored a 6 on the VEI scale, which, like the Richter scale for earthquakes, grows logarithmically—an increase of 1 on the index equates to 10 times as much ash and rock. That means Pinatubo—which had a measurable cooling effect on the planet's climate, enough to temporarily offset global warming—was only about 1 percent as powerful as Toba likely was.

Just as paleontologists can reconstruct extinct animals by searching the fossil record, paleoclimatologists can reconstruct past climates by sampling bubbles of air trapped tens of thousands of years ago in Arctic ice. Ice cores and other geologic evidence recovered from Greenland indicate a sudden and lasting drop in global temperatures around the time of Toba, with the eruption as a prime suspect. It's not conclusive—over the past few million years, the Earth has swung between ice ages and warmer periods, and around the time of Toba the planet was entering into a prolonged glaciation.[12] But ice ages take thousands of years to develop, and what could have happened after Toba would have been immediate and drastic.

In a 2009 computer simulation of Toba's climate effects, Alan Robock of Rutgers University estimated that global temperatures would have fallen by 18 degrees on average for several years after the eruption,[13] and potentially by as much as 30 degrees. Precipitation would have been fallen by 45 percent, and vegetation cover would have shrunk dramatically, with broadleaf evergreen trees and tropical deciduous trees dying out.[14] Imagine a winter that lasted for years, like something out of *Game of Thrones*, shriveling life on land. According to Michael Rampino, a geologist at New York University, up to three-quarters of plant species in the Northern Hemisphere might have perished after Toba, along with countless animals.[15]

For our hunter-gatherer ancestors, who at the time were still mostly concentrated in Africa and parts of southern Asia, what would it have felt like to live through Toba's volcanic winter? Imagine waking up day after day, year after year, the dim sun obscured behind a film of sulfate aerosols that never lifted. That 18-degree drop in temperatures would have felt like moving from New York City to Anchorage, Alaska. Warmth would become nothing more than a memory, as the plants and animals you depended on for survival died out in the lingering chill. The weakest members of your community—the sick, the old, and the very young—would follow. Though climate models indicate that temperatures returned to close to normal within a decade, it must have felt in that prolonged darkness and cold like the end times had come.

They nearly did. In the 1990s, Rampino and Stanley Ambrose, an anthropologist at the University of Illinois, as well as the science journalist Ann Gibbons, put forward a theory that the effects of the supereruption might have indirectly reduced the total human population at the time to as few as four thousand people. This was an existential risk that actually played out. "Toba might have caused such dire environmental conditions that it would have led to a die-off," said Rampino. "It's a reminder that we are at the mercy of geology."

As the existential risk scholars Nick Bostrom and Milan Ćirković write in the introduction to the book *Global Catastrophic Risks*, Toba "is perhaps the worst disaster that has ever befallen the human species."[16] It is likely the closest we have ever come to extinction before or since.

We know that the population of *Homo sapiens* went through a bottleneck around the same general time as the Toba eruption because we can read the evidence in our genes. Unlike most of our fellow primates, and even other mammals, human beings across races and nations show an unusual degree of genetic uniformity. The science writer Charles Mann has noted that two human beings from two different parts of the world might share all but 0.1 percent of their DNA. Compare that to two different strains of the simple bacteria *E. coli*, which might share only 95 percent of their genes,[17] even though their genome is perhaps a quarter the size of the human genome. That genetic similarity—along with evidence that human populations suddenly exploded about fifty thousand years ago[18]—indicates that contemporary human beings descend from a relatively small group of ancestors. Rampino and Ambrose believe that the extreme if short-term climate change caused by the Toba supereruption left just a small group of survivors—our forebears.

We don't know how they made it, though Ambrose believes communities that were able to work together cooperatively had a better chance of survival. "You might think that in an apocalyptic event that people would be stealing from each other," Ambrose told me. "It's not true. In a stable environment, when population density is high, you don't need to rely on your neighbors to get the next meal, and you may even need to defend yourself from them. But in a situation where it's the equivalent of a small lifeboat and everyone needs to cooperate, selfish people will be weeded out. You end up with a population that is more sharing and caring."

We're lucky that the survivors pulled together. You and I and everyone we know—those who came before us and everyone who might come after us—are here because the human beings who lived through Toba found a way to survive the eruption and its long, cold aftermath. Without their resourcefulness, the human story would have ended in its earliest chapters. Extinction was a possibility.

In fact extinction has been the rule over the 3.8 billion years that life has existed on Earth, not the exception. Some species go extinct due to disease, some due to predators, some through the process of evolution, as one species gradually gives way to a successor better suited to the environment. This is what scientists call the background rate of extinction, which amounts to an estimated one to five species per year dying out[19]—biological business as usual. But five times over the course of the planet's history, life has been washed away by mass extinction events, waves of death that resulted in the loss of half or more of the species existing on Earth at the time. These were the actual end times, played on repeat.

The demise of the dinosaurs from the Chicxulub asteroid is now so well known that when we imagine a mass killer, we picture a flaming rock screaming across the sky. Over the course of the Earth's history, however, volcanoes have been the cause of far more extinctions than anything from space. That may even include the dinosaurs as well—some scientists believe that the Chicxulub impact might have accelerated the ongoing eruptions of sprawling volcanoes called the Deccan Traps, in what is now India, and that the two catastrophes would have combined to cause the icy global cooling that ultimately killed off life.[20] But we know for certain that the most severe mass extinction event the Earth has ever experienced—one where an estimated 75 percent of life on land and 90 percent of life in the ocean went extinct, an event so terrible that it has been called the "Great Dying"—was due to volcanic activity.[21] And not just any volcanoes—these were eruptions that make Toba look like a popgun.

The Siberian Traps are now large plateaus and rolling grassland that cover a broad expanse of northern Russia. But beneath the

grass is roughly a million square miles of hardened basalt rock, the
cooled remains of an eruption that turned the Earth inside out. About
300,000 years before the start of the Great Dying, colossal amounts of
lava began pouring through the surface, enough to cover the continental United States in magma half a mile deep.[22] Unlike Toba, though,
this wasn't a single powerful explosion, but rather a flow—the technical term is "Large Igneous Province," or LIP. Seven hundred thousand
cubic miles of lava flowed for hundreds of thousands of years. When
the magma running underground reached a massive coal basin in
modern Siberia, it sparked a raging inferno that released vast amounts
of carbon into the atmosphere. That tremendous pulse of greenhouse
gases pushed climate change into overdrive, particularly in the oceans,
where temperatures rose by as much as 18 degrees.[23]

Life simply couldn't adjust. In a blink of geologic time, Earth
went from hospitable to hostile. As the science writer Peter Brannen
tells us in his excellent book *The Ends of the World*, the fossil record
went silent. And it took 10 million years for life on Earth to begin
to recover.[24]

If Toba marked the moment when humanity was nearly driven to
extinction, the Great Dying nearly ended the story of life altogether.
And both began with a volcano—a reminder that, as the historian
Will Durant wrote, "Civilization exists by geological consent, subject
to change without notice." No natural force on Earth puts humans
at greater existential peril than a supervolcano.

There's almost no chance that the Earth will experience another LIP
like what caused the Great Dying—at least not over the next 10–100
million years. Supervolcanoes like Toba, however, are known to erupt
more frequently. Toba isn't even the planet's most recent supereruption. About 26,500 years ago, the Taupo volcano in New Zealand hit

8 on the VEI scale, spewing 260 cubic miles of volcanic material into the air,[25] making it a little more than a third the size of Toba.

About 50,000 years separate the Toba and Taupo eruptions. That's almost eternity on a human scale. It's long enough to rerun the entire history of civilization multiple times over. But geologic time runs at a very different speed than human time. To geologists, 50,000 years ago is like yesterday. Or to be more precise, if you think of the entire history of the planet as a single day, it's just a second ago.

Geologists can comprehend the endless sweep of Earth's multibillion-year history—but most of us are not geologists. We are confined to the brief human time horizons of our own experience. What happened in the deep past or what could happen in the far future has little bearing on our lives as we live them, the seven or eight or nine decades we count ourselves lucky to get. It's as if we are born, grow up, and die all on the same spot, ignorant of the vastness that surrounds us. "Think about it in terms of a mayfly that lives and dies in a single afternoon in a creek on the mountains," said Jake Lowenstern, a geologist with the U.S. Geological Survey. "A mayfly doesn't really believe in nighttime or hailstorms or winter or ice. They never see anything but light and water, so why would they believe in anything else?

"We're the same way."

Just because the mayfly doesn't live to see hailstorms or winter doesn't mean they don't exist, and just because we've never experienced a supereruption doesn't mean one couldn't strike in our future. Humans aren't mayflies—through the practice of science, we have the ability to extend ourselves into the deepest corners of the past and the farthest reaches of the galaxy. But human psychology is a very different matter. Our species has been fortunate enough over the past ten thousand or so years to develop in a stable environment that—despite the usual earthquakes and hurricanes and droughts and floods—hasn't been marred by a megacatastrophe that might have set back civilization to zero, or even threatened human extinction. Our inability to see beyond the narrow boundaries of human history—the availability heuristic I mentioned earlier—means that we suffer from a blind spot that can leave us vulnerable to an array of extremely rare but extremely lethal existential risks, including supervolcanoes.

We assume that our experience is normal, that yesterday was like today and that tomorrow will be the same. But there's nothing special about our experience so far, either as individuals or as a species. Over the 4.5-billion-year life span of our planet, our experience is just an infinitesimal and random sampling, no more representative of the Earth's full potential than a few snatched seconds would be of your entire life. So to protect ourselves from supervolcanoes and other existential threats, we need to understand that our experience isn't normal—and we need to see beyond it. We need to see like a geologist.

So how vulnerable are we today to a supereruption? It helps first to understand the checkered history of the word "supervolcano." It is not exactly a scientific term. "Supervolcano" wasn't used in a geology publication until 2002, in an article by R. B. Trombley, who duped people into believing he was a trained geologist and could predict when volcanoes would erupt. (Trombley was uncovered as a fraud after the eruption of the Icelandic volcano of Eyjafjallajökull in 2010.) The term took off in 2005 after a BBC/Discovery docudrama about a future eruption at Yellowstone National Park called, of course, *Supervolcano*. The film and the title proved popular enough that geologists were basically forced to begin using it whether they wanted to or not. Professional volcanologists still prefer "supereruption" to "supervolcano," though in practice they mean the same thing.

There are about twenty known supervolcanoes on Earth,[26] and if any of them were to erupt at full power, the effects would be spectacular, and spectacularly destructive. The authors of a 2005 report for the Geological Society of London put it in perspective this way: if a supervolcano were to erupt in London's central Trafalgar Square, it would produce at least enough magma to bury all of Greater London and everything in it to a depth of 700 feet.[27] But what really sets a supervolcano apart from anything else the Earth can throw at us— tsunamis, earthquakes, hurricanes—is that they are the only natural disaster that can truly go global. What happened after Toba could happen again—ash and sulfur from a supereruption would reach the stratosphere, and depending on the location of the volcano and its eruptive strength, it could spread around the globe, blocking incoming sunlight and casting the world into winter. A supervolcanic eruption in Trafalgar Square would be very bad news for London, but the entire world would feel it.

Because the last supervolcano blew tens of thousands of years ago, scientists employ the tools of paleoclimatology and computer modelling to reconstruct what might have happened in the aftermath of a supereruption. But those models are imprecise, which leaves room for debate—including about Toba. While Alan Robock's models projected double-digit temperature declines after Toba, Claudia Timmreck of the Max-Planck Institute for Meteorology in Germany argued in a 2010 study that cooling from the supereruption might have been much less, perhaps just five or six degrees on average.[28] That would have still made for abnormally cold weather— roughly the difference between a winter in Boston and one in Washington, D.C.—but Timmreck concluded that the cooling would not have been "large enough to severely affect the species survival of modern humans."

Scientists have also tried to piece together archaeological evidence from around the time of the Toba eruption, in hopes of better understanding how the aftermath affected humans. In a 2018 study published in *Nature,* researchers excavated a cliff near the town of Mossel Bay on South Africa's southern coast and discovered a layer of microscopic glass shards called cryptotephra that could be traced back to Toba.[29] That was clear evidence that ash from the supervolcano had reached as far as southern Africa, more than 5,400 miles from Toba in Indonesia. Just as important were the bones, tools, and other signs of human activity found above and below the cryptotephra layer. In this settlement at least, prehistoric humans seemed largely unaffected by the distant volcano. In fact, signs of human occupation in the settlement seemed to grow in the years following the supereruption.

In villages in southern India, the University of Cambridge archaeologist Sacha Jones found stone tools above and below a layer of Toba cryptotephra—more evidence that human settlement in the region hadn't been overly disrupted by the blast. (Jones told me she discovered the Toba deposits when she found out that Indian villagers were mining and selling the ash—which she described as "bright white, gritty, and abrasive"—as a skin exfoliant.)

The new research doesn't necessarily mean that our ancient ancestors simply shrugged off Toba. Settlements like the kind Jones found in India may have served as refugia, safe zones where animals—including human beings—ride out environmental catastrophes. But conditions beyond those refugia might have been much harsher. "We will find that Toba had very varied effects on different regions of the world, and that they could be especially negative if the population was quite vulnerable to begin with," Jones told me. "If you're doing well, then maybe you're fine. But if you're vulnerable, if your resilience is not so good, you're in trouble." Any attempt to reconstruct a world 74,000 years gone will inevitably be imprecise, whether the tools employed are climate computer models or the shovels and spades of an archaeologist. And any attempt to predict exactly how a supervolcano would disrupt our world today would be almost as imprecise. That's an occupational hazard of dealing in existential risk. We look to past analogues to try to forecast how a future event will unfold, but existential risks by definition are on a level that we have never known.

While we haven't experienced a supereruption in recorded history, we have come close, and more recently than you might know. What happened afterward could help us understand why a supervolcano would represent an existential threat like no other.

———————

Tambora might have once been among the tallest peaks in Indonesia, though we'll never know for sure, because on the evening of April 5, 1815, this mountain on the Indonesian island of Sumbawa detonated. In his book *Tambora: The Eruption That Changed the World*, the historian Gillen D'Arcy Wood writes that "huge plumes of flame issued from the mountain for three hours, until the dark mist of ash became confused with the natural darkness, seeming to announce the end of the world."[30] As the eruption unfolded over days, whole villages were vaporized by flame, ash, and hurricane-strength winds, instantly killing some 10,000 people living in the immediate vicinity of the volcano.[31]

The force of the eruption sent tsunamis racing across the sea. Even those living far from Tambora could hear the rumble of the eruption as it continued, on and off, for days. Sir Stamford Raffles, the British imperialist and the founder of modern Singapore, reported that a detachment of troops had been mobilized because officials assumed the booming sound in the distance could only be cannon fire. Raffles wrote that local Indonesians assured him that the sound came not from a military clash, but from a battle between the devil and the souls of their departed ancestors.[32]

What they had all witnessed was a VEI 7 volcano eruption. That is one ranking below a supervolcano, but it still sufficed to make Tambora the most powerful eruption in known history. Nearly ten cubic miles of ash, gases, and magma were blasted into the atmosphere, and the mountain itself lost more than two-thirds of its estimated height.[33] In the days that followed the eruption a blanket of toxic volcanic ash draped over the island of Sumbawa, suffocating rice paddies and poisoning wells. Many of the islanders who survived the initial eruption succumbed to sickness and starvation in the following weeks, raising the estimated death toll to 100,000 and making Tambora not only the strongest but the deadliest eruption known to man.[34]

But the deaths didn't stop there. As Tambora's ash cloud spread across Southeast Asia, darkness fell over the region. For weeks after the eruption, British colonial officials were forced to conduct business by candlelight, even during the day.[35] The cloud—higher than 27 miles[36]—spread by stratospheric winds until the entire globe was affected. As happened after Toba, the ash and sulfuric gases in the atmosphere reduced the amount of sunlight reaching the Earth's surface, leading to global cooling. Average temperatures fell by 2.7 degrees between 1810 and 1819, making this decade the coldest in the historical record.[37] A drop of a little less than three degrees Fahrenheit may not seem like much of a change—you'll experience a much greater range of temperatures in the course of a typical day—but that average masks extremes that yielded famine and death.

In New England, 1816 was called the "Year Without a Summer," or "Eighteen Hundred and Froze to Death."[38] Rainfall levels were abnormally high in much of Europe, and sodden harvests rotted in the fields. The ceiling of the Earth's atmosphere actually dropped, chilling temperatures and causing further havoc with the weather.

Albany, New York, experienced snow in June, and in Europe late spring snows fell brown and red, tainted by the ash cloud. The very sunset was altered, as the ash filtered the light into the fiery reds and oranges immortalized in the works of contemporary painters like the English artist J. M. W. Turner.[39] Wood writes of the Scottish meteorologist George Mackenzie, who kept daily records of cloudy skies over the British isles in the first few decades of the nineteenth century. "In 1816, there wasn't a single sunny day," Wood told me. "Not a single one."

Lord Byron was inspired by the Tambora gloom to write the poem "Darkness," with the lines:

The bright sun was extinguish'd ...
Morn came and went—and came, and brought no day.[40]

Of even greater literary value was Mary Shelley's novel *Frankenstein*, written during that summer that wasn't in 1816. Mary, her lover (and future husband) Percy Shelley, as well as Byron, his lover Clair Claremont, and his doctor John William Polidari were forced by the terrible weather to seek indoor pastimes at Byron's villa on Lake Geneva in Switzerland. Byron proposed that the group entertain each other by composing ghost stories, and Mary Shelley's *Frankenstein*—which would become a metaphor for another existential risk, the invention of artificial intelligence—was the result.

But if the well-off were inconvenienced by the endless cold and rain, the poor died in droves. Epidemics of typhus raged through Europe, with 100,000 people in Ireland dying from the disease.[41] Crop yields in the British Isles fell by as much as 75 percent. In Switzerland—today one of the richest countries in the world—families starved en masse. In China peasants sucked on white clay to stave off hunger pangs, and some even sold their children as slaves, in hopes that their offspring would at least be given something to eat. There were worse fates than bondage, after all. "From Indonesia to Switzerland, people killed their children and themselves, rather than have their children face slow death from starvation," said Wood. No wonder historians have characterized the years from 1816 to 1819 as the last great crisis of hunger to afflict the Western world.[42]

Beyond the art, and the death, Tambora reshaped global history. Migration to the United States accelerated—the number of European immigrants arriving at American ports in 1817 was more than double any previous year. Desperate farmers in New England abandoned their lands and moved west in search of a better climate, with Vermont alone in 1816 and 1817 losing some 10,000–15,000 emigrants, twice what had been the usual number.[43] The sheer misery of the volcanic famine was so great that it may have helped encourage laissez-faire European governments to start actively responding to natural disasters. "It was the beginning of the realization that a responsibility of government is to care for citizens in times of crisis," said Wood. "We take it for granted but that principle did not exist at all."

The history of the Year Without a Summer makes for horrifying and fascinating reading, but what matters is this question: would our world suffer the same fate if another VEI 7 volcano of the size and scale of Tambora were to erupt—let alone a supervolcano?

In some ways we'd clearly be better off. In 1815 most of humanity still made their living as subsistence farmers, which left them vulnerable to the failure of a single harvest. Global trade was limited, so there was no easy way to ship food from places that were less affected by volcanic cold to those in dire need. There was no Red Cross, no World Health Organization, no world anything. The institutions and the wealth that cushion the world today—or at least the developed world—from even the worst natural disasters did not exist.

But as rich and as technologically advanced as we are, there are some aspects of life in 2019 that would make us more vulnerable than our ancestors to the shock of a global volcanic eruption. Global trade could help cushion the blow, but it depends in part on aviation, and as we saw after the 2010 eruption of Eyjafjallajökull in Iceland, even a minor volcano with a VEI score of just 4 can bring air travel to a halt should it strike at the wrong time and the wrong place. Volcanic ash can be as toxic to jet engines as it is to lungs, and Eyjafjallajökull grounded airlines in Europe for days, costing the global economy nearly $5 billion.[44]

The wealth that globalization has created has also led to specialization, magnifying the misery caused by any lengthy disruption in trade. My great-great-great-great grandfather in Ireland would have known how to grow his own food, because he had to. I do not. "Even an eruption on the scale of Tambora could cause real problems," said Bill McGuire, a British volcanologist and natural disaster expert. "This is especially worrying as such events may happen as frequently as every several hundred years."

Human history has repeatedly been marked by volcanic eruptions around the size of Tambora. In 1257, the Samalas volcano erupted on Lombok island in Indonesia, dampening global temperatures and helping to usher in what is known as the Little Ice Age, a period of unusual cooling that lasted from the fourteenth century into the 1800s. As happened after Tambora, cold weather and heavy rains ruined harvests in Europe; archeologists in England discovered mass graves with more than ten thousand skeletons dating back to the year after the eruption. One English monk in 1258 recorded that the year was marked by "such unendurable cold, that it bound up the face of the earth, sorely afflicted the poor, suspended all cultivation." Some historians have even suggested that the climatic change triggered by

Samalas hastened the decline of both Byzantium in Eastern Europe and the Mongol Empire in Asia.[45]

Scientists have connected even older volcanic eruptions to reductions in the flow of the Nile River, which in turn may have triggered political revolts and war in Ptolemaic Egypt.[46] In 2018 the journal *Science* named AD 536 the "worst year to be alive," largely because a massive volcanic eruption in Iceland that year caused temperatures to fall even more than they did after Tambora, leading to widespread famines, starvation, and even the terrible Plague of Justinian, an infectious disease that killed up to half the population of the eastern Roman Empire and sped its decline.[47] "You have political disturbances that seem to come out of nowhere, but are connected to volcanic eruptions," said David Pyle, a volcanologist at the University of Oxford. "That suggests there could be quite dramatic impacts in terms of the changes wrought by volcanoes."

In 2018, Alan Robock, along with Stephen Self and Chris Newhall—two veteran volcanologists who created the VEI scale—publicly warned that the world was utterly unprepared for Tambora- and Samalas-scale volcanos.[48] Just as few in Indonesia or Sri Lanka feared the threat of major tsunamis before December 26, 2004, because none had occurred in memory, the world has done little to prepare for a VEI 7 volcanic eruption because we have no ready experience of one. But Tambora erupted a little more than two hundred years ago—beyond our living memory, but virtually concurrent in geologic terms. VEI 7 eruptions seem to happen on average every few hundred years, which means it may not be long before another occurs. Yet the threat barely registers. "We have to overcome that human tendency to not want to deal with problems until they are painfully obvious," Newhall told me.

What Tambora and the other eruptions of its class teach us is that volcanoes can have global effects, ones that continue well after the volcano itself has fallen silent. A volcanic change to the climate can cause starvation, massive refugee flows, even political revolts. We can't prevent a volcano from erupting—at least not yet. But we can control how our society responds to the shock of a catastrophe, be it a supervolcano or another existential threat. "Tambora was an object lesson in how human systems affect the severity of human outcomes that result from a climatic crisis," said Wood. "The extremity of the natural event met the vulnerability of the human community. The result was disaster."

Will we fare better when our time comes? And what happens if that test is even bigger than what followed Tambora?

———————————

Michael Poland is the scientist-in-charge at the USGS's Yellowstone Volcano Observatory, and he wanted me to know this much: he is not trying to cover up the end of the world. As well as being America's first national park, Yellowstone is home to the most notorious supervolcano in the world. The volcanic system has produced three supereruptions over the past 2.1 million years, and Yellowstone is still highly active, as the park's dazzling array of geysers demonstrate on a daily basis for the more than four million people who visit each year.[49] So when Yellowstone experienced a swarm of hundreds of minor earthquakes in the summer of 2017, it drew the attention of the media and conspiracy theorists alike. It was Poland's job as Yellowstone's top volcanologist to assure anyone who asked that, no, the earthquakes did not mean that a world-ending supereruption was nigh. Which isn't to say that he convinced everyone. "There's always a significant part of the population that wants to believe the worst," Poland said.

In the case of Yellowstone, the worst would be like nothing the planet has experienced for tens of thousands of years. Yellowstone sits atop a volcanic hot spot, a long plume of magma that rises up from deep inside the Earth's mantle, like water being squeezed up a straw. That magma fills a shallow cauldron of molten rock just beneath the surface, a process that causes the ground in Yellowstone to rise and fall perceptibly, as if the Earth itself were breathing. When that magma reservoir fills to capacity, Yellowstone can blow. In the center of the park is a caldera, like the one that now forms Lake Toba in Indonesia, measuring 28 miles by 47 miles.[50] It is the hole left by Yellowstone's last supereruption some 640,000 years ago, which disgorged an estimated 240 cubic miles of volcanic material. Lava alone spread across some 1,700 square miles during that eruption, an area larger than the state of Rhode Island. Deposits of volcanic ash from the eruption have been found as far afield as present-day Iowa, Louisiana, and California.[51]

The magma reservoir beneath Yellowstone is filling again even now, and one day America's first national park could become its first supervolcano.[52] First would come a swarm of increasingly intense earthquakes, a sign that magma was rushing toward the surface. The pressure would build until, like champagne in a bottle given a vigorous shake, the magma would burst through the ground in a titanic eruption that would discharge the toxic innards of the Earth to the air. It would continue for days, burying Yellowstone in lava within a forty-mile radius of the eruption.

These earthquakes should fortunately give visitors and workers weeks of warnings, permitting time to evacuate the park. (That's why the Yellowstone Volcano Observatory is more properly an earthquake observatory.) Yellowstone as we know it, though, would cease to exist. But as with past supereruptions, the true existential threat would come from the ash, lava, and volcanic gases that would shoot upward with a force equal to a thousand Hiroshima-sized nuclear bombs,[53] sufficient to reach a height of fifteen miles or more.

The USGS volcanologist Larry Mastin modeled a Yellowstone supereruption and found that the northern Rockies could be buried in up to three feet of volcanic ash, smothering large swaths of Colorado, Wyoming, and Utah. Depending on the prevailing weather patterns, much of the Midwest would receive a few inches of ash, which would fall like black rain, plunging the region into darkness. Even the coasts—where the majority of Americans live—would likely see a dusting as the vast ash cloud spread out like an unfolding umbrella in the days following the eruption.[54] "What Yellowstone could produce looks almost like a hurricane, something that is almost continental in size," said Mastin.

Raging lava may be the volcanic showstopper, but ash is how a supereruption kills. Volcanic ash is not like the burnt-out embers of a campfire, or the bits that fall from the end of a cigarette. It is hard and abrasive, made up of tiny, jagged pieces of pulverized rock and glass. The ash is corrosive to living tissue, and in the aftermath of a Yellowstone supereruption, hospitals would be choked with victims coughing up blood as the silicate in the ash slashed at their lungs.

"Drowning in a liquid form of concrete" is how Marie Jones and John Savino put it in their book *Supervolcano*.[55] Those who avoided the ash in the air would still struggle to find uncontaminated water. Volcanic ash isn't soluble, so if ash-laden sewage enters a treatment plant, it will act as a wrecking ball, shredding mechanical equipment and stopping up biological reactors.

Accumulated volcanic ash is also heavy, and in the states closest to Yellowstone, buildings would collapse beneath the weight, just as snow from a blizzard can cave in roofs. But collapse would require only four inches of accumulated volcanic ash, far less than what will fall in the aftermath of a supereruption.[56] Ashfall will bring down power and telephone lines and ruin power plants. It could even fully collapse the grid by destroying the transformers that are the backbone of the electricity distribution system. While downed power lines can be restored in a matter of hours, it can take as long as two years to order and install a replacement for a single damaged electrical transformer.[57] If Yellowstone's ashfall managed to knock out a significant share of America's 2,100 electrical transformers, vast swaths of the country could be in the dark for a very, very long time.

Even as we grappled with a prolonged and widespread blackout, our food system might be crippled. More than a few inches of ashfall can kill crops in the fields. If Yellowstone were to erupt during the spring or summer, an entire season's worth of midwestern corn or soybeans could be wiped out in a few days. Depending on how much poisonous fluoride is contained in the ash, pastureland could contaminated by as little as 1/32nd of an inch of ash. Any livestock that fed on the grass could die horribly from gastrointestinal blockage, while those that survived would eventually die of starvation or thirst.[58] Even the earth itself would be poisoned. It could take generations before soil buried in more than four inches of ash could be restored to fertility. Oh, and aviation across the country would probably be grounded for weeks or longer, although with the blackouts and the food shortages and the air pollution, canceled flights would be the least of our concerns.

A FEMA estimate pegged the total damage to the United States from a Yellowstone supervolcano at $3 trillion, some 16 percent of the country's GDP.[59] That would be almost twenty times more than Hurricane Katrina, the most expensive natural disaster in American history, which cost $125 billion and resulted in the deaths of nearly two thousand Americans.[60] But it's guesswork to put a dollar figure on something as catastrophic as a Yellowstone supereruption, in part because it would be the first truly continental-scale disaster this country has ever experienced. In every past catastrophe—hurricanes, earthquakes, floods—most of the United States remained untouched, which meant safe parts could divert aid to and take in refugees from affected regions. But no corner of the continental United States would be exempt from the effects of a supervolcano. "If Yellowstone erupts, with the ash fallout, you're looking at the eradication of most of the Midwest and the U.S.A.," said John Grattan, an archeologist and disaster expert at Aberystwyth University in Wales. "It will be incredibly dramatic, something beyond the scale of government."

It will also be felt well beyond the United States. Half of the world's cereal grains are produced on North American farmland, and losing much of the Midwest farm belt, even temporarily, could mean global famine.[61] Take $3 trillion or more out of the U.S. economy and you have a recipe for a financial crash that would make the Great Depression look like a mild bump.

The most lasting effect of a Yellowstone supervolcano, however, would be on the global climate. In 2006, Claudia Timmreck and Hans-F. Graf modeled the climatic effects of a Yellowstone-like supereruption. They reported that aerosols from the volcano could spread globally if the eruption occurred during the summer. Over the short term, as the cloud of aerosols from Yellowstone blocked sunlight, global average temperatures could plunge as much as 18 degrees and not return to normal for as long as a decade.[62] Rainfall would decline significantly; along with the cooling, that might be enough to trigger a drastic die-off of tropical rain forests. The planet's most productive wildlife habitats would be annihilated and billions of tons of carbon that are stored in those trees would be released into the atmosphere. That would accelerate global warming—though not, unfortunately, in time to counter the volcanic winter.

The world has less than two and a half months of estimated food reserves,[63] so we could quickly find ourselves in a race against global starvation. The 800 million people who are already hungry might be the first to die,[64] but no part of the planet would remain unscathed. It would be, as Hans-Peter Plag wrote in a 2015 report for the European Science Foundation, "the greatest catastrophe since the dawn of civilisation."[65] There's a reason why a 2017 survey by Oxford's FHI listed supervolcanoes as the single natural existential threat that poses the highest probability of human extinction.[66] "This is absolutely an existential risk," said Oxford's David Pyle. "The sheer knockout effects, the cascade of disruption, would be mind-blowing."

You may now be asking yourself: what are the chances that this mind-blowing, knockout catastrophe is going to happen? This is where geologic time scales work in our favor. The probability of a supereruption at Yellowstone in any given year is 1 in 730,000, or 0.00014 percent.[67] We're closer to Yellowstone's next supereruption than we are distant, but that day—should it come—is likely still tens or even hundreds of thousands of years away. It's also possible that the threat from Yellowstone could be receding. As the North American tectonic plate slowly shifts, the magma hot spot that powers the Yellowstone volcanic system is beginning to encounter thicker and colder rocks that take more energy to burn through, which would forestall or even eliminate the chance of an eruption.

Yellowstone is one of the most closely watched volcanic systems in the world, so if it begins to show signs of building toward a supereruption, USGS volcanologists like Michael Poland will almost certainly notice, hopefully with years to spare. But there are many other potential supervolcanoes around the world, and not all have their own observatory. That's why volcanologists in a 2005 report advocated for the creation of a kind of Intergovernmental Panel on Climate Change for volcanoes—a global scientific body that would meet regularly to report on the latest research in volcanic risk.[68] This would hopefully place the threat of supervolcanoes squarely on the public radar, while putting more pressure on governments to create a global volcano monitoring system—something that very much does not exist today. Given that the world already spends more than $50 million to track incoming asteroids, it makes sense to spend at least that much on what is a greater threat. (The 2005 report came about in part because volcanologists were envious of the resources being made available to asteroid defense in the post-*Armageddon* 1990s. Scientific competition can have its upsides.)

Hans-Peter Plag calculated that such a monitoring system would cost perhaps $370 million per year. For comparison, the United States currently spends $22.3 million on its volcano hazards program.[69] Plag believes that sufficient pre-eruption warning could cut potential fatalities in half by giving humanity time to evacuate anyone near a supervolcano and adapt to the deprivations of a volcanic winter. Employing the same kind of existential risk math we used for asteroid defense, Plag estimates that the cost from expected fatalities from a rare supervolcano eruption still comes to between $1.1 billion and $7 billion per year, which would make the benefits of a global volcano monitoring system as much as 18 times greater than the expense,[70] simply because of the protection it would provide against the worst of the worst eruptions.

"I believe we are very naive to think that because we can handle small disasters, it means we can handle big ones," Plag told me. "There's no reason to think that because you can handle a cold, you can handle pneumonia. But that's where we are on volcanoes."

We can't yet prevent a volcanic eruption the way we can theoretically deflect an incoming asteroid. But I wanted to know if that could change—which is how I ended up connecting with Brian Wilcox. Wilcox is a longtime NASA engineer who served on an agency task force focusing on planetary defense. For NASA, planetary defense means asteroids and other things from outer space. When Wilcox ran the numbers, however, he became convinced that volcanoes were a much greater existential threat than asteroids—and a more likely one. "The probability of a two-kilometer-sized [1.2-mile] asteroid hitting the Earth, a big enough collision to cause major global cooling, is 2 to 10 times less likely than a supervolcano that could have a similar effect," he told me.

But while the mechanics of asteroids are simple—at least for a NASA engineer—the mechanics of volcanoes are not. "With volcanoes there's a lot of guesswork about what is actually going on," Wilcox said. "We don't know what fraction of the magma chamber in Yellowstone is melted, and we don't know the percentage of volatile gases in it that could trigger cooling." (The amount of cooling that might occur after a supervolcano blows depends not just on the sheer size of the eruption, but also on the amount of sulfur that would be released in the debris cloud. More sulfur means more cooling.)

That's an excellent argument for spending more money on volcanic science and monitoring, which Wilcox supports. But NASA simply wouldn't be NASA if it couldn't come up with an ambitious (and expensive) engineering project that might just save the planet. So in 2015 Wilcox and a group of colleagues from NASA's Jet Propulsion Laboratory published a paper with the modest title of "Defending Human Civilization from Supervolcanic Eruptions."[71]

As Wilcox sees it, volcanoes are essentially heat generators. Yellowstone's geysers are what happens when water seeping into the churning magma reservoir beneath the park is superheated and explodes through the surface, a process that Wilcox compares to a coffee percolator. But not all of the heat is released, and what remains builds up in the reservoir, melting more rock and sulfur, adding more magma, until an eruption becomes inevitable.

Wilcox and his colleagues suggest that the most direct way to defuse a supervolcano would be to simply cool its magma reservoir down so that it never reaches the eruption stage. That could be done directly by increasing the amount of water that reaches the reservoir. It would likely head off a supereruption—and would make the geysers of Yellowstone even more spectacular—but you would need a Great Lake's worth of liquid to make a dent in the park's vast volcanic system.

Wilcox instead suggests drilling some six miles down into the supervolcano and constructing tubes that can carry water around the magma reservoir. That water would leach heat from the magma, gradually cooling down the system and pulling it back from the eruption point. The entire project would cost about three and a half billion dollars, but it would have the side benefit of generating carbon-free geothermal electricity at a competitive rate. (Geothermal power plants use steam produced by hot water found underground to drive electrical turbines.) "The thing that makes Yellowstone a force of nature is that it stores up heat for hundreds of thousands of years before it goes kablooey all at once," said Wilcox. "It would be good if we drained away that heat before it could do a lot of damage."

There are, it's safe to say, a few caveats. Current drilling technology is barely sufficient to reach the necessary depths of six miles, and even then, the holes made are less than a foot across, which would limit the amount of water that could be delivered to the magma reservoir.[72] The extreme heat and corrosive gases beneath Yellowstone would eat away at drilling equipment. Oh—and this one might be important—the whole process is so slow that it would take *tens of thousands* of years to properly cool off the supervolcano. As USGS's Jake Lowenstern told me about Wilcox's plan, "It all seems a bit fanciful."

To be clear, Wilcox's concept is a thought experiment, not a NASA-approved doomsday strategy. There's little chance of it ever being put into practice, but it's worth pondering, because doing so demands that we step outside our brief human time frame and think like a geologist. As Wilcox told me, "If you want to defend human civilization on that time scale, you need to act on that time scale."

Time is what makes supervolcanoes such a confounding existential threat—time and uncertainty. While astrophysicists can calculate the orbital paths of potentially hazardous asteroids decades into the future, we know far less about what's happening beneath our feet than we do about outer space. But what we're learning should worry us. Simply based on geologic evidence, scientists had estimated that supereruptions occur on average about every 100,000 years.[73] But in a 2017 paper, Jonathan Rougier, a professor of statistical science at the University of Bristol, provided new calculations predicting that supereruptions occur every 5,200 to 48,000 years, with a best-guess value of about 17,000 years.[74] Given that the last VEI 8 eruption was more than 26,000 years ago, it might seem as if we're overdue. We're not necessarily—the Earth doesn't follow a regular timetable. But it does mean, as Rougier put it in his paper, that human civilization has been "slightly lucky not to experience any supereruptions" so far.[75]

We may have been luckier than we can possibly know. Scientists estimate the risk of natural catastrophes like supervolcanoes or asteroid impacts largely by looking back to the past and observing how many times such events have occurred. But it is by definition impossible for us to observe a catastrophe so great that it would have destroyed the human race while we existed, or if it happened before we came on the scene, have interrupted the chain of life that led to the rise of *Homo sapiens*. That's not because such megacatastrophes are impossible; rather, it's because if they had occurred, we humans would not be here to do the observing. Our own existence now casts what Nick Bostrom, Anders Sandberg, and Milan Ćirković called in a 2010 paper an "anthropic shadow" over our planet's past.[76]

This matters to us because while our current existence guarantees that a supervolcano—or any other existential threat—could never have wiped out humanity in the past, we're offered no such protection in the future. The anthropic shadow only falls backward. We may look at our planet's geologic record and feel confident that supervolcanoes and major asteroids are so rare that we need not worry about them. But that confidence could be fatally misplaced.

Sooner or later—and on a geologic time scale at least, much sooner—we will face a supereruption. It is in the nature of the planet we live on, the most geologically active in the solar system. Yet of all the existential risks we'll explore in this book, natural and man-made, it is the one for which we've done the least to prepare. Volcanoes have caused mass extinction on this planet before; in fact, they are the serial killers of life. We have been lucky—our existence after 3.5 billion years of evolution is nothing short of a cosmic miracle, even if it was a miracle that had to happen. But that luck may not hold.

NUCLEAR

The Final Curtain on Mankind

The chief asset of a missile range is its emptiness. At maximum size—its dimensions shrink and grow like a shadow depending on what its owners the U.S. Army is testing on a given day—the White Sands Missile Range in central New Mexico covers about as much land as the state of Connecticut.[1] Nearly all of it is vacant, devoid of buildings or roads or even the few animals that live in the flat, dry scrub the Spanish called the Jornada del Muerto, the Journey of Death. It's a name that seems almost too perfect, like that of the Sierra Oscura, the Dark Mountain, which loomed to the east as I drove up on an April morning to Stallion Gate, the northern entrance to the missile range and the one closest to Trinity Site.

It was to see Trinity that I had come to New Mexico and the White Sands Missile Range. We think of Hiroshima as the dawn of nuclear weapons, but it was here at Trinity Site that the very first nuclear bomb was detonated, the work of more than 100,000 people—from ditch diggers to dozens of past and future Nobel Prize–winning scientists—employed by the U.S. government's Manhattan Project.[2]

Hiroshima, Nagasaki, the Cuban Missile Crisis, the North Korean standoff, and whatever might come next—it can all be traced back to what happened here in New Mexico on July 16, 1945 at 5:30 a.m., when a successful atomic test resulted in an explosion more destructive than anything humans had caused before.

Trinity, however, inaugurated more than just the nuclear age. It ushered us into the era of man-made existential risk, an era that we live in still. Before Trinity, the world could end, and nearly did, as mass extinction events repeatedly erased most of life on Earth. But the cause each time was natural: supervolcanic eruptions, a collision with an asteroid, sudden and drastic climate change. After Trinity, though, human beings could be the authors of our own annihilation. All the existential risks that would follow—anthropogenic climate change, biotechnology, artificial intelligence—flow from what happened at Trinity, a hinge point in human history. So you can bet I wanted to see the place where it all began.

Nearly seventy-five years after the test, though, there was little evidence of the sheer physical power of what happened that morning in 1945, nor of its significance. Places where history was made are memorialized and meant to be visited—think Gettysburg National Military Park, Independence Hall in Philadelphia, even sites of scientific achievement like the Thomas Edison National Historical Park in New Jersey, where the lightbulb was invented. After World War II the Interior Department tried to create a national monument at Trinity, but its efforts were continually frustrated by the military, which wanted to retain White Sands to test its growing inventory of missiles. There was also the problem of Trinity's mixed legacy—the greatest of the technical and scientific accomplishments that helped win the war for the Allies, and the birthplace of the first weapon of mass destruction. It wasn't until 1975 that Trinity was finally declared a national historic landmark—a few ranks down from a national historical park—but even now it remains closed to most visitors, save for two Saturdays a year in October and April, when the army runs an open house and cars line up for miles in the desert sunlight, waiting to enter through Stallion Gate.

I was fortunate enough to schedule a private tour of Trinity Site the day before the April open house in 2018. My guide was Jim Eckles, a welcoming and talkative Nebraska native in his sixties who had worked for years as the missile range's public affairs officer and wrote a history of the Trinity test not long after retiring. Eckles knew how to dress for the New Mexico desert, even in the spring—a broad hat that shaded his entire face and white hair, and sunglasses with narrow oval lenses. He also knew how to conduct a tour of a historical site that didn't necessarily have much to see. "The starkness you see out here is important," he told me. "It's not a house where George Washington stayed. It's a bomb test site, and that emptiness is the right way to memorialize what happened here."

By sheer area Trinity is mostly a parking lot, albeit one that usually sits empty, waiting for those two days when it will fill with RVs dusty from the seventeen-mile drive from Stallion Gate. That lot holds a mangled metal carcass called Jumbo, a 200-ton steel tube originally designed to contain the priceless plutonium in the Trinity bomb—the Gadget, as the scientists called it—in the event of a malfunction during detonation. In the end Jumbo wasn't used, and it was eventually ordered blown up by Lieutenant General Leslie Groves, the military commander of the Manhattan Project—in part, Eckles told me, because Groves was worried that someone would ask why the army had spent $12 million on a custom-built steel tube no one actually needed. But eight 500-pound bombs only managed to blow off both ends of the tube—which perhaps shouldn't have been surprising, given that Jumbo had survived the actual Trinity nuclear blast. "It turns out that it's really hard to destroy a big thing made of steel," Eckles said. Now Jumbo sits rusting in the Trinity parking lot, a footnote to human fallibility of a different sort.

In fairness, though, everything done at Trinity was being done for the first time, which meant some guesswork. Before the test, some of the senior Manhattan Project scientists organized a betting pool with a one-dollar entry fee, guessing the size of the explosion. Official estimates before the test suggested that the expected strength of the bomb would fall between 500 and 7,000 tons of TNT.[3] This was at a time when the most powerful explosives in the U.S. arsenal had about 10 tons of TNT equivalent power.[4] Then there was the question of whether the Gadget—the product of more than three years of effort—would even work. Groves himself thought there was only a 40–60 percent chance of success.[5]

The conditions the night before the test were atrocious—heavy rain and lightning, which is not ideal when there are miles of electrical cable snaking through the open desert and about thirteen pounds of highly radioactive plutonium, wrapped in wires and screws and silver and gold and high explosives, all mounted on top of a hundred-foot steel tower.[6] But there was no time to wait for better weather. Harry Truman—who had been president for barely three months and had only been told about the Manhattan Project when he assumed the office after Franklin Roosevelt's death in April—was due to meet the day after the scheduled test with Soviet premier Joseph Stalin. A successful test of the atomic bomb would be a powerful card for the new president to play as he charted the endgame of the war with his current ally and soon-to-be adversary. Trinity may have heralded a new era of man-made existential risk, but as ever it was immediate politics that drove the decisions on the ground, not any reckoning with truly long-term consequences. With one exception.

In 1942, the Hungarian-American physicist Edward Teller—who would go on to develop the more powerful hydrogen bomb and become one of the inspirations for Stanley Kubrick's film *Dr. Strangelove*[7]—ran some calculations and concluded that an atomic bomb might just possibly create enough heat to ignite the atmosphere and the oceans, causing a global inferno and the end of the world. When Robert Oppenheimer, the scientific leader of the Manhattan Project, told the physicist Arthur Compton about Teller's figures, the older man reportedly responded in horror. "This would be the ultimate catastrophe!" Compton recalled in an interview after the war with the author Pearl Buck. "Better to accept the slavery of the Nazis than run a chance of drawing the final curtain on mankind!"[8]

According to Buck, Compton told Oppenheimer that if additional calculations showed that the odds of igniting the atmosphere with a nuclear explosion were more than approximately three in one million, all work on the bomb should stop.[9] Fortunately, about six months before the Trinity test, Teller and the Polish-American nuclear scientist Emil Konopinski produced a report on the subject, titled "LA-602: Ignition of the Atmosphere with Nuclear Bombs."[10] They concluded that it would be virtually impossible for even a much larger bomb than what would be tested at Trinity to create such a runaway reaction and end the world.

"Virtually impossible" was good enough for wartime, so the work on the bomb continued. After the war, Oppenheimer told a Senate panel that "before we made our first test of the atomic bomb, we were sure on theoretical grounds that we would not set the atmosphere on fire."[11] At the time, though, not everyone may have been fully convinced. The night before the test, the Italian-American physicist Enrico Fermi offered to take wagers on whether the bomb would indeed ignite the atmosphere—and if so, whether it would merely destroy New Mexico or the entire world.[12]

Fermi's antics irritated the military commanders overseeing the Manhattan Project, who worried that his joke—and it was mostly a joke—might spook the soldiers securing the Trinity site. But as Richard Rhodes writes in his magisterial history, *The Making of the Atomic Bomb*, "a new force was about to be loosed on the world; no one could be absolutely certain—Fermi's point—of the outcome of its debut."[13]

The LA-602 report was a footnote in the history of the Manhattan Project, but it holds a special place in existential risk studies. It marked the first time humans had tried to figure out in advance whether their actions could bring about the end of the world. "It's the first technical assessment of an anthropogenic existential risk, rather than a religious one, or one related to a natural hazard," said Jason Matheny, the former director of the Intelligence Advanced Research Projects Activity (IARPA), perhaps the closest entity the U.S. government has to a successor to the Manhattan Project. The debate over the existential dangers of biotechnology, the dueling visions about the threat from artificial intelligence, even this book—they can all be traced back to LA-602, a report that wasn't even declassified until 1973.

As it turned out, the Trinity test demonstrated that for atomic weapons at least, it wasn't what would happen if they went wrong that we should fear most. It's what would happen when they went right.

———————

With twenty minutes to go before ignition, Sam K. Allison of the University of Chicago began the world's first countdown over a loudspeaker. As Allison reached the last few seconds, a local radio station began broadcasting on the same wavelength, overlaying Tchaikovsky's *Nutcracker Suite* upon the falling numbers. Twenty miles away, observing the test site from Compania Hill, Edward Teller began making everyone nervous—or more nervous, at least—

by offering to pass around suntan lotion he had brought.[14] At 5:29:45 a.m., the Trinity bomb detonated.

From the parking lot Eckles and I walked a few hundred feet to Trinity Site itself. It was fenced in from the desert. There was bright yellow grass, surrounded by the greener brush that extends out to the Sierra Oscura. If I had a Geiger counter I would have detected the slightest uptick in radiation—another echo of the bomb[15]—though not anywhere near enough to cause harm. Near the center of the site is the actual and first Ground Zero, the spot where on July 16, 1945, a hundred-foot steel tower stood, topped by the Gadget. (The tower itself was called Zero; the ground at the foot of the tower was named Ground Zero, which is where the term originates.) Today Ground Zero is memorialized with a stone obelisk mined from nearby volcanic rock—black and brown stone born of fire, as Eckles told me, to represent the ultimate fire. A plaque states the facts of what happened that day in the simplest terms: *Trinity Site—where the world's first nuclear device was exploded on July 16, 1945.*

Standing before the obelisk, I recalled that a few weeks after the Trinity test, Lieutenant General Groves had been driven out to the site on an observation trip. Victor Weisskopf, a nuclear physicist who had worked on the Manhattan Project, remembered Groves looking at the crater left by the first atomic test, and the general remarking: "Is that all?"[16] I now understood what he meant.

Beyond the obelisk, though, I could see black-and-white historical photos arranged against the fence. Between portraits of Manhattan Project scientists and images of the Gadget itself was a frame-by-frame series of the milliseconds after the Trinity bomb exploded.

At 0.006 seconds there is a bubble of perfect light, as if the dawn itself had blossomed suddenly out of the desert ground. The heat of the blast is thousands of times hotter than the surface of the sun, and the light in that single moment was a dozen times brighter.[17] At 0.025

seconds, the bubble head keeps rising, while a fringe of fire spreads across the ground. It will carve a crater half a mile across, and suck up hundreds of tons of sand into the blast interior. Later the silica in the sand will adhere to radioactive particles and rain back to the surface as something wholly new: bright green trinitite, also known as Alamogordo glass. For decades after, tourists will collect shards of trinitite as souvenirs from the site, even though removing it is technically illegal.

At 0.053 seconds, that perfect bubble begins to lose its clarity, becoming diffuse and unfocused, as if overwhelmed by its own energy, while the inferno at the surface expands, gouging out the earth below. At this point every living thing within a radius of a mile is dead, or will be soon. At .10 seconds, the blast looks like nothing less than a halo ringing the head of some Renaissance painting of Christ, as the exposure itself begins to degrade. The atomic heat has made the air glow luminous, as the force of the shock wave expands outward, shredding the matter in its path. Everything is ravaged, everything is burned. And at 15 seconds after detonation comes the familiar image of the mushroom cloud, what the art historian John O'Brian called the "logo of logos in the 20th century,"[18] a symbol that would shadow humanity for decades to come. That mushroom cloud—like nothing seen on Earth before—is the result of the intense heat at the heart of the blast, causing the air to rise in a column, before it spreads out in a cap.[19]

Less than a minute after the explosion, Enrico Fermi stood up and released slips of paper into the air. He estimated from their deflection in the blast wave that the Trinity explosion had released the equivalent of 10,000 tons of TNT. Fermi was off, but his impromptu experiment proved far more accurate than the Manhattan Project's conservative pretest estimates. Trinity's destructive power was close

to 21,000 tons of TNT. Isidor Isaac Rabi, a Manhattan Project physicist who'd come late to the test and took the last available bet in the
pool, won with a guess of 18,000 tons.

Entire books can and have been filled with the testimonies of the
men who were there at Trinity. There is the scientific reaction and
the military one, the religious and the poetic. But of all the words
spent in witness I prefer those of Rabi, who had filled the tense night
before the Trinity test playing poker:

> A new thing had been born; a new control; a new under-
> standing of man, which man had acquired over nature. . . .
> Then, there was a chill, which was not the morning cold;
> it was a chill that came to one when one thought, as for
> instance when I thought of my wooden house in Cambridge,
> and my laboratory in New York, and of the millions of people
> living around there, and this power of nature which we had
> first understood it to be—well there it was.[20]

News of the successful Trinity test reached President Truman,
who was touring a bombed-out Berlin on the way to the Allied
conference in Potsdam, Germany. He wrote in his private diary that
the atomic bomb "seems to be the most terrible thing ever discovered,
but it can be made the most useful."[21]

In one way Truman was wrong. The atomic bomb had not been
"discovered," in the sense of simply being found somewhere, but had
been called into being by the decisions of politicians and generals and
the hard work of thousands of scientists. But in another way Truman
was more correct than he could know. Before the Manhattan Project was launched, before the decision to test and then use the atomic
bomb, scientists who made pioneering discoveries in nuclear physics

in the decades leading up to the war had been laying the groundwork for Trinity, without fully knowing what their work might lead to. Many dismissed an atomic bomb as impossible.

Not all, though. Leo Szilard was one of the few physicists who from the start knew that the likely outcome of the groundbreaking atomic research of the 1930s would be nuclear weapons. The Hungarian-born Szilard—who earned his PhD in Germany, which he fled following Adolf Hitler's rise to power in 1933—drafted the famous letter sent by Albert Einstein to President Roosevelt in 1939 warning that nuclear bombs were possible and urging the United States to launch what would become the Manhattan Project. And it was Szilard who, in the months leading to the Trinity test, became the most prominent voice urging Washington not to use the bomb, should the test prove successful. His call went unheeded.

Three weeks after the test, the B-29 bomber *Enola Gay* took off from an air base on the Mariana Islands and flew toward Japan. A version of the Trinity device was nestled in its bombing bay. At 8:15 a.m. local time on August 6, at 31,000 feet over the southwestern Japanese city of Hiroshima, the *Enola Gay* released its payload. Robert Lewis, the copilot on the flight, later recorded the dissociation he felt as the bomb fell: "The bomb was now independent of the plane. It was a peculiar sensation. I had a feeling the bomb had a life of its own now that had nothing to do with us."[22]

The bomb detonated 1,900 feet above the city. Within a millisecond, the heat was so intense that as far away as 2.3 miles from Ground Zero, the temperature of a person's skin could be raised to 120 degrees.[23] Within minutes, 9 out of 10 people inside a half-mile radius of Ground Zero were dead, their bodies burned away to black char. The suffering of those who survived defies description, though it remains seared in the memories of victims like Setsuko Thurlow, who was a thirteen-year-old girl in Hiroshima the morning the bomb was dropped. What she witnessed was still vivid in her mind seventy-two years later, when she co-accepted the Nobel Peace Prize for the International Campaign to Abolish Nuclear Weapons (ICAN). "When I remember Hiroshima," Thurlow told the audience in Oslo, Norway, "the first image that comes to mind is my four-year-old nephew, Eiji—his little body transformed into an unrecognizable melted chunk of flesh. He kept begging for water in a faint voice until death released him from agony."[24]

Some seventy thousand people likely died as a result of the initial blast, heat, and radiation, and thousands more would die from injury and radiation-induced cancer in the months and years that followed.[25] As high as the numbers were, though, it was not the death toll alone that set Hiroshima apart. America's incendiary bombings of Tokyo with conventional weapons in 1945 had killed even more people, but that had required the work of 300 planes dropping 8,000 bombs over the course of two nights.[26] Hiroshima had needed but one plane, and one bomb, a bomb of unimaginably concentrated destruction.

After the atomic bombings, it was said to Leo Szilard that it was a tragedy for scientists that their discoveries were used for destruction. No, Szilard replied, it is not the tragedy of scientists. It is the tragedy of mankind.

Those words were prophetic, and they'll rain over the rest of this book like Alamogordo glass. Scientists move civilization forward through their pursuit of knowledge, but Trinity demonstrated that their pursuit can inadvertently create the conditions for our own doom. Existential threats can be brought into the world not by those who wish to end it, but by those who hope to better it. Intentions don't matter for the fate of the world—results do. As Richard Rhodes wrote of the Manhattan Project's legacy: "The scientific method doesn't filter for benevolence. Knowledge had consequences, not always intended, not always comfortable, not always welcome."[27]

There were no pictures of Hiroshima at the Trinity site the day I visited. Like so much else at Trinity, its presence is felt through its absence, just as the site itself is a reminder of a threat we'd rather forget. "All those missiles and all those weapons are still out there, and we blithely go along in our everyday lives," Jim Eckles told me, as we stood under the New Mexico sun. "Every minute of the day. It's in the back of our minds, and maybe Trinity, it brings it forward just a little bit. Let's meditate on that, that this is still a threat."

Eckles looked at the emptiness that makes Trinity Trinity. "It doesn't really change much for anybody, but at least there's this reminder here."

Once the destructive power of the nuclear bomb had been demonstrated at Hiroshima, and three days later at Nagasaki, other nations rushed to arm themselves with this terrible new weapon. The Soviet Union tested its first bomb in 1949, Great Britain in 1952, France in 1960, China in 1964. The thermonuclear or hydrogen bomb, hundreds of times more powerful than the fission device tested at Trinity, was developed by the United States in 1952. The Soviet Union followed with its own thermonuclear weapon the next year.

These new weapons of mass destruction were at first brandished with little care. Atomic bombs were tested—which at the time meant exploded above the ground—almost once a week on average during the 1950s and '60s.[28] Because the U.S. military in Europe was outnumbered by the Soviets by as much as ten-to-one during the initial stages of the Cold War, Washington was dependent on the threat of its nuclear arsenal to repel a conventional attack[29]—and was quick to threaten its use. President Dwight Eisenhower considered using nuclear weapons during the end stages of the Korean War in 1953, over the Taiwan crisis in 1958, and over a dispute with the Soviets about the fate of Berlin in 1959. President Lyndon Johnson was prepared to preemptively strike China to prevent the communist government in Beijing from developing its own nuclear arsenal. Eisenhower even signed an agreement delegating the authority to use nuclear arms to generals and admirals outside of Washington, in the event that the U.S. capital had been destroyed, and those generals and admirals in turn delegated that power down the chain of command until junior commanders aboard oceangoing warships had the power, on their own, to launch nuclear weapons in a crisis.[30]

Nuclear arms were initially considered another weapon—powerful, of course, but not necessarily special or civilization-threatening. There wasn't yet a full understanding of what a global nuclear war would actually mean for the human beings living through it. This was the age of backyard fallout shelters, when *Life* magazine could run a cover in 1961 of a man in what the editors identified as a "civilian fallout suit," cowering beneath the headline "How You Can Survive Fallout: 97 out of 100 people can be saved."[31] The Federal Civil Defense Administration (FCDA), created in 1951 by President Truman, pumped out a steady stream of dubious advice for surviving the unthinkable. That included "Bert the Turtle," an animated character who showed children in song what to do in the event of a nuclear strike: "There was a turtle by the name of Bert and Bert the turtle was very alert; when danger threatened he never got hurt. He knew just what to do. He ducked! And covered!"[32] The U.S. Postal Service printed 60 million change-of-address labels and sent them to regional offices, in case a nuclear war left tens of millions of American refugees who nonetheless wanted their magazines forwarded.[33] At one point the Eisenhower administration had a plan to dig trenches alongside public highways so that if Americans were caught out in their cars during a nuclear strike, they could try to survive by abandoning their automobiles, lying down in the trenches, and covering themselves with dirt.[34]

Ducking and covering is actually good advice should you have the grave misfortune to be caught near a sudden nuclear strike. "If you're talking about a kiloton-range weapon, there's actually quite a lot you can do to change your odds of survival," said Alex Wellerstein, a historian at the Stevens Institute of Technology who studies the history of nuclear weapons. "But if you're talking about an exchange between nuclear powers with thousands of weapons, there isn't much you can do." The idea that 97 out of 100 Americans could survive a full-scale nuclear war through ducking and covering was plainly absurd. These plans for what came to be known as civil defense represented what the sociologist Lee Clarke has called "fantasy documents"—exercises that were done to give both the citizenry and the bureaucracy a sense of control, however fantastical, over the uncontrollable.[35]

But we shouldn't be too quick to judge. Again and again we find that the biggest barrier to combating existential risks—including nuclear war—is a mental one, for the human mind rebels against the sheer scale of extinction.[36] *Homo sapiens* evolved to live in small groups, and so we feel acutely the grief of small-scale loss, that of a friend or a relative or even a stranger. But there is no individualizing the deaths of billions of people, perhaps even our entire species. And so we choose to deny it.

Our empathy actually erodes as potential death tolls grow, thanks to a psychological process called scope neglect. Few researchers have studied this more closely than the University of Oregon psychologist Paul Slovic. Slovic has found that sympathy can begin to fade as soon as we're presented with two needy people, rather than one[37]—what he calls the "arithmetic of compassion." Counterintuitively, instead of concern and our willingness to act rising as the size of a potential catastrophe grows, it can actually contract. In one telling experiment, Slovic told volunteers that they could spend $10 million to save 10,000 people from a disease that killed 20,000 people per year, or spend the same amount of money to save 20,000 people from a disease that killed 290,000 people per year. The study subjects preferred to save 10,000 people from the disease that killed 20,000 people—even though the same amount of money could have saved twice as many people if it had been used to treat the more deadly disease.[38]

The psychic numbing that Slovic identifies makes it that much more difficult to come to grips with existential risks of any sort—including the ones, like nuclear war and climate change, that result directly from our own actions. Rather than being motivated to prevent global catastrophes, we prefer to ignore them. And if we can't accept those risks, we can't do anything about them. In a 1982 address to the American Psychological Association, the behaviorist B. F. Skinner argued that there was "something in the very nature of human behavior" that blocks us from working to prevent huge catastrophes we have not yet experienced—which applies to every existential risk covered in this book. "Instead of our worry increasing as the size of the consequence increases, it degrades," Slovic told me. "Our attention is scarce, and so is our ability to worry. We treat these catastrophes as if they're impossible unless there's an experiential aspect that lends an air of reality to it."

But the bomb seemed to—and still does—defy reality. Like all existential risks, nuclear war really was unthinkable to most people, as in it couldn't be thought of. This was true even among the few who could testify to that reality: the survivors of Hiroshima and Nagasaki. The psychiatrist Robert Jay Lifton, who studied the aftermath of the bombings, found that survivors were deeply confused about what had happened to them. They had no previous model that could help them grasp the experience of a nuclear bomb. The survivors, Lifton writes, "wondered whether it was a huge 'electric short,' a form of 'Buddhist hell,' or 'the end of the world.'"[39]

Denialism was embedded in the very language used by nuclear strategists. A nuclear war was referred to as a "nuclear exchange," as if Washington and Moscow were returning unwanted Christmas gifts, not potentially annihilating the world. Firing off intercontinental missiles that would kill hundreds of millions of people was termed a "nuclear expenditure," as if it were a line item in a budget. The explosive power of a nuclear warhead was called its "yield," the same word used to describe the product of farmland or the interest from a financial investment.

There was the occasional bit of nuclear terminology that said exactly what it was—like "megadeath," the word for one million deaths from a nuclear strike. "Megadeath" was the invention of Herman Kahn, a systems theorist at the RAND Corporation, a Santa Monica, California–based think tank that was the nerve center of Cold War strategizing. Kahn was the author of 1960's *On Thermonuclear War*, a book that if nothing else delivered precisely what its title promised. He identified different levels of American deaths from a nuclear war, ranging from two million to 160 million, and how long it would take for economic recuperation—at least according to his calculations. To Kahn there were, as he put it in his book, "Tragic but Distinguishable Postwar States."[40] (If that phrase sounds familiar, it's because Stanley Kubrick lifted it almost verbatim for *Dr. Strangelove*.) To Kahn it was thinkable to fight a nuclear war, and thinkable to believe you could win it—which was why the title of his next book was *Thinking About the Unthinkable*.

Yet it wasn't hawks like Kahn who truly understood what nuclear war would mean, but rare objectors like Daniel Ellsberg. Today Ellsberg is best known for leaking in 1971 what became known as the Pentagon Papers, a secret Defense Department study showing that successive presidential administrations had systematically lied to the American public about the country's involvement in Vietnam. But years before he was both celebrated and vilified for his activism on Vietnam, Ellsberg was a strategist at the RAND Corporation working on theories of nuclear deterrence—just like Herman Kahn. Ellsberg took the possibility of nuclear war seriously, so seriously in fact that he didn't bother paying into RAND's generous retirement plan because he assumed he would be killed in a nuclear conflict before he could ever collect.[41] Before his time at RAND, Ellsberg had been an infantry platoon commander with the U.S. Marines. Far from being a peacenik, Ellsberg was a red-blooded Cold Warrior who believed in the strategic importance of America's nuclear arsenal.

What changed Ellsberg forever was the moment he was shown how the world could end. He was working as a consultant for the Office of the Secretary of Defense in the spring of 1961 when he saw the answer to a question President John F. Kennedy put to the Joint Chiefs of Staff: "If your plans for general [nuclear] war are carried out as planned, how many people will be killed in the Soviet Union and China?" The answer was stark: 325 million people. But that was just the beginning. Drilling down further, Ellsberg found out that the Pentagon expected an additional 100 million deaths in the communist states of Eastern Europe, perhaps another 100 million from radioactive fallout in the friendly nations of Western Europe, and an additional 100 million in adjacent nations like Finland, India, and Japan. Altogether the Pentagon estimated that roughly 600 million human beings would die, at a time when the global population was 3 billion. And those numbers assumed, somehow, that the United States would escape any nuclear retaliation from the Soviet Union in a nuclear war—almost certainly a fantasy.[42]

More than fifty years later, as he spoke to me from his home in Berkeley, California, I could still hear the horror in Ellsberg's voice as he recounted what he read that day in 1961. It wasn't just the numbers that appalled him, but the fact that this was the Pentagon's *only* nuclear war plan. In response to any armed conflict with the Soviets, even a small conventional one, even a mistake, the war plan called for the firing—the expenditure—of America's entire nuclear arsenal. There was no plan B, no room for a measured response. The strategy included targeting Chinese cities even if Beijing had nothing to do with the conflict. "It was insane," Ellsberg told me. "No human could ever imagine doing such a thing in the history of our species—and here they were doing it, planning it. It goes beyond ordinary concepts of crime or even sin. It transcends any human concepts to be planning to kill hundreds of millions of people."

This was Ellsberg's moment of conversion, the first step on the path to becoming who he is today—an eighty-eight-year-old former national security insider who has spent decades fulminating against nuclear policy and government secrecy. In 1969, as Ellsberg began photocopying what would become the Pentagon Papers, he also copied those far more secret nuclear war plans, the ones marked "For the President's Eyes Only." He planned to leak these as well. That never happened—after the *Washington Post* began to publish the Pentagon Papers and it became clear that his arrest was only a matter of time, Ellsberg gave the nuclear war plans to his brother in upstate New York for safekeeping. Ellsberg's brother buried them in a nearby trash dump, but the documents were destroyed in a freak tropical storm. It wasn't until the publication of his book *The Dooms-day Machine: Confessions of a Nuclear War Planner* in 2017 that Ellsberg was able to fully describe what he had seen.

Ellsberg was and remains a person of rare moral courage, one who was willing to go to jail for years, even the rest of his life, to fight against the madness of war. Yet even today he wonders why the revulsion he felt toward America's nuclear strategy was so rare, and why so many of his colleagues—the country's best and the brightest—were willing to go along with a blueprint for mass murder. "It didn't take a lot of thinking to say that this is madness and absolutely intolerable," he said. "Yet they all recoiled from the opposition, from the Joint Chiefs, from Congress, and from the military-industrial complex, which involves so many jobs, profits, campaign donations, and everything else."

As with other existential risks, the desire for short-term political and economic gain sabotages attempts to focus on long-term threats. We now know that much of what America's nuclear war strategy was based on was false. The plan Ellsberg saw in 1961 called for bombing China because Beijing and Moscow were seen as communist comrades, yet by then the two countries had all but split. That same year RAND estimated that the Soviets possessed hundreds of intercontinental ballistic missiles—yet it later turned out that the Russians in fact had just *four* missiles at the time.[43] When Kennedy was elected in 1960—after running a campaign accusing the Eisenhower administration of letting the United States fall behind in the arms race— America had nearly twelve times as many nuclear warheads as the Soviet Union did.[44] But it was in the political and economic interests of some to hype the Soviet nuclear threat, even though doing so arguably brought the world closer to a ruinous war.

We now know, nearly sixty years later, that the missiles never left their silos, that the Cold War never went nuclear. The world didn't end. But this wasn't because of a surfeit of wisdom on either side of the Iron Curtain. We were lucky—and few people know just how lucky we were than another national security insider turned renegade: William Perry.

If there's an important post in America's national defense establishment, chances are that William Perry has held it. He worked as a civilian expert in electronic intelligence in the 1960s, served as undersecretary of defense for research and engineering—where he was instrumental in developing the stealth technology that helped the United States win the Cold War—and ended his career in government service as President Bill Clinton's defense secretary from 1994 to 1997. He served on the University of California's board of governors for the laboratory at Los Alamos—where the first nuclear bomb was developed—and is currently the head of the board at the *Bulletin of the Atomic Scientists*. Even at ninety-one years old his voice still exudes authority, and his words demand attention in capitals around the world.

What makes Perry special, however, is that he is one of the last living American statesmen who saw with his own eyes just how close we came to nuclear annihilation. And what he came to understand was that the real threat of nuclear war wasn't from military competition, but from the way that simple misunderstandings and technical errors could spiral out into planetary catastrophe. It wasn't the war in nuclear war that was so dangerous—it was the nuclear, the fact that thousands of megatons of explosive power kept on a hair trigger made any mistake irrevocable.

In the fall of 1962 Perry was working as director of Sylvania's Electronic Defense Laboratories, in the San Francisco Bay area. He spent his time calculating missile trajectories and nuclear yields. One day that October Perry received a call from a friend in the CIA asking him to fly into Washington for a consultation—immediately. That was how Perry got involved in what would become known as the Cuban Missile Crisis. In D.C. Perry pored over reconnaissance photos of Soviet missile sites in Cuba and helped write technical reports for President Kennedy and his staff. As the standoff between Washington and Moscow became increasingly tense—with Kennedy instituting a naval quarantine of Cuba and contemplating an invasion of the island if the Soviets wouldn't remove the missiles—Perry became convinced that each day would be the last day of his life. And things were even worse than he knew. "Kennedy's assessment was one chance in three of nuclear war," Perry told me. "It was at least that in my judgment, because there were possibilities of that war starting from circumstances he wasn't even aware of."

Washington didn't know it at the time, but tactical nuclear weapons—small-scale atomic bombs that could be employed on the battlefield—had already been placed in Cuba. Nor did they know that Soviet submarines operating off the island were armed with nuclear-tipped torpedoes. Commanders on the ground in Cuba had been given the authority to use the tactical nukes in the event of an American invasion, and Soviet submarine captains had been given permission to fire their nuclear torpedoes without explicit commands from Moscow. Had Kennedy ordered an attack—as his hawkish military advisers were urging—those weapons would have been used, which would have quickly escalated to full-scale nuclear war between the superpowers.

This set the stage for the single moment in the modern age when the human race may have come closer to extinction than it ever has before or since. On October 27, 1962, as part of the U.S. naval quarantine of Cuba, American destroyers and the aircraft carrier USS *Randolph* managed to corner the Soviet submarine *B-59*. The U.S. ships began dropping small depth charges—underwater explosive devices—around the sub. The American commanders weren't trying to sink the sub but rather to force it to the surface, an intention they had made clear to Soviet military leaders in Moscow.

What the Americans didn't know was that the sub had been out of touch with Moscow for days. Its batteries were depleted, and without power to cool the sub or clean the air, temperatures inside *B-59* rose to more than 113 degrees and the sailors began suffering from carbon dioxide poisoning. When depth charges began exploding around the sub, the crew had every reason to believe that World War III had begun. An exhausted Captain Valentin Savitsky gave the orders to prepare the sub's nuclear torpedo for firing. A successful hit on the *Randolph* would have vaporized the aircraft carrier, which in turn could have put the U.S. nuclear war plan for total retaliation into play. Thousands of American warheads would have been on their way to targets in the Soviet Union, China, and other nations. The Soviets would have responded. The insanity Daniel Ellsberg glimpsed would have become real, and would have consumed us all.

The decision to launch a nuclear weapon on board the Soviet sub had to be authorized by three officers. Ivan Maslennikov, the deputy political officer, said yes. But Vasili Arkhipov, Savitsky's second in command, refused. He convinced Savitsky to instead bring the sub to the surface, where a U.S. destroyer ultimately allowed the ship to return to Russia.[45] That same day, Soviet premier Nikita Khrushchev sent a letter to the White House proposing that the USSR would dismantle its missiles in Cuba in return for the United States removing medium-range ballistic missiles from its NATO ally Turkey, which bordered the Soviet Union. After a day of deliberation, Kennedy accepted the offer, though the missiles in Turkey weren't moved until months later, to avoid the appearance of a quid pro quo. The crisis—this crisis, at least—was over, but if not for Kennedy's prudence and Arkhipov's courage under fire, a nuclear war may well have begun. And then, as Ellsberg told me, "you and I would not be having this conversation."

Vasili Arkhipov holds a special place of honor in the field of existential studies. There is an Arkhipov Room at Oxford's FHI,[46] and fifty-five years to the day after his actions aboard sub *B-59*, Arkhipov was posthumously honored with the inaugural Future of Life award from the Cambridge, Massachusetts–based Future of Life Institute (FLI). As FLI president Max Tegmark said at the ceremony, "Vasili

Arkhipov is arguably the most important person in modern history, thanks to whom October 27, 2017, isn't the 55th anniversary of World War III."[47]

But the Cuban Missile Crisis is only the best known of many, many times when World War III was almost triggered by accident, as the writer Eric Schlosser has shown in his magisterial book, *Command and Control*. Perry himself lived through one when he was serving in the Department of Defense in 1979 and was awakened in the middle of the night by a watch officer at NORAD who said his monitors were showing two hundred Soviet intercontinental ballistic missiles (ICBMs) en route to the United States. It turned out to be a computer error. Less than a year later, on June 3, 1980, military computers showed thousands of Soviet missiles headed toward the States. Then–national security adviser Zbigniew Brzezinski was about to recommend a counterattack until he was told at the last minute that the alarm had been generated by a faulty computer chip—one that cost all of forty-six cents.

Perhaps the closest the world came to nuclear war after the Cuban Missile Crisis was in 1983. President Ronald Reagan entered office promising to confront the Soviet Union. He modernized the U.S. nuclear arsenal, asking Congress for billions for civil defense efforts and calling for a missile shield—the Strategic Defense Initiative— that would render nuclear weapons "impotent and obsolete." Reagan genuinely wanted to reduce the risk of nuclear holocaust, but to the Soviet Union he seemed to be preparing to fight and win a nuclear war. The USSR was on edge, and by the summer of 1983 its forces had a "shoot to kill" order if the U.S. military crossed into Soviet territory. That tension led to tragedy on September 1, 1983, when a Soviet pilot, convinced a South Korean passenger airliner was actually on an espionage mission for the United States, shot it down, killing 269 people, including 63 Americans.

It was against that dire backdrop that the Soviet early warning system on September 26, 1983, reported the apparent launch of
several ICBMs from the United States. Lieutenant Colonel Stanislav
Petrov was on duty that night, and his job was straightforward: regis-
ter the missile launch and report it to Soviet military and political
command. An ICBM takes half an hour to reach its target,[48] which
meant Petrov had only minutes to authenticate the apparent attack in
time for the Soviets to launch a counterattack—and given how close
the superpowers were to all-out war, they may well have. Yet Petrov
judged that the United States would not launch a first strike with only
a handful of missiles, so he instead reported a system malfunction.
And then he waited. "Twenty-three minutes later I realized that nothing had happened," Petrov told the BBC in 2013. "If there had been a
real strike, then I would already know about it. It was such a relief."[49]

Petrov, too, has a room named for him at the Future of Humanity
Institute. Nick Bostrom has said of Petrov and Arkhipov that "they
may have saved more lives than most of the statesmen we celebrate
on stamps."[50] This is almost certainly true. But what made Petrov's
and Arkhipov's heroism necessary—and what made the many close
calls of the Cold War so dangerous—is inherent in the nature of
nuclear deterrence.

During the Cold War, and even today, the nuclear powers had a
policy of mutually assured destruction, which meant that each was
restrained from attacking the other because they knew they would be
attacked and destroyed in turn. In one sense this worked perfectly—
fear of nuclear war kept the Cold War from going hot, and the second
half of the twentieth century proved far less violent than the first. But
the side effect of nuclear-enforced peace was the creation of existential risk for the entire species. Every year, every day, every moment,
global catastrophe could strike at the push of a button. The very speed
of a nuclear exchange, the doctrine of total retaliation, meant that

any mistake could result not in a battle, not even a war as we had known war before, but the end of the world. "Today, every inhabitant of this planet must contemplate the day when this planet may no longer be habitable," President Kennedy told the United Nations in 1961. "Every man, woman, and child lives under a nuclear sword of Damocles, hanging by the slenderest of threads, capable of being cut at any moment by accident or miscalculation or by madness."[51] And we live under that sword still.

Something else happened in 1983 that drove home the existential nature of the nuclear threat. As the top-secret plan that Ellsberg saw showed, military and political leaders on both sides knew that nuclear war would result in mass death the likes of which the world had never experienced. But even at the height of the Cold War—when the strategy of "overkill" meant that the two superpowers had built up massive nuclear arsenals capable of destroying the other many times over[52]—hundreds of millions of people would still survive the missiles, enough to hopefully rebuild civilization after the nuclear fires were quenched. Radioactive fallout was a further threat, but not one capable of killing off everyone who survived. As Herman Kahn might write, there were "tragic but distinguishable" differences between hundreds of millions of deaths and human extinction.

Beginning in the 1970s, however, scientists began to examine what thousands of nuclear explosions would do not just to human beings, but to the environment they depended on. Early reports suggested that nuclear war might destroy much of the atmospheric ozone layer for years, damaging crops and leading to a spike in skin cancers. Dust and soot from the explosions and fires, meanwhile, could result in what one report termed "minor changes in temperature and sunlight."[53] But as scientists continued to refine their climate models, those changes in temperature and sunlight began to look less minor and more catastrophic.

In 1983 a paper was published by Richard Turco, Brian Toon (whom I referred to in chapter 1), Thomas Ackerman, James Pollack, and Carl Sagan. Known as *TTAPS*—an acronym of the coauthors' surnames—the paper used meteorological models, derived in part from the study of volcanic eruptions, to predict the global effects of nuclear war on sunlight and temperature.[54] *TTAPS* made the case that the dust and smoke generated by nuclear fires could reduce sunlight levels by more than 90 percent, while global temperatures could fall by as much as 27 to 45 degrees. As with asteroid impact and supervolcano eruptions, such rapid and drastic cooling could make farming impossible, even in those regions spared by the missiles. There would be no escape from what the *TTAPS* authors memorably termed "nuclear winter."

Nuclear winter seized the public's attention because it blew away the position—built up over the years by strategists on both sides— that nuclear war could be survivable, let alone winnable. New Soviet leader Mikhail Gorbachev was shaken by the argument, as was President Reagan, who said that a nuclear war "could just end up in no victory for anyone because we would wipe out the Earth as we know it."[55] The stakes of a nuclear conflict were now plausibly infinite. "A nuclear war imperils all of our descendants, for as long as there will be humans," Sagan wrote in 1983.[56] "Extinction is the undoing of the human enterprise."

The nuclear winter theory had its share of detractors, including some scientists who were skeptical of the *TTAPS* conclusion. Climate models of the early 1980s were crude, and as the authors concluded, the problem of nuclear war is "not amenable to experimental investigation,"[57] which was a technical way of saying that the only way to be sure of what would happen after a nuclear war is to have a nuclear war. Some of the *TTAPS* authors themselves later said that their findings had been somewhat overblown by the media. But while scientists have disagreed on just how wintry nuclear winter might get—some argue that the aftermath of a war would be more akin to "nuclear autumn"—there is no doubt that the climatic effects would be meaningful, even devastating. A 2007 study by Alan Robock forecast cooling by as much as 36 degrees in the core farming regions of the United States, with even more drastic temperature drops in Russia.[58] Robock is a climate scientist as well, but he told me, "I'm much more scared of nuclear war than I am of global warming because nuclear war can be instant climate change."

The threat of nuclear winter made a difference in public opinion and political policy. In 1986—the year that Turco, the lead author on the *TTAPS* paper, was awarded a MacArthur genius prize[59]—the number of nuclear warheads globally reached a peak, and then began to decline.[60] The Cold War ended; mutually assured destruction relaxed. In 1996, William Perry, who had become one of the first U.S. defense secretaries of the post–Cold War era, joined his Russian counterpart Pavel Grachev to plant sunflower seeds at a field outside a Ukrainian missile base, a public event meant to symbolize the end of the nuclear threat.[61] When Perry came into office the Doomsday Clock was set to seventeen minutes to midnight, as far from annihilation as it has ever been before or since.[62] "I thought the problem was over," Perry told me.

It isn't.

———————————

January 13, 2018, was a warm and humid morning in Honolulu, and Robert de Neufville was sleeping in. De Neufville is the director of communications for the Global Catastrophic Risk Institute, a U.S. think tank that focuses on existential threats, and at the time he had been hard at work on an important project. Along with his colleagues Tony Barrett and Seth Baum, de Neufville was trying to answer a vexing question: how likely was a nuclear war in the future?

It was de Neufville's job to dig through historical archives and uncover every time the world had brushed close to a nuclear conflict, whether because of military tensions or simple accident. History mattered. As near as the world had come to a nuclear holocaust, again and again, more than seventy years had passed since Nagasaki, and no nuclear weapon had since been used in war. We were lucky, clearly—all the near misses demonstrated that. But de Neufville and his colleagues also wondered whether there was some obstacle or complication that made an actual nuclear war unlikely, no matter how close we seemed to come. Knowing that would help us understand how much danger nuclear weapons might pose in the future.

It was about 8:10 a.m. that morning when de Neufville's girlfriend woke him up and showed him the message on her cell phone: "BALLISTIC MISSILE THREAT INBOUND TO HAWAII. SEEK IMMEDIATE SHELTER. THIS IS NOT A DRILL."[63] "There was a moment of confusion," he told me. "I thought, 'Well, this could be it. It's not totally impossible.' I don't know if I was scared, exactly. But in the back of my mind there was a thought: 'We might all be about to die.' And I think a lot of Honolulu residents thought that. 'We might all be about to die this morning.'"

De Neufville knew that he and his girlfriend should try to find shelter, somehow, against an incoming missile, but instead he found himself thinking about friends who were staying with him. "They had come out from the mainland," he said. "They had two kids, ten and twelve. And I thought, 'They were in California, and now maybe they're going to die.'" He thought about his cats. "My cats are going to die," he said. "Literally, do I try to save the cats?"

The best strategy in the event of an imminent nuclear explosion is to shelter in place, away from windows, preferably in the center of a building. But de Neufville's house was made of wood. He considered herding everyone into the sewers to provide protection against the coming blast wave, but he realized there wasn't time to get the crowbar he would have needed to pry open a manhole. As he debated, the minutes ticked by. "What really struck me was that I'd actually thought before about what we would do in this situation," de Neufville said. "But it turns out that when it happened, I didn't really know what to do."

So, eventually, de Neufville did what most of us would probably do in the same situation: he went on Twitter. And there he saw a tweet from Hawaii congresswoman Tulsi Gabbard: "HAWAII - THIS IS A FALSE ALARM. THERE IS NO INCOMING MISSILE TO HAWAII. I HAVE CONFIRMED WITH OFFICIALS THERE IS NO INCOMING MISSILE."[64] Thirty-eight minutes after the original alert—longer than it would have taken an ICBM to arrive—the state of Hawaii sent out a correction.[65] It was indeed a false alarm.

It's still not clear exactly what happened. Initial reports suggested that a worker at the Hawaii Emergency Management Agency had accidentally activated a real-world alert code instead of a test missile alert—and then for good measure, had clicked "yes" when the computer asked him to confirm his choice.[66] The worker himself has said that he thought the alert was real, while Hawaiian officials later put out that the worker—who was fired soon after the incident—froze after he pushed the wrong button.[67] Regardless, the damage was done.

Being awakened by a phone alert warning that a nuclear missile was inbound would shake anyone, but the political atmosphere at the beginning of 2018 made it easy to believe that the possibility of nuclear war—once consigned to the Cold War past—was very real.

North Korea had spent the first half of 2017 completing a flurry of ballistic missile tests, and in September had conducted its sixth-ever test of a nuclear device—one that the regime claimed was its first hydrogen bomb. President Donald Trump had responded by

threatening "fire and fury," and on January 2, 2018, he tweeted that
his nuclear button was "much bigger and more powerful" than North
Korean dictator Kim Jong Un's.[68] At the beginning of December,
Hawaii sounded the nuclear attack warning siren for the first time
in three decades.[69] No wonder everyone was ready to panic.

Nuclear tensions between the United States and North Korea
ebbed somewhat after Trump and Kim met for a summit in Singapore just six months later. But as of early 2019 Pyongyang had yet to
give up a single nuclear warhead, even as Moscow was once again
flexing its nuclear might and the United States had embarked on a
major upgrade of its own atomic arsenal. Make no mistake: the existential threat of nuclear war is still great, in some ways as great as
it was during the darkest days of the Cold War. "I couldn't imagine
how we could go back to a Cold War again and a nuclear arms race
again," said Perry. "But I was wrong."

That might seem hard to believe. Years of arms control treaties
have vastly reduced the number of nuclear warheads in the world,
from a high of nearly 70,000 in 1986 to around 14,500 now.[70] The
United States and Russia control 93 percent of those warheads, and
while relations are clearly strained, the two major nuclear powers
are no longer avowed ideological enemies. No one has even seen a
mushroom cloud since China undertook the last aboveground atomic
test in 1980—every nuclear test since has been done belowground,
including those by North Korea.[71] Duck-and-cover drills are a thing
of the Cold War past—in 2018 New York City began retiring the
once-ubiquitous yellow fallout shelter signs, in part because it had
been decades since anyone had bothered to take care of the facilities.[72]

Beneath those positive signs, however, the nuclear status quo
was eroding. During much of the Cold War, there were only five
declared nuclear powers: the United States, the Soviet Union, the
United Kingdom, France, and China. Today India and Pakistan—

enemies that in seventy years have fought three wars and numerous skirmishes—possess growing nuclear arsenals, along with North Korea. Israel has never officially acknowledged its status but is widely known as a nuclear power. By withdrawing from the Obama-era deal that constrained Tehran's nuclear ambitions, President Trump may have inadvertently opened the door to a future Iranian nuclear weapon, though there is no evidence yet that the country has resumed development. And while preventing the spread of nuclear weapons has long been a goal of Republican and Democratic presidents alike, Trump has signaled in the past that he isn't too bothered by the thought of American allies like Japan or South Korea, or even Saudi Arabia, developing nuclear arms of their own.[73] The math is simple: the greater the number of countries that have nuclear weapons, the harder it is to keep the nuclear peace, and the more likely that accidents will occur.[74]

Far from acting to stabilize the nuclear balance, in recent years Russia and the United States have reversed years of nuclear arms cuts, and both nations are now embarking on expensive—and dangerous— nuclear modernizations and expansions. Moscow's position is reversed from what it was at the start of the Cold War—Russia's conventional military is hopelessly outmanned and outgunned by America, which spends nine times more than Moscow does on defense.[75] So just as the United States once did, Russia is increasingly relying on its willingness to threaten use of its nuclear arsenal to offset its conventional military deficit.

Russian president Vladimir Putin has raised the possibility of employing tactical nuclear weapons in the event of a conflict with NATO, in the apparent hope that the West would simply back down rather than escalate a border skirmish to a full-scale nuclear war. Moscow is also developing and deploying new nuclear warheads and launch systems,[76] including superweapons like a "Doomsday" torpedo—an autonomous stealth submarine armed with a giant nuclear warhead that could reportedly render much of the North American east coast radioactive.[77] Nor is Putin shy about showing off the power of Russia's nuclear arsenal. "Despite all the problems with the economy, finances, and the defense industry, Russia has remained a major nuclear power," Putin said in his annual address in March 2018. "Nobody really wanted to talk to us about the core of the problem [with the West], and nobody wanted to listen to us. So listen now."[78]

The United States is listening. President Barack Obama was awarded the 2009 Nobel Peace Prize in part for his commitment to nuclear nonproliferation, and within thirteen months of his inauguration the United States and Russia had signed the New Strategic Arms Reduction Treaty (New START), the latest in a series of nuclear arms control agreements. Yet by the end of his second term—even as Obama became the first sitting president to visit Hiroshima, where he called for a "world without nuclear weapons"—Washington had begun work to upgrade its nuclear arsenal.[79]

Since Obama left office those efforts have only accelerated. According to the Defense Department's 2018 Nuclear Posture Review, the Trump administration has planned for an expansion and enhancement of American nuclear weapons systems that will cost an estimated $1.2 *trillion* over thirty years.[80] In February 2019, the United States announced that it would suspend the Intermediate-Range Nuclear Forces Treaty, a deal signed in 1987 between Reagan and Gorbachev to eliminate their stocks of ground-launched ballistic missiles capable of traveling from 300 to 3,400 miles, what are known as tactical arms. The Trump administration's stated reasoning was that Russia had been effectively cheating on the treaty for years. That was true, but by unilaterally withdrawing from the treaty rather than punishing Russia over its violations, the United States effectively closed the door to future arms control treaties while raising the possibility of a mini arms race between the two countries to develop low-yield, tactical nukes.[81]

The Nuclear Posture Review also stated that the United States reserved its right to respond to nonnuclear strikes—like a major cyberattack—with nuclear weapons. This is no longer a bipolar Cold War. Where once U.S. nuclear policy was wholly dedicated to countering a massive launch by the Soviets—hence the overkill plans that Daniel Ellsberg saw—American presidents must now negotiate a multipolar world of varying dangers from varying competitors with varying aims. Cyberattacks are on the rise, but so is the threat of chemical or biological terrorism, or regional conflicts involving U.S. allies like the bloody war Saudi Arabia has prosecuted in Yemen. If a dictator like Syrian president Bashar Assad seems ready to use lesser weapons of mass destructions on his own people, should the United States threaten to respond with a nuclear attack? What about a cyberattack on a major U.S. corporation that might be linked back to China or Russia? Nuclear war just isn't as simple as it was in the megadeath days of Herman Kahn.

What this all means—the changing nuclear posture, the abandonment of arms control treaties, the new weapons—is that the barriers to nuclear war are falling. Tactical, low-yield nukes are seen as "gateway drugs" to full-scale conflict, and expanding the range of attacks the United States might choose to respond to with atomic weapons blurs what should be very sharp lines around nuclear war. Any rational leader would blanch before launching an all-out nuclear attack that could ultimately kill billions of people. But firing off a single tactical nuke might seem a lot closer to ordering conventional air strikes—something American presidents rarely hesitate to do when confronted with a range of threats that all fall short of the existential. Once the nuclear seal has been broken, though—even by what seems like a comparatively minor bomb—no one knows what will happen next. "Once you start using nuclear weapons," said Joe Cirincione, the president of the Ploughshares Fund, an antinuclear nonprofit, "that is not a stable situation."

The question of stability brings us to the most unstable factor of all: Donald Trump. The most powerful person in the world is the president of the United States, and above all else that is true because they retain sole control over America's nuclear arsenal. If the president wishes to launch a nuclear strike—and the United States has never abandoned its right to wield nuclear weapons first in a conflict—no one can stop him or her. Not Congress, not the secretary of defense, not the Joint Chiefs of Staff. The order for a launch goes from the president to an officer at U.S. Strategic Command, and from there to the crews manning the ICBM silos and bombers and nuclear submarines.[82] While military officers are technically required to refuse "unlawful" orders—and General John Hyten, the commander of U.S. Strategic Command, has said he would push back against what he termed an "illegal" strike[83]—in practice it wouldn't be difficult to devise legal rationales for almost any nuclear attack the current occupant of the White House desires. All the president really needs is the support of a majority of American voters to put them in office—and as we've discovered a couple of times now, they may not even need that.

Absolute presidential control of the nuclear arsenal did not begin with Donald Trump. The order to drop nuclear bombs on Japan, though verbally approved by President Truman, was actually drafted by General Groves of the Manhattan Project and signed by Truman's secretary of war, Henry Stimson. But when Truman raised the possibility of giving military commanders the power to use nuclear weapons in the Korean War, much as they might decide how to employ any other armament, it led to a public outcry. The Truman administration eventually released a statement confirming that the president had to authorize the use of a nuclear bomb.[84]

It took some time to work out the chain of command, but by the Kennedy administration it had become policy that the president and only the president could order the use of nuclear weapons. Which, it should be noted, was definitely a good thing. General Douglas MacArthur, the commander of American and UN forces during the Korean War until he was fired by Truman, later told a biographer that he wanted to drop dozens of atomic bombs on North Korea and its ally China.[85] In 1968, General William Westmoreland activated a plan to move nuclear weapons to South Vietnam before he was overruled by President Johnson.[86] If the Pentagon under President Trump has come to be seen as a moderating force, that has most assuredly not always been the case.

While presidential control of nuclear arms does keep potentially world-ending weapons in civilian hands, however, it matters quite a bit who those hands belong to. During his last, blighted days in office, when he was drinking heavily, President Richard Nixon once boasted that "I can go into my office and pick up the telephone, and in 25 minutes 70 million people will be dead."[87] He raised the possibility of using nuclear weapons during an offensive by North Vietnam in 1972 and had to be talked down by Henry Kissinger. By the time of President Reagan's last year in office, his aides were reportedly worried enough about his declining mental state—he would die suffering from Alzheimer's in 2004—that they discussed invoking the Twenty-Fifth Amendment, which calls for the vice president to take over in the case of the president's incapacity.

Trump is a teetotaler, and while some experts have raised concerns that he might be suffering from early-stage dementia, he seems in better shape than Reagan was by the end. But Trump has given the world plenty of reason to worry about whether he can be trusted with the nuclear codes. He has directly threatened North Korea with a nuclear attack, welcomed a renewed nuclear arms race with Russia, and at one point in the summer of 2017 reportedly told his military that he wanted a nearly tenfold increase in America's nuclear arsenal, returning it to a level not seen since the Cold War. His Twitter account is an electrocardiogram of his tempestuousness— and tempestuousness is not a quality we should welcome in a person who wields as much power as Trump does.[88]

In the fall of 2017 a congressional committee held a hearing to examine presidential authority to use nuclear weapons.[89] The very fact the hearing was held—the first one on the subject in forty-one years—was notable. But so was the outcome. Witness after expert witness told the Senate Foreign Relations Committee the same thing: there is virtually nothing preventing Donald Trump, or any president, from launching a nuclear war at their whim. Congress could act to restrict the president's power—and a pair of Democratic congressmen introduced legislation to do just that—but it's unlikely to pass, and of course, the president would be well within his authority to veto it.

Is Donald Trump himself an existential threat to the planet? Perhaps—but if he is, it is because the architecture the United States has built in the decades since Trinity permits him to be. The entire nuclear command-and-control system is built for surprise and for speed, the speed of missiles and rockets—not the speed of rational human decision making. It's why accidents and miscalculations around atomic weapons are so uniquely dangerous. If the leader of an atomic power has reason to think their country is under attack from nuclear missiles, they have just minutes to decide whether that attack is authentic and whether a response—a response that could mean the end of humanity—must be launched. But should that leader hesitate, a real nuclear first strike risks wiping out their country's ability to retaliate. End of war, end of country, end of leader.

When it comes to the existential threat of nuclear weapons, it's less the finger on the trigger than the trigger itself, as Elaine Scarry, an English professor at Harvard University and the author of *Thermonuclear Monarchy*, a book that sharply critiques the presidential monopoly on nuclear weapons, told me. "People are particularly concerned about Trump, but what they don't realize is the insanity, the obscenity of the system itself," she said. "The obscenity doesn't depend on any unlikeable characteristics, even if you take the person you like best in the world. You just have to run the odds for how many times you have to come close before it can really happen, and then it's over in a day."

We can at least begin to change that system. William Perry, who is spending the final years of his life raising the alarm around nuclear weapons, wants Washington to take one finger off the nuclear trigger and end the outdated "launch-on-warning" policy for its land-based ICBMs. Originally instituted to ensure that America's missile silos couldn't be destroyed by a surprise Soviet attack, the policy has long outlived whatever usefulness it had. Launch on warning is why the president has only a few minutes to authenticate a possible attack and respond, which raises the risk of mistakes. The United States holds a nuclear triad—the ICBMs, but also bombers and submarines— and the other two legs would be able to retaliate even if land-based missiles and the U.S. command structure were somehow eliminated by a first strike.

There's no time to wait. Martin Hellman, an electrical engineer and cryptologist at Stanford University, has run the odds and compares the probability of nuclear war to a game of Russian roulette. Every nuclear crisis—every point of geopolitical tension between two atomic powers, every accidental close call with nuclear weapons that could lead to an exchange—represents a pull of the trigger. Eventually, inevitably, we'll come upon the chamber with the bullet.

In a 2009 paper, Hellman put the annual probability of a "Cuban Missile–type Crisis" producing a nuclear war at 0.2 percent to 1.0 percent.[90] That seems reassuringly low, but every year that passes compounds those odds, so much so that Hellman estimated that there was a minimum 10 percent chance that a child born in 2009 would suffer an early death in a nuclear war. Other experts have come up with higher or lower estimates, but the point is that as long as nuclear weapons exist, they will remain an intolerable, existential risk. As the report from the Canberra Commission on the Elimination of Nuclear Weapons, an international panel on the future of atomic warfare, put it in 1997: "The proposition that nuclear weapons can be retained in perpetuity and never used—accidentally or by decision—defies credibility."[91]

Like any child of the 1980s, there were nights when I went to sleep fearing that I would be killed in a nuclear war—or worse, that I would somehow survive one. I was thirteen when the Cold War was said to end, and with it, so did the fear. The sword that President Kennedy said was suspended above our necks had suddenly been lifted. So it enrages me to know that my son has been born into a world where that fear is very real again, where the sword is again poised to fall.

While supervolcanoes are the most dangerous natural existential risk, nuclear weapons, I believe, remain the single most significant existential risk we face right now, today, man-made or otherwise. Not asteroids, not disease, not artificial intelligence, but the old nuclear nightmare. There is no defense against them should they be used. They are weapons of war but primarily murder civilians. They have the capacity to ruin the entire planet, even end our species. And after years of hopeful progress, we now somehow find ourselves in a moment of renewed peril. "Each of us knows, at least most of the time, that we are mortal individuals, that our friends and family are mortal," Daniel Ellsberg told me. "What we should know is that our species is mortal as well. We have achieved the capacity to extermi- nate ourselves. And I've been using my life, the best I can, to stop that. Because I think it is worth it. Civilization with all its ills is worth struggling to preserve." And we can preserve it, for just as the abil- ity to end this world with nuclear weapons is in our power, so is the ability to save ourselves.

4

CLIMATE CHANGE

What Do We Owe the Future?

From thirty thousand feet in the air, the Greenland ice sheet seems invincible, nearly 800 trillion gallons of water locked safely away in a deep freeze. But if you get closer, you can see the cracks begin to emerge, veins of the purest blue meltwater running between folds of ancient ice. They are cracks in the foundation of the world we've built.

I was in Greenland in the summer of 2008 as part of a tour put on by the Danish government, which was set to host a major UN climate change summit in Copenhagen the following year. Our group spent a couple of days at a scientific camp in Greenland's white and frozen interior, where climate researchers were digging a mile and a half deep into the ice to find clues about the state of the Earth's past climate. The ice cores mined by the scientists contained tiny bubbles of ancient air that could reveal the temperature, the concentration of greenhouse gases, and even the ambient dust from the year the layer was formed. (Similar ice cores, as we saw in chapter 2, helped scientists reconstruct the climatic effects of supervolcanoes like Toba.) The ice cores functioned like tree rings for global warming, and by unlocking their secrets, scientists could better understand how Greenland's titanic ice sheets would respond to future global warming. And if you're part of the 40 percent of the world's population that lives within sixty miles of a coastline, there's no more important question.[1]

That work was the past of climate change, however. I had come to Greenland to see its future. On one of our last days the group took a helicopter tour of Jakobshavn Glacier, Greenland's largest, a more than thirty-five-mile-long tongue of slithering ice on the territory's west coast. Outlet glaciers like Jakobshavn are the drainpipes of the Arctic—as the glacier flows toward the ocean it breaks apart and melts, steadily adding to sea level rise. And as global warming has accelerated, so has Jakobshavn Glacier, which now flows at more than 145 feet a day on average, nearly three times its speed during the 1990s.[2] While a 2019 study found that over the last couple of years Jakobshavn had actually begun growing again, that's due to temporary natural cooling of nearby ocean water, and may actually indicate that the glacier could be even more vulnerable to warming in the future.[3]

Our helicopter landed on a rocky outcrop overlooking the glacier. Days on Greenland's ice cap at the scientific camp—and only days, as the sun shines through midnight during the high Arctic summer—had made it seem as if ice were something eternal and unchangeable. I could walk outside the tent I shared, behold white stretching out in every direction to the horizon, and know that there were thousands of feet of ice packed beneath my feet.[4] But what I saw below me now was a speckled river of ice and rock and water churning toward the ocean. And then I heard the sound. The first stage was a low rumble, like the march of a far-off army coming closer. What followed was an echo of thunder that rolled and rolled, for ten, twenty, thirty seconds, before a boom, boom, boom, each low note resounding like an orchestra's timpani drum. Then almost a sigh, a release of tension and pressure, as if the air were being let out of the Earth, followed by a still and cold silence. This was the sound of the glacier itself giving way, calving off icebergs the size of mansions to slide into the water, where they would begin to melt and add to the seas. This was the sound of climate change in motion.[5]

Climate change is unlike any of the other existential risks in this book. It's neither purely natural nor solely man-made, but rather the product of an interaction between the mechanics of the planet and our own actions as an industrialized species. While the other risks are largely binary—a nuclear war or an asteroid impact either occurs or it doesn't—climate change is already happening, gathering force with every year that humans burn fossil fuels, cut down forests, and otherwise add billions of tons of greenhouse gases to the atmosphere. While natural risks like supervolcanic eruptions are surpassingly rare and the probabilities of coming man-made risks like artificial intelligence are all but unknowable, scientists can forecast the next several decades of global warming with chilling accuracy. And what they're learning is frightening.

The planet's average temperature has already warmed by about 1.6 degrees since the late nineteenth century, and it is almost entirely certain that the primary cause is man-made greenhouse gas emissions.[6] That warming has accelerated, and the five hottest years on record have all occurred this decade. Global sea level has risen by eight inches over the past century, a process that is also speeding up. Arctic sea ice has shrunk by a 12.8 percent per decade on average, going back to the beginning of satellite data records in 1979,[7] in part because the far north is warming so quickly that in the *winter* of 2018 temperatures at the North Pole occasionally rose above freezing.[8] One hundred fifty-seven million more people were exposed to heat wave events in 2017 compared to 1990,[9] while the annual number of global weather disasters between 2007 and 2016 was 46 percent higher than during the 1990s.[10] The year 2017—which saw Hurricanes Harvey and Maria devastate American territory—was the costliest year on record for extreme weather, with an estimated $320 billion in losses worldwide. The fall of 2018 was marred by devastating forest fires— fueled in part by years of climate-change-related drought and heat— that tore through California, roasting dozens of people to death in their cars and homes.[11]

What's set to come will be much worse. A shocking 2018 report from the UN's Intergovernmental Panel on Climate Change (IPCC)—the gold standard for climate science—found that if global greenhouse gas emissions continued at their current rate, the atmosphere would warm by as much as 2.7 degrees above preindustrial levels by 2040. Some 50 million people globally could be exposed to increased coastal flooding by that year, leading to refugee flows that would dwarf what we experience today.[12] By 2050 nearly a third of the world's land surface could turn to virtual desert.[13] Should warming reach 3.6 degrees—which seems likely, barring a political and economic revolution—the effects would multiply.[14] And until we can stop emitting carbon—and even after that for a time—climate change won't stop, either. This is a disaster that rolls on and on.

If it sounds bad, it is. But bad is one thing—existential is an entirely different category, and despite all we know and all we've experienced, it's not certain into which category climate change will fall. This is another factor that sets climate change apart from other potential existential risks—the uncertainty of its effects. The chance of a supereruption, a major asteroid impact, or even a global nuclear war may be minuscule, but we can be confident that should they occur, the results will be catastrophic, up to the point of threatening human extinction. With climate change, though, matters are reversed. The future of the climate depends largely on our actions—

on how our political, economic, and scientific systems will respond in the decades to come, whether we'll invest in zero-carbon energy technology that cuts carbon or even invent ways to suck greenhouse gases directly from the atmosphere. But it also depends on how the climate system will respond to the carbon dioxide and other greenhouse gases we're pouring into the atmosphere. The uncertainty is in how bad things will get, whether it will be uncomfortable, catastrophic—or existential.

There is a chance that the climate system will prove less sensitive to the billions upon billions of tons of CO_2 that we're certain to pour into the atmosphere, and therefore produce less warming than expected in the decades to come. Perhaps temperatures will rise sharply, but sea ice and tropical rain forests will show resilience to the heat. It's possible, in the way that rolling double sixes at the craps table is possible, in the way that winning at Russian roulette is possible—which doesn't mean it's advisable to play.

But uncertainty runs the other way. The climate system may turn out to be highly sensitive to elevated levels of CO_2 and spit out even more warming than models currently project. Arctic sea ice and rain forests could prove less resilient to climate change, leading to global feedback effects that further accelerate warming. The resulting heat waves could become so intense that major swaths of the globe would become uninhabitable to human beings. The migrations and misery that would follow could destabilize what political order remains, leaving us more vulnerable to other existential risks.

Scientists long assumed that the Greenland ice sheet I visited—parts of which have survived for a few million years—would remain mostly frozen for thousands of years more, even as the climate continued to warm. Yet more recent research warns that the Greenland ice sheet, and the even larger one in Antarctica, may be approaching or even passing a tipping point, putting them on a path to total melt one day.[15] That could mean sea level rise of ten feet or more over the next 50 to 150 years, enough to swamp cities like London, New York, and Shanghai by 2100.[16] What I witnessed on the Jakobshavn Glacier in 2008 could have been the beginning of the end of the world—at least as we know it. Or it might have been the melting of a warm summer day.

We won't know for sure until the future arrives. That's another way that climate change confounds us. Unlike the conventional air pollutants that cause smog, which can be washed from the air in a matter of days, greenhouse gases like CO_2 linger in the atmosphere for decades, even centuries. Man-made climate change has a cumulative effect that worsens as time passes. What we're experiencing today is only a fraction of the warming that will ultimately result from the carbon we've emitted since the first coal-fired engines were ignited. The rest will be felt by our descendants, our poisoned legacy to them. Given how self-centered and shortsighted human beings tend to be—political leaders very much included—that delay dissipates the motivation to act on climate change now.

And there's one final characteristic that sets climate change apart from other potential existential risks. We can fight nuclear war by pushing for disarmament. We can defend the Earth against asteroids with very large nuclear weapons. We can combat engineered pathogens by regulating genetic engineering technologies. In each case—unless we're the ones launching nuclear missiles or releasing a deadly new virus—the existential threat is outside of us, and largely beyond our individual actions. But with climate change, virtually everything we do on earth—eat, drive, fly, work, watch television, have children—contributes to the threat, a threat built out of each of those actions, multiplied a billion, trillion times. As *New Yorker* writer Elizabeth Kolbert has written, when it comes to climate change, "we are the asteroid."[17]

This is the conundrum of climate change as an existential risk. The economists Gernot Wagner and Martin Weitzman put it this way in their book *Climate Shock: The Economic Consequences of a Hotter Planet*: "There's always a small chance that any particular final temperature wouldn't cause any damage. There's also the small chance that it would cost the world."[18] The challenge we face—in all existential risks but in climate change especially—is to keep that uncertainty and that responsibility from crushing us.

The early indications are not good.

———————

A year and a half after my trip to the Arctic, I traveled to Copenhagen for the 2009 UN climate change summit. The annual meeting hosts the more than 190 countries that are party to the United Nations Framework Convention on Climate Change (UNFCCC), the environmental treaty that launched international climate diplomacy more than twenty-five years before. If it was in Greenland that I witnessed climate change with my own eyes, it was in Copenhagen that I began to understand that we would not solve climate change—not in the way we hoped, at least.

It was the second UN climate summit I'd attended as the environment correspondent for *Time* magazine. The first, two years earlier, had been held on the considerably more pleasant Indonesian island of Bali, where if nothing else the sweltering tropical temperatures were proof that we all really were working on global warming. But Bali had been a bust. In one of his earliest acts in office, then-president George W. Bush withdrew the United States from the Kyoto Protocol, the first global deal mandating specific reductions in greenhouse gas emissions. For the remainder of their eight years in office, Bush's team acted as spoilers, blocking most meaningful efforts to create a stronger international climate pact and ensuring that summits like Bali were largely exercises in futility. Though I did enjoy witnessing stuffy UN technocrats sweating in the Indonesian batik shirts that were included in the summit's formal dress code.[19]

But the Copenhagen conference was going to be different. Climate change had become a global priority. Everyone, it seemed, from activists to corporations, wanted to be seen as good and green. Most promising of all, there was a new U.S. president, Barack Obama, who had come into office determined to restore American leadership on climate change. Obama would personally attend the conference, along with more than one hundred other world leaders.[20] By the time they were scheduled to arrive in the last days of the summit, diplomats would have hammered out a successor treaty to the expiring Kyoto Protocol, one that would have committed the world—including both the United States, which had left Kyoto, and major developing nations like China, which had been exempt from the treaty's mandated cuts—to reducing carbon emissions. Copenhagen was set to be the place where we turned the tide on climate change—or so I was told again and again that week, as I trudged through the city's snowy streets from event to event.

It didn't work out that way. By the time Obama had arrived for the final official day of the summit, the talks had all but collapsed. The major sticking point was the responsibility of those big developing nations. UN climate talks had always divided the world into developed and developing nations. In the UN's phrase, those two blocs had "common but differentiated responsibilities" toward climate change action. The common part was that global warming was a global problem with global causes and global effects, and therefore every country had a common responsibility to do something about it. The differentiated part recognized that nations that had industrialized first—the bloc of developed countries—had historically contributed much more to warming, simply because they had been emitting carbon at greater levels and for a longer period of time. That meant they had a different responsibility to cut carbon before developing nations.

It sounded fair, and it was, if you were comparing the United States with its 16.5 tons of CO_2 per person to, say, Rwanda with its 0.1 tons of CO2 per capita.[21] But the bloc of developing nations included not just desperately poor countries but major industrializing powers like China that were growing explosively, with all the carbon emissions that growth demanded. China had passed the United States as the world's top carbon emitter as early as 2005, and by the Copenhagen summit in 2009 China's emissions were 65 percent greater than America's.[22] New administration or not, U.S. diplomats still insisted that a new climate deal had to include limits on those major developing nations, but their representatives refused to accept any restrictions on their nations' right to emit greenhouse gases in the future, pointing to the fact that developed countries had only become developed by treating the atmosphere as a free carbon dump.

Each side had a point: developed nations like the United States had emitted the most carbon historically, but major developing nations like China were projected to produce the bulk of future carbon emissions[23]—the only kind of carbon emissions, after all, that a treaty could try to restrict. The result was paralysis. As the days and nights in Copenhagen ticked on without progress, it seemed possible that the entire UN climate system, which had been in place since the original UNFCCC treaty had been negotiated at the Earth Summit in Rio de Janeiro in 1992, would disintegrate. And if that happened, there was no plan B for the planet.

It is to President Obama's credit that he helped salvage the day. While reporters stalked the halls of Copenhagen's Bella Center, trolling for rumors and leaks, Obama personally negotiated with other world leaders, at one point crashing a meeting between China, Brazil, and India.[24] Together they finally managed to hammer out what became known as the Copenhagen Accord—an agreement, essentially, to agree about the importance of fighting climate change and allow individual countries to make their own pledges to mitigate carbon emissions in the future.[25] It sidestepped the common but differentiated responsibilities question by allowing nations to decide their own responsibilities, while putting in place a global system that would regularly check the progress of those pledges.

The Copenhagen Accord fully satisfied no one—especially not the most ardent environmentalists or the representatives of island nations that would be swallowed up by rising seas—but it kept the game going. Copenhagen ultimately helped pave the way for a better outcome six years later at the 2015 UN climate summit in Paris, where all major emitters—including China, which now emits more CO_2 than the United States and the European Union (EU) combined[26]—made voluntary pledges to reduce carbon emissions and work to keep global temperatures from rising more than 2.7 degrees above preindustrial levels.[27]

By 2015 I had been promoted to become *Time*'s international editor in New York, so I missed the UN climate summit in Paris that year, which is a shame because I hear the French capital is lovely in the fall, and because it might have been nice to experience just a measure of optimism as a climate journalist. Obama was certainly optimistic after Paris. "We came together around the strong agreement the world needed," he said from the White House after the summit concluded. "We met the moment."[28]

But I think it's the mood of Copenhagen in 2009—dark, despairing, wintry—that better captures the state of the global fight against climate change, even years later. That last night in Denmark, after Obama left on Air Force One—to get ahead of a blizzard bearing down on the east coast,[29] one of the many weather ironies of an ice-cold global warming summit—it fell to the delegates at the Conference of Parties to debate the finer points of the Copenhagen Accord, vote on the thing, and let us all go home. And yet the talks stretched on for hours upon hours, through the night and into the following morning, while bleary-eyed reporters watched. Final agreement remained elusive as the proceedings bogged down in UN legalese. I remember one delegate, from Saudi Arabia of all places, finally exploding in frustration: "This is without exception the worst plenary I have ever attended, including the management of the process, the timing, everything."[30]

And this was at a conference where almost everyone attending believed that climate change was real and must be addressed. The point was hammered home again and again: the time to act was now. We could delay no longer. Our very future was at stake. Yet inside the Bella Center, inside the corridors of power, where it mattered, no one really wanted to do all that much to slow climate change. Not if it carried any political or economic risk. Not if it could cost them their job, or restrict their citizens in almost any way. Climate change was important, sure—but not *that* important. Not if you judged political and business leaders—and ordinary people, too—on what they did, not what they said.

Copenhagen represented the death of a specific kind of environmental dream: that we would come together and in a single broad stroke legislate the planetary threat of climate change out of existence. We would not save the Earth at a conference, after all. That was true as well of 2015's Paris Agreement, which allowed countries to pick and choose their own carbon emission targets, and which unlike the Kyoto Protocol was nonbinding. Just how nonbinding became clear less than a year after the agreement was signed, when a climate-change-denying Donald Trump became president. True to a pledge he made during his campaign, on June 1, 2017, Trump announced his intention to withdraw the United States from the Paris Agreement.[31]

Even before Trump, however, all the world's good intentions weren't nearly good enough to prevent dangerous global warming. Tally up the pledges made by countries at Paris—including the United States—and experts say that greenhouse gas emissions in 2030 will exceed the level needed to keep eventual global temperature rise below 3.6 degrees by 12 to 14 billions tons of CO_2.[32] A 3.6-degree increase has long been considered a rough red line for dangerous climate change—while not precise, scientists believe that anything greater than that is more likely to lock in devastating environmental damage, including enough sea level rise to eventually erase entire island nations.[33] The actual gap between what we need to do and what we will actually do is likely even greater—as of 2017 not a single major industrial country was on track to meet its Paris pledges.[34] After three years of remaining roughly flat, global carbon emissions rose 1.6 percent in 2017 and an estimated 2.7 percent in 2018, reaching a historic high.[35] While clean renewable energy sources like solar are growing rapidly, so is overall demand for energy—and the vast majority of that energy demand is still being met by fossil fuels, especially in developing countries.[36] Polluting fuels like coal and oil still make up about 80 percent of global energy consumption, the same rough percentage they did more than thirty years ago,[37] before climate change had begun to penetrate the public consciousness.

Sixteen years ago, a climate scientist named Ken Caldeira calculated that the world would need to add about 1,100 megawatts' worth of clean-energy capacity—roughly what's produced by a single nuclear power plant—*every day* between 2000 and 2050 to avoid dangerous climate change.[38] In 2018 Caldeira checked on our progress and found that we were falling more than 85 percent short of where we needed to be.

In an article that same year in *MIT Technology Review*, the journalist James Temple also calculated that at the pace we're on, it would take not three decades to remake our energy system, but nearly four centuries.[39] In a 2018 report from the IPCC, scientists concluded that net greenhouse gas emissions had to reach *zero* by midcentury in order to have a decent shot at keeping temperature increase below 2.7 degrees, which means we have at most fourteen years of current emissions levels before that becomes impossible.[40] It has been more than a century since the basics of the greenhouse effect were discovered by the Swedish chemist Svante Arrhenius, and nearly thirty years since the 1992 Earth Summit in Rio de Janeiro. Yet we're being borne backward against the current of our carbon emissions. Optimism in the light of these facts is ignorance.

Part of the problem is that despite the apocalyptic rhetoric, we treat climate change not as the existential risk it may be, but as the inconvenience we hope it will be. This is true not only of our world leaders, but also many of the environmentalists, activists, and scientists who passionately care about climate change. Ted Nordhaus, the executive director of the Breakthrough Institute, an environmental and energy think tank with a sometimes contrarian bent, has noted a disconnect between the apocalyptic rhetoric of hard-core climate activists and what they actually do with their own lives. While they warn that climate change is an existential threat and that the world will quite literally end if we don't take radical actions immediately, they don't act as if they're in a fight to the death. They fly to international climate conferences like the one I attended in Copenhagen; they eat meat (if organic and grass raised); and they oppose nuclear power, which emits virtually zero carbon emissions, because it conflicts with conventional environmental opinion. That's not how you would act if you truly believed that a climate asteroid was on its way. "For all the discourse about climate change being an existential threat," Nordhaus told me, "when you look at what people actually do, it doesn't look like that at all."

That might seem like hypocrisy, but it's more accurately a measure of how deeply challenging climate change is, how embedded the causes of global warming are within our daily lives. It's easy to blame climate-change-denying politicians or big oil companies—and there's no doubt they play an outsize role in opposing commonsense climate solutions, even to the point of spreading outright lies. But the real fault lies in ourselves, and in the systems we've built.

Copenhagen taught me one thing: we will not, as countries or as individuals, make the radical changes in our lifestyles and our energy use that would be needed to fully eliminate the risk from global warming using current technology. That means we will almost certainly blow past the various red lines climate scientists have set. We won't keep atmospheric carbon concentrations below the 450 parts per million that climate diplomats have aimed for, not when levels have already passed 410 ppm and are rising by the year. We won't keep global temperatures from rising above the 2.7-degree limit that the Paris Agreement calls for, and we'll probably miss the 3.6-degree limit established by the original UN climate agreement. Without an ethical or political or technological revolution, seas will rise, forests will burn, superstorms will strike, tens of millions of people will be displaced, and who knows how many people will die. Whatever we might say about it, whatever our fears are, our ultimate attitude toward climate change has been this: we hope it won't be that bad. But hope isn't a policy—not when the world might be at stake.

We can't say we weren't warned. All the way back in 1861, the Irish physicist John Tyndall established that gases like CO_2 could warm the planet. By the 1950s, scientists were noticing a pronounced global warming trend, and began to build crude computer models of the climate. In 1965, a group of experts now known as the President's Council of Advisors on Science and Technology (PCAST) reported to President Johnson that growing levels of atmospheric CO_2 produced by the burning of fossil fuels would almost certainly cause significant changes in the climate, changes that "could be dele-terious from the point of view of human beings." In the decades that followed there would be millions of hours and hundreds of millions of dollars spent further refining climate science, but the basics were already established.

So was the political reaction: don't do much. The idea of delib-erately reducing fossil fuel emissions wasn't even suggested to John-son in that 1965 report. Instead his advisers thought that spreading reflective particles over 5 million square miles of ocean, in order to repel one percent of incoming sunlight away from the Earth, might do the trick. Nothing came of it, though this was one of the first times the possibility was raised of engaging in what is now called geoen-gineering, the attempt to intentionally shape the Earth's climate through direct technological intervention—a last-ditch option we'll return to later. It was also an idea typical of the high nuclear age, a time when scientists were far less humble about tinkering with the machinery of the planet. In 1971, the Soviet Union even attempted to use three underground nuclear explosions to build a massive canal in an effort to reconstruct the Volga River basin. It didn't work, though the effort had the unexpected side effect of convincing the United States to launch a climate change research program of its own.

From the beginning, everything about the way climate change worked as a physical process made it easy to ignore politically. One ton of CO_2 burned will on average lead to approximately 0.0000000000027 degrees Fahrenheit of warming, according to a 2009 study by scientists at Concordia University in Montreal.[41] There are other greenhouse gases, like methane, and thousands of other factors—water vapor, ocean heat circulation, El Niños—that influence changes in the climate. But at its foundation, man-made climate change really is that straightforward. And that's a problem, because carbon is everywhere in the modern world.

We expel carbon every time we breathe, and when we burn plant life to make room for farms or cities or anything else we need space for, carbon is released into the atmosphere. The fossil fuels like coal and oil that have made the modern world go since the dawn of the Industrial Revolution are mostly carbon. And unlike other potential pollutants— the radiation from nuclear explosions, for example—the sources of fossil fuel emissions are everywhere, from everyone. The reason Johnson's scientific advisers didn't even consider calling for a reduction in fossil fuel emissions in 1965 is that at that point, reducing the supply of fossil fuels would have been like reducing the supply of air. We've made progress—renewables contributed almost half of the growth in global energy generation in 2017, and are poised to keep growing as technologies improve and prices drop.[42] But not nearly enough.

Perhaps the best way to understand the unique challenge of climate change is to compare it to another environmental threat that once imperiled the world, and which is now largely forgotten: the destruction of the ozone layer. In the 1930s, the chemicals known as chlorofluorocarbons (CFCs) were developed as refrigerants. The chemicals quickly found their way into air conditioners, refrigerators, and aerosol sprays, in part because they were nontoxic to humans, unlike earlier artificial refrigerants. But in the early 1970s scientists discovered that when CFCs reached the stratosphere, they triggered a chemical reaction that eroded the ozone layer. When ozone is stripped away, more of the sun's harmful ultraviolet radiation reaches the Earth's surface. That radiation—the same radiation we use sunscreen to shield our skin from—can damage DNA, increase the chance of skin cancer, and harm animals and plants. Without the ozone layer, life as we know it would not exist.

This was bad, existentially bad, as the University of California, Irvine chemist F. Sherwood Rowland realized during his work on CFCs in the early 1970s. Rowland told his wife one day, when she inquired about his research, "The work is going very well. But it may mean the end of the world."

That it didn't was thanks in part to his efforts, and those of Paul Crutzen and Mario Molina, who shared the 1995 Nobel Prize in Chemistry with Rowland for their research connecting CFCs and ozone depletion. But science alone wasn't enough. In the 1980s governments began to take steps to ban CFCs from use, efforts that accelerated after a shocking hole in the ozone over Antarctica was discovered in 1985. By 1987 the international community had adopted the Montreal Protocol, a global agreement to phase out ozone-depleting substances, including CFCs.[43]

The Montreal Protocol was the first global treaty to be ratified by every country in the world, and it has been called "the single most successful international agreement to date" by former UN secretary-general Kofi Annan. Thanks to the protocol, depletion of the ozone layer was eventually halted. By 2050 the ozone layer should be mostly restored, preventing more than 1.6 million skin cancer deaths in the United States alone.

The Montreal Protocol represents the ideal response to man-made existential risks: recognize the science, come up with a technical and political solution, implement it. And do it quickly—a little more than a decade passed between Rowland, Crutzen, and Molina's work on the ozone-depleting effects of CFCs and the adoption of the Montreal Protocol.

This is exactly what environmentalists wanted to do with climate change at Copenhagen—and they failed. Now all but the most optimistic environmental activists have given up hope that we could ever adopt a binding global treaty that would limit CO_2 emissions as clearly as the Montreal Protocol did for CFCs. And the primary reasons lay not in political failures or public apathy, but in the "super-wicked" nature—an actual scientific term—of climate change.

While CFCs were largely limited to refrigerants and aerosol sprays—a small part of the economy[44]—fossil fuels are everywhere. While the effects of ozone depletion could be seen immediately and undeniably—most of all in a giant ozone hole that had opened up over the South Pole—the effects of climate change are time delayed and muddled, easily lost amid the normal swings of extreme weather. While many of the major companies manufacturing CFCs had begun work on effective substitutes years before the Montreal Protocol was adopted[45]—which helped ensure that it cost the United States just $21 billion to comply with the treaty[46]—there is no simple and inexpensive technical substitution for oil, coal, and natural gas.

The political leaders of 1987 weren't wiser than those of 2009—it was under President Reagan, no one's idea of an environmentalist, that the United States negotiated the Montreal Protocol. And it wasn't that the companies that made billions off CFCs necessarily cared about the planet more than oil and coal companies—DuPont, the main manufacturer of CFCs, resisted the science linking the chemicals to ozone depletion for years. It was simply much, much easier to phase out CFCs than it has been to eliminate fossil fuels. And that will not change anytime soon.

———————

You may never have heard of Vaclav Smil, a professor emeritus at the University of Manitoba, but no less than Bill Gates believes he's one of the smartest people in the world. The Microsoft founder once wrote: "I wait for new Smil books the way some people wait for the next Star Wars movie."[47] Smil has written dozens of books, but he is best known for his work on energy, especially the social and technological transition from one source of energy to another. That is essentially what the campaign against climate change amounts to—accelerating the transition from carbon-intensive fossil fuels to carbon-free sources like solar.

Many environmentalists believe that the only thing preventing that transition is a lack of political will. In 2008, former vice president Al Gore called on the United States to make the transition to a carbon-free electricity system within ten years. More recently the Green New Deal championed by progressives like Democratic representative Alexandria Ocasio-Cortez called for moving the United States to 100 percent renewable electricity by 2035.[48] "This goal is achievable, affordable, and transformative," Gore said in 2008. "The future of human civilization is at stake."

Smil might agree with the latter statement—but not the former. And the reason why boils down to a simple factor: energy density. Fossil fuels have it—renewable energy sources, for the most part, do not.

From our time as hunter-gatherers to our current lives as office workers, humans have repeatedly transitioned from one source of energy to another. Every time that transition took place, we shifted to an energy source that has greater density, meaning it can release more power per unit of fuel. From human muscle power to harnessing animals, from burning wood to burning coal, from refining oil to splitting the atom through nuclear fission—with each transition we earned more bang for our energy buck. Energy transitions take a long time, much longer than you might expect—almost the entire nineteenth century passed before ancient biofuels like wood were replaced as the major global energy source by fossil fuels, mainly coal.[49] That's largely because when an energy source becomes dominant, an entire infrastructure is built to produce it and service it, and that infrastructure takes time to tear down and replace with something new—even if that something new is superior.

When it comes to the full transition to renewable energy, that means we could be in for a long wait. Wind and solar only began to be a measurable part of the global energy mix in the 1980s.[50] Even today—despite years of rapid growth—non-hydro renewables account for less than 4 percent of global energy consumption.[51] A 2018 report by the International Energy Agency found that as fast as renewables are growing—displacing fossil fuels like coal while they do so—it's not nearly fast enough to prevent dangerous, perhaps even catastrophic levels of warming.[52] This is exactly what Smil has predicted.

And the renewable energy transition must overcome another obstacle, one that previous transitions didn't face. Instead of shifting to a fuel source that has greater energy density than the current incumbent, the renewable revolution demands that we move to a less dense energy source. Wind and solar and biofuels require more land and produce less energy per unit than fossil fuels like oil and coal, which is a problem, since land is one of the few resources on this planet that are truly scarce. Renewables are also intermittent—wind turbines only produce power in the breeze, and solar panels only produce electricity when the sun shines—while fossil fuels can be stored and burned anytime and anywhere. Scientists are hard at work developing technologies that could economically store the energy produced by wind and solar so it can be used around the clock, but a barrel of crude oil is already a stable, movable battery in liquid form, one that is especially useful for powering transportation, which accounts for 14 percent of global greenhouse gas emissions.[53] "Give me mass-scale storage and I don't worry at all. With my wind and [solar] photovoltaics I can take care of everything," Smil told *Science* in 2018. But, he added, "we are nowhere close to it."[54]

Smil believes the transition away from fossil fuels must happen and will happen, both because he is worried about climate change and because while we've gotten better and better at digging them out of the Earth, fossil fuels are ultimately finite in a way that the wind and the sun aren't. Fossil fuels also have dire costs besides climate change, like the air pollution they create that hastens the deaths of millions of people per year.[55] (The smoggy skies of New Delhi or Beijing—or Los Angeles on a bad day—are largely the side effect of burning oil and coal.) There are security and performance benefits to switching to renewables, especially solar, which doesn't require a sprawling and vulnerable electrical grid.

All of these factors—in addition to the drive to reduce carbon emissions—explain why more solar power capacity was added globally in 2017 than fossil fuels and nuclear power combined.[56] The price of solar power has fallen by an incredible 99 percent over the last four decades, driven by a mix of government policy, technological improvements, and simple experience.[57] But the basic barriers Smil identified haven't been broken. "We are an overwhelmingly fossil-fueled civilization," he has written. "Given the slow pace of major resource substitutions, there are no practical ways to change this reality for decades to come."[58]

It's not for lack of trying—not exactly, at least. According to a 2017 report from Stanford University, the world is spending about $746 billion a year on renewable energy, energy efficiency, and other low-carbon technologies.[59] More money is probably earmarked to fight climate change than every other existential risk in this book—combined. We're just not spending *enough*, if enough is defined as forcing an energy transition that is fast enough to eliminate the risk of dangerous climate change. The Stanford report calculated that if we want to keep global carbon concentrations below 450 ppm, the private sector needs to start spending $2.3 *trillion* a year, approximately three times what we spend now and more than the global budget for defense.[60] And many scientists and environmentalists believe we actually need to bring carbon concentration down to 350 ppm to ensure our safety, closer to the level the Earth experienced during the preindustrial age, which would surely require trillions upon trillions of dollars more.

Could we do it? Sure—if we decided to mobilize the entire global economy and population around the singular goal of cutting carbon as fast as possible. That's what the United States did during World War II, only with the aim of defeating Japan and Germany. Washington nationalized much of the steel, coal, and transportation industries, and required car companies, for example, to stop making vehicles for domestic consumption and instead churn out planes and tanks. It worked because the government forced business to do its will, with the support of virtually the entire American population, a unity that is unimaginable today. Meat, sugar, gasoline, and more were strictly rationed. The United States became the arsenal of democracy and won the war chiefly because it could outproduce its enemies—though the labor of some physicists at Los Alamos played a part, too.

Bill McKibben, the author and environmentalist who helps lead the 350.org climate activist movement, cited the example of World War II in a 2016 article for the *New Republic*.[61] McKibben wrote: "If Nazis were the ones threatening destruction [from climate change] on such a global scale today, America and its allies would already be mobilizing for a full-scale war." But World War II lasted only a few years, after which the government loosened the reins and the private sector got back to the business of business. A war against climate change, by contrast, would be open-ended. And Nazi Germany represented a clear existential threat to the world, at least the world as we valued it. Can we say climate change is an existential threat in the same way?

That the natural world is degraded from what it once was is indisputable. If Christopher Columbus were to arrive in the Americas today aboard his *Nina, Pinta,* and *Santa Maria,* he would find 30 percent less biodiversity than in 1492.[62] The global population of vertebrates has declined by an estimated 52 percent between 1970 and 2010.[63] The current extinction rate is 100 to 1,000 times higher than it has been during normal—meaning non-mass extinction—periods in biological history,[64] with amphibians going extinct *45,000 times* faster than the norm.[65] One point eight trillion pieces of plastic trash, weighing 79,000 tons, now occupies an area three times the size of France in the Pacific Ocean[66]—and this Great Pacific Garbage Patch is expected to grow 22 percent by 2025.[67] And of course there is the climate change that has happened and the climate change that is to come.

Yet amid all this natural loss, human beings, on the aggregate, have largely thrived. That's definitely true compared to the days of Columbus—the economist Angus Maddison estimates that between 1500 and 2008, global average per-capita gross domestic product (GDP) multiplied by more than thirteenfold.[68] Much of that gain has occurred in recent years, as globalization helped lift more than a billion people out of extreme poverty in the developing world since 1990 alone[69]—the same years when environmental damage, including the first signs of climate change, began compounding. Nor is this simply a story of GDP. The Human Development Index (HDI) is a composite statistic developed by the Pakistani economist Mahbub ul Haq to track life expectancy and education, as well as per capita income. Graph every country for HDI since 1990 and you'll see—with the occasional ups and down of individual nations—a decidedly rising trend that shows no signs of reversing.[70]

Perhaps a better way to understand that picture is to look at those years when economic growth temporarily halted, and rolled backward: the 2008 global financial crisis and its immediate aftermath.

The Great Recession led to a loss of more than $2 trillion in economic growth globally, a reduction of nearly 4 percent.[71] By one estimate the crisis cost every American nearly $70,000 in lifetime income on average.[72] But that's just money. The real human cost of the financial crisis was in broken families, sickness, even death. One study found that the crisis was associated with at least 260,000 excess cancer-related deaths around the world, many of them treatable.[73] As unemployment rates rose, so did suicide rates—a correlation that has been seen in past financial crises.[74] Nothing we've experienced with climate change or any other environmental threat compares—yet—to the sheer suffering that was inflicted on the world when the wheels of growth temporarily stopped. "We're still dealing with the political and economic blowback from the 2008 financial crisis, and everything about it makes climate change seem like a walk in the park," Ted Nordhaus told me.

This picture—a degrading natural world poised against a generally improving humanity—has a name: the environmentalist's paradox. In 2010, a team of researchers led by Ciara Raudsepp-Hearne at McGill University tried to figure out what could explain it. They came up with four hypotheses. It could be that humanity only appears to be better off—but the numbers belie the conclusion that improved human well-being is an illusion, including those, like the HDI, that measure more than just raw economic growth. It may be that the remarkable improvements in food production over the past century are so important that they simply outweigh any environmental drawbacks, however severe. For most of human history deadly famine was just one bad harvest away. That's no longer the case for most of the world.

It may be that technological advances and increased wealth could make us less dependent on a healthy ecosystem. That third hypothesis—a brand of technoptimism—is implicit in climate economics. The Stern Review, put out by the British economist Lord Nicholas Stern in 2006, is one of the most exhaustive explorations ever completed on how climate change might damage the global economy. It's also one of the most pessimistic. Yet even the most extreme scenario that Stern outlines has climate change triggering a 32 percent decline in economic output relative to the baseline in 2200. That sounds like a lot, and it is—but since Stern also assumes that economic per capita growth will continue through all of those years, humans at the dawn of the twenty-third century will still be nine times richer per person than they were when the Stern Review was published.[75] In other words, climate change will make us poorer than we would be in its absence—but we'll keep getting richer, much richer, than we are today.

If that really is how climate change will unfold, then it is not an existential risk. The natural world will continue to suffer and degrade, and we'll be affected—some of us far more than others, especially those who have done the least to contribute to the problem—but humanity on the whole will be okay. More than okay, at least by our generally grim historical standards.

But there's a fourth possible answer to the environmentalist's paradox: the worst is yet to come. In this hypothesis, the material benefits that humanity has enjoyed over the last few decades have been purchased on overextended carbon credit, and just as previous credit bubbles have inevitably led to painful economic contractions, so it will be with our subprime environmental loans. In this scenario the desire of 7-plus billion human beings for an American-style middle-class lifestyle, with all the energy and meat and pollution that entails, is not sustainable, environmentally or economically. The bill will come due, and when it does, no degree of innovation will save us from collapse. Climate change in this scenario is very much an existential risk. Not only are there environmental limits that we can't innovate our way around, but there are also tipping points for the climate—and we exceed them at our existential peril.

The uncertainty at the core of the environmentalist's paradox is emblematic of global warming science more generally. Why provide only one answer when you can have four—or more? Climate scientists try to predict the future by constructing models that anticipate both how humans will react to warming—by reducing carbon emissions or substituting energy sources—and how the climate itself will respond to those efforts. Those models spit out an array of different scenarios for the future—more than a thousand in the IPCC's most recent assessment[76]—that are further whittled down into projections of a range of possible future climates.

Skeptics often seize on the uncertainty as evidence that climate change isn't real or isn't something we need to worry about. They're dead wrong—the physical basics behind the science of man-made climate change are well established, and existing climate models have been confirmed by real-world warming over the past several decades. But in part to avoid giving skeptics more ammunition, scientists tend to emphasize the most likely, middle-of-the-road projections for future warming, rather than the extremes.

In its 2014 assessment of climate science, the IPCC reported that without significant emissions reductions, we're most likely to experience warming in the range of 7 degrees above preindustrial levels by the end of the century.[77] Seven degrees Fahrenheit would be almost twice the 3.6-degree rise that the UN climate system was originally formed to prevent, all the way back in 1992. It would be a climate warmer than anything *Homo sapiens* has ever experienced before, warmer than it was during the Pliocene epoch more than 3 million years ago, when sea levels were as much as eighty-two feet higher than they are today.[78] Would that represent an existential threat? Perhaps—although some of the worst impacts, like the coastline-reshaping rise in sea level, would still take centuries more to play out, giving us time to adapt, perhaps in ways we can't begin to anticipate now. "Will the planet still be around then, too?" Gernot Wagner asked himself when I visited his office in Cambridge, Massachusetts. "Yes. Will society as we know it? No. The rich will be fine. They'll buy a second air conditioner and fly their private jet to Aspen. The poor as usual will suffer extraordinarily more than the rich. But is it an existential risk for them? No."

The uncertainty in climate science runs both ways, however. The most likely outcome by the end of the century may be around 5.4 degrees of additional warming—or less if we either get much better at reducing carbon emissions or the climate system turns to be less sensitive than we generally expect.[79] But there's about a 10 percent chance, as Wagner and Martin Weitzman describe in their book *Climate Shock*, that the uncertainty in those projections could skew in the other direction, and warming might be far worse: as much as 10.8 degrees by 2100 or more, with a 3 percent chance of 18-degree warming.[80] That's the shock in *Climate Shock*—the shock of a very nasty surprise, one that might be enough to snuff out human life as we know it.

What Wagner and Weitzman are describing is a "fat-tailed risk." It's easy to picture. If you distribute the probability of an event over a line graph, you'll usually get what's called a bell curve—close to zero probability at either extreme, and a high probability in the middle, which shows up as a bell shape on the graph. Think of a random day in your own life: you're highly likely to, say, eat lunch, but very unlikely to win the lottery or get hit by a bus—extremely good or extremely bad events.

Graph the probability of future global warming, however, and you end up with a curve that has a fat tail at the end, which means the chance of extreme warming—of an extremely negative scenario— is something like 10 percent. How worried should we be about a 10 percent chance of something? It depends on the context. If you get a hit once out of every 10 times you go to the plate in Major League Baseball, let's just say you'd better be a really good pitcher. But if there were a 10 percent chance that the bridge in front of you would collapse, you wouldn't cross it. And if you thought there were a 10 percent chance your home would burn down in any given year, you'd pay a lot for insurance—or find somewhere else to live.

That 10 percent probability is itself an estimate, and as we learn more about the climate, that fat tail may grow fatter or thinner. Ken Caldeira and his colleague Patrick Brown of the Carnegie Institution for Science took the most popular climate models, then examined how well they've predicted the amount of warming we've seen up to the present. The simple idea is that if a model accurately predicted the past, it's more likely to accurately predict the future. Unfortunately for us, the models that best predicted the past are also the ones that predict greater levels of warming. Their study finds that the most likely warming is about 1 degree greater than the raw models have suggested—raising the chances that a climate shock could be waiting for us.

What would that shock feel like? In August 2018, a team of scientists led by Will Steffen of the Australian National University and Johan Rockstrom of the Stockholm Resilience Centre published a grim prophecy for the planet. Even if the world managed to meet the carbon emission reduction goals laid out in the Paris Agreement—which we are very much not on track to do—they calculated that there is a chance that climate change could still continue unimpeded. Instead of a global temperature rise of around 3.6 degrees—terrible but probably endurable—the planet would keep warming, perhaps as high as that 10.8 degrees that Wagner and Weitzman warned us about. Already hot parts of the planet would become uninhabitable. Sea levels could rise 30 to 200 feet above where they are now. "You're going to eliminate most of the coastal megacities on the planet,"

Steffen told me by Skype from his home in Australia, where recent summers have become so hot that fruit has cooked in the trees.[81] "We would lose terrestrial biomes, lose rain forests, lose ice." We would be living in what Steffen and his colleagues called "Hothouse Earth"— assuming we could live there at all.[82]

Mainstream climate economics generally treats the global environment as a linear system—so much carbon added to the atmosphere equals so much warming equals so much economic damage. But many scientists believe that the linear system could break down if we pass certain environmental tipping points, creating feedback processes that would make Hothouse Earth inevitable. Steffen pointed to the Arctic sea ice that caps the planet. While it exists, the white ice and snow reflects sunlight, cooling the planet the way a white T-shirt keeps you cooler in the summer, but as it disappears it has been replaced by dark ocean water that absorbs sunlight, further warming the Earth. The hotter the Earth gets, the faster the ice melts, which in turn opens up more dark water, accelerating climate change. And that's just one tipping point. The Hothouse Earth paper identifies more than a dozen, some of which are connected to each other, so they could fall like dominoes as the climate warmed.

"The dominant paradigm is that the level of climate change is solely determined by the level of human carbon emissions," said Steffen. "That's true at lower levels of emissions. But we are saying that there is a risk that warming as low as minus 2 degrees Celsius [3.6 degrees Fahrenheit] could allow these intrinsic feedbacks to take control of the trajectory of warming."

Is the climate system riddled with such tipping points, like land mines on a battlefield? We don't know for sure—and that's the biggest climate question of them all.

What's true for climate change is true for all the existential risks surveyed in this book, only more so. We're forced to grapple with uncertainty in a contest where the price of getting it wrong, of failing to do enough to forestall catastrophe, may be infinite. And the human brain does not do well with uncertainty, especially the kind of fat-tailed uncertainty presented by climate change. As Weitzman told me, "there is a human tendency to say that the probability is effectively one, or the probability is effectively zero." Instead of treating that 10 percent probability of catastrophically high warming as if it has a one-in-ten chance of actually coming true—which would demand that we do much, much more—we act as if it's impossible. "With these worst-case scenarios, it's as if we can't seem to afford to act on them," said Weitzman.

Part of what holds us back is the distributed, globalized causes of man-made climate change. A ton of carbon emissions emitted at any one place in the world has a warming effect on the entire world. To prevent nuclear war we have to focus our efforts on just a few nuclear-capable nations, but every country contributes to climate change, as indeed almost every individual does through their use of energy. Ever have a dispute at work where no one will take responsibility because everyone is implicated? That's climate change times 7.7 billion—the ultimate collective action problem.

While the causes of climate change are distributed around the world, the effects of warming are primarily distributed in the future. There is a time lag built into the machinery of climate change. Even if we somehow halted all man-made greenhouse gas emissions today— no more coal, no more oil, no more gasoline—the climate would continue to warm by an additional 0.5 degrees simply because of the greenhouse gases emitted in the past, going all the way back to the Industrial Revolution.[83]

The real victims every time we turn the ignition on our cars or switch on the air-conditioning are not us, the people of the present generation, but our children, our grandchildren, and their descen-

dants. With every other existential risk—a nuclear war, a bioengineered pandemic, even hostile aliens—the people of today would
be the first victims, even if the greater tragedy is the loss of the far
larger number of human beings who would have lived in the future.
But the worst consequences of climate change, those that might truly
qualify as existential, won't hit home until many or most of us are
long gone. That raises a vital question: what do we actually owe the
people of tomorrow?

When you think about yourself—at least while doing so inside the
narrow metal tube of a machine that uses functional magnetic reso-
nance imaging (fMRI)—a certain part of your brain, called the
medial prefrontal cortex, or MPFC, will light up like Times Square
on New Year's Eve. If you think about a family member, the MPFC
will still light up, though less robustly. And if you think about other
people whom you feel you have no connection to—like, say, the
inhabitants of the South Asian island nation of the Maldives, which
will likely one day be erased by climate-change-driven sea level rise—
the MPFC will light up even less.

You don't need a $3 million MRI machine to know that human
beings are self-centered creatures.[84] But as Jane McGonigal, the
research director of the Institute for the Future, noted in a 2017 article for *Slate*, what's strange is that if you think about your own self
in the future, you'll see less activation in the MPFC than when you
imagine your present self. The further out in time you imagine that
self, the weaker the activation is.[85] As McGonigal writes: "Your brain
acts as if your future self is someone you don't know very well and,
frankly, someone you don't care about."[86] And if we view our own
selves in the future as virtual strangers, how much less do we care
about the lives of generations yet to be born?

There are sound economic reasons for privileging the present over the future—the foremost being that, whatever the actuarial tables might say, the future isn't guaranteed to anyone. The question is how much we should value the present over the future. Economists have a figure for this—the "social discount rate," which quantifies how much present-day value declines as we move into the future. It's essentially the inverse of an interest rate, which calculates how much an investment will grow in the future. An array of factors go into determining the social discount rate, like how much interest banks charge on loans and the expected rate of return from investments. The Obama administration, for instance, used a discount rate of 5 percent for its analysis of the social cost of CO_2 emissions—how much economic damage each ton of carbon dioxide is estimated to cause.[87]

Economists also discount the future in part because they expect it to be richer—much richer—than the present. The temporary dip of economic crashes notwithstanding, that's been the case historically, especially in the modern era. Global GDP increased from about $1 trillion in 1900 to over $40 trillion in 2000—and then more than doubled over the next eighteen years.[88] We invent new technologies and we become more productive, and for mainstream economists, there's no reason to think that the engine of growth will run out of fuel anytime soon. And so economists on the whole tend to be less worried about problems like climate change that will mostly be felt in the future. At the current rate of growth, the global economy will be twice as large in about twenty years—and a world that is twice as rich should be able to handle disasters that are twice as severe. Or at least that's the hope.

Discounting makes sense over relatively short time horizons, like the decision of a business to take out a loan. But when we begin to look into the further future—future on the scale of climate change, many decades and even centuries from now—discounting can spit back results that seem confounding. The philosopher Derek Parfit, whom we met back in the introduction, wrote that "at a discount rate of five percent, one death next year counts for more than a billion deaths in 500 years."[89] What that means is that the economic reasoning of discounting concludes that it would be more worthwhile to save one life next year, rather than save more than a billion lives five centuries from now. To put that in monetary terms, with a 5 percent discount rate, it would only be worth spending about $2,200 today in order to prevent $87 *trillion* in damages—the size of the total world economy now—in 500 years. Make it 700 years and it would only be worth spending 13 cents today. That's how much we discount the far future.

An easier way to understand the queasy moral math of discounting is to turn the tables and imagine ourselves not as the people of the present, looking into the future, but as the future of the past. I'm indebted to Nick Beckstead, a program officer at the Open Philanthropy Project—a Silicon Valley–based nonprofit that supports evidence-based charities—for the following analogy. Let's say the ancient Romans had been told by their ancient Roman scientists that an act that seemed to benefit their society in the short term— like burning cheap but polluting coal, for instance—could expose generations two thousand years in the future to existential risk? From our perspective in the present, it would have been insane for the Romans to do something that would risk our quality of life and even our existence, even if it seemed to benefit them in their time. Yet if the Romans employed the same discounting analysis mainstream economists employ for climate change, the math would tell them to go ahead—the value of even all of human civilization a couple thousand years into their future would be practically nil to them, just as the value of humans hundreds of years into our future is nil to us—at least with a 5 percent discount rate.

But that 5 percent discount rate wasn't chiseled on a stone tablet and handed down by Adam Smith. We can choose our own discount rate, and that choice is a rough reflection of how much we value the future, which is another way of saying how much we care about climate change. "If you want to trivialize climate change, you want a high discount rate, which will legitimately trivialize it," said Weitzman. "And if you want climate change to matter and to get people to do something about it, you're going to have to have a story that presumes a low discount rate."

You'll recall that the 2006 Stern Review on climate change predicted much greater economic costs from warming than most other models. A main reason is that Stern chose an unusually low discount rate: just 1.4 percent. This was in part because Stern was more pessimistic than the average economist about how much wealthier future generations would be, and in part because he took an ethical stand that we, the people of the present, should value our distant descendants almost as much as we value ourselves. Stern used that discount rate to argue that it made economic sense to spend as much as 1 percent of global GDP now to avert dangerous climate change in the future—a figure he later raised to 2 percent of GDP,[90] which today would amount to well north of a trillion dollars a year.

We can let the economists argue over the validity of Stern's choices—and they certainly did. Weitzman, for one, was critical of where Stern set the discount rate in his models, but he also acknowledges that conventional discounting fails in the face of the fat-tailed risk of catastrophic climate change. If one degree of warming equals the loss of 1 percent of GDP growth, and two degrees equals 2 percent, then we could probably count on future economic growth more than compensating for whatever global warming throws at us, however bumpy the ride may be. But if there are hidden tipping points, if the relationship between temperature rise and economic damage is not linear but rather exponential, then we have a very different story.[91]

How we choose to value future generations will help decide what we should do now about climate change. And that should worry us. We know from neuroscience and psychology that we don't instinctually care about the future, and we know from economics that it doesn't seem to pay to care about the future. But ethically, what *should* our relationship to our future descendants be, those great-great-great grandchildren whose quality of life, perhaps their very life, will depend in part on what we do today? It's a question that directly matters for climate change, but it applies to every existential risk threatening the loss of that future. So we need to speak not just to the economists but to the ethicists.

Most ethicists would say that it is wrong to automatically value the life of someone close to you in space over the life of someone far away. We often do that in practice—we tend to care more about our countrymen and neighbors than we do about foreigners—but that doesn't justify our chauvinism. In fact, the growth of internationally focused charities like the Gates Foundation and the general increase in foreign aid over time[92] show that as humanity has matured, we have taken a larger interest in the well-being of those separated from us by space, in part because global media allows us to witness their lives. So why don't we take a greater interest in the well-being of future generations separated from us by time?

We do, to a certain extent. I love my son more than I thought was possible, and my wife and I will sacrifice some of our present enjoyment for his future well-being. (It began with sleep.) We all love our children—in the particular. But while the children may be our future, they're not a very useful stand-in for the farther future that matters for climate change and other existential risks. A better question would be: was I ready to sacrifice for my son when he was just a possibility? Will we sacrifice for our hypothetical children and grandchildren and great-grandchildren, whom we may never know and who may never know us?

For most of us, our revealed preferences indicate that the answer is no, as the collective lack of sufficient action on climate change demonstrates. The problem is that, as the Yale futurist and sociologist Wendell Bell has written, "a present sacrifice for the welfare of the future appears to be a one-way street."[93] We experience the sacrifice in the here and now, and people we will never meet enjoy the benefit. So instead the present is essentially "colonizing the future," in the words of the social philosopher Roman Krznaric, treating it "as a distant colonial outpost where we dump ecological degradation, nuclear waste, public debt and technological risk."[94]

Because they don't exist yet, these people of the future have no voice, no way to lobby for their needs. The United Nations exists in part to give representation to everyone living on Earth now, no matter where they are, but as the New York University ethicist Samuel Scheffler told me, "There is no trans-temporal United Nations. The only way we're going to work for the future is if we are motivated to act on their behalf."

Scheffler is the author of books with dire titles like *Death and the Afterlife*, but for an ethicist he's actually something of an optimist. In his spare and neat office off New York's Washington Square, he explained to me that we're more connected to future generations than we may appear. Much of what we do in the present day only has meaning if there is a future that continues beyond our own deaths. "If I'm trying to find a cure for cancer, and in fact the human race is going to die out in thirty years, that activity really won't be very valuable," he said. We will never know the far future, but we depend on its existence. Take it away and you take away something essential about being human. You lose value in the present day—perhaps all value.

To demonstrate what it would feel like to lose the future, Scheffler pointed to the example of a novel, *The Children of Men*, by the English writer P. D. James. (It later became a brilliant 2006 film by the director Alfonso Cuarón.) The book is set in England in the year 2021, in a world where some years before, global sperm counts mysteriously dropped to zero. Humans can no longer reproduce, and science is helpless. The last people—known as Omegas—were born twenty-six years before the start of the novel. As a result, the human race is slowly dying out.

If you were forced to pick an extinction event, you could do worse than the one in *Children of Men*. No one is killed violently in a super-volcanic eruption or an asteroid strike. No one watches their children die of an engineered superflu. The climate is much as it is today. The world loses no money, nothing that it already possesses—except, of course, the future. And as James deftly shows, the future turns out to be everything.

Society slowly collapses. People retreat into puerile pastimes, and the pursuit of justice and art and truth—the pursuit of the timeless—withers away. "Without the hope of posterity," writes James in the voice of her narrator Theo, an Oxford history professor with no students left to teach, "for our race if not for ourselves, without the assurance that we being dead yet live, all pleasures of the mind and senses sometimes seem to me no more than pathetic and crumbling defences shored up against our ruins."[95] It is a world that may live for a time, but is already dead.

In *Children of Men*, the end of the world is simply happening, and humanity is helpless to do anything about it. But with climate change, and with the other existential risks in this book, we may well be choosing extinction, through our actions or our inactions. This is another reminder of how important our moment is for everything that might come after us—or might not. Previous generations may not have cared much more about the future than we do now, but there was also much less they could do to affect the future, negatively or positively. Our thought experiment notwithstanding, there was nothing the ancient Romans could have done or not done that would have led to the end of the world. But we have that power. That is an awesome and terrible responsibility—to provide the best and the safest space for the future in an age of escalating existential threats. But it's one that we too often ignore.

We're asleep to the future—and Alexander Rose wants us to wake up. Rose is the executive director of the Long Now Foundation, a San Francisco–based collective dedicated to fostering long-term thinking. And I mean really long—Rose and company, which includes the musician Brian Eno and *Wired* magazine cofounder Kevin Kelly—are pondering the next ten thousand years of humanity, longer than civilization has so far existed. The group holds talks and conferences, and it even has its own bar in San Francisco's Fort Mason—the Interval at Long Now, where, true to its focus on time, you can get six different versions of an Old-Fashioned. It holds a series of long bets—wagers that will play out over the course of decades, like one betting that the human population will be smaller in 2060 than it is today. The Long Now Foundation is building a library of human knowledge meant to be archived over the next ten millennia, with much of it placed on "Rosetta disks"—palm-sized disks of solid nickel that contain thirteen thousand pages of information engraved in a microscopic font.

The members of the Long Now Foundation believe that whereas we once looked to the far future with anticipation—dreaming of moon colonies and rocket packs—we've since become consumed with the ever-present now, fearful that the human story might be coming to an end but helpless to do anything about it. Rose, whom I met for a coffee one bright March afternoon at the Interval, is realistic about the existential risks that humanity faces—including climate change—but he worries that our anxiety keeps us from dreaming about tomorrow. "A lot of people think of ourselves as being at the end of a ten-thousand-year story of human civilization," he told me. "But we're arguing that we should think of ourselves as being in the middle of a twenty-thousand-year story. How would we act differently, how would we feel about ourselves and about civilization?"

The Long Now's grandest project is a symbolic one: a timepiece. The 10,000-Year Clock, as it's called, is being built inside a mountain in western Texas. It will be some two hundred feet tall, a giant mechanical computer that will mark out the days and centuries over the next ten thousand years. Each time the clock's chimes ring over those ten millennia, it will play a tone it has never played before. It will be a monument, but not to its creators—the clock will contain no references to the individual leaders of the Long Now. Its point is its endurance—if we can build a clock that can ring for ten thousand years, perhaps humanity can last that long as well. And just as the first photographs of Earth from space helped us realize that we all shared the same fragile blue marble, the clock is meant to restore the idea of the future to a civilization stuck on the present. "The only message that we are trying to get across is that there are long-term things that matter, and you get to decide what those are for you," said Rose, who serves as the project manager of the clock.

Planning for the far future doesn't have to mean obsessing over it, and it doesn't automatically mean sacrificing the good of this generation to preserve those who will come next. Rose and most of the other members of Long Now are optimistic about the future, which is part of the reason they're building the clock in the first place—because they believe there will be humans alive in ten thousand years to hear it chime.

To those of the Long Now, what we owe the future, more than anything else, is our respect. We now have the beginning of the ability to predict the future, and it is easy to become convinced—as many environmentalists are—that we know exactly what to do today to ensure the best tomorrow, if only we have the will and the political power to do it. But in truth we see the future as if through a glass darkly, and we only need to look to the past to see how quickly political, cultural, and scientific certainties can be overthrown. Instead of trying to control the future with our choices, we should endeavor to give the people of tomorrow the maximum amount of space safe to operate. That means we shouldn't bequeath them a climate so ruined that a flourishing civilization is all but impossible, or biodiversity so drained by extinction that the Earth is left irrevocably impoverished. But we also shouldn't lock the future into certain choices around technology because we might find them questionable or even repellant today. (Among the sci-fi-like projects that the Long Now has been involved with is Revive & Restore, which aims to resurrect extinct species like the woolly mammoth using cutting-edge genetic engineering—one novel solution to the extinction crisis.[96]) Instead we should embrace the power of science to repair what we've lost and ensure that our descendants are as wealthy and as strong as possible, because we have faith that they will be smarter and fairer than us, just as we are smarter and fairer—for the most part—than our ancestors. "We trust that future," said Rose, "so try not to make decisions that limit our choices or the choices of future generations."

Conservationists want us to preserve the past as much as possible—the climate as it was, the environment as it was. But while I don't share Rose's degree of confidence that the future is necessarily going to be better and smarter than the past, I agree that we must act as if there will be a long term, and we have to embrace the change and disruption that will come with it, as scary as it is. Our history has been one of growing, making a mess, inventing new technology or politics that can clean up that mess, and then repeating the whole process over and over again. Long before humans began burning fossil fuels, we had already unrecognizably reshaped the planet through the creation of farms and cities.[97] It's a tightrope walk—and we may have already passed the point of no return for the climate—but it's the only way we know to go forward. There's a line from Long Now cofounder Stewart Brand—who went from being a 1960s radical to a *Whole Earth Catalog* hippie to the godfather of modern Silicon Valley—that sums up our existential challenge around climate change and other technological risks: "We are as gods—and have to get good at it."[98]

I don't believe the solution to saving ourselves from climate change is embracing total sacrifice in the present, in part because I don't think human beings can be convinced to do so, and in part because I agree with Rose and Long Now that the best way to help the future is to ensure that we keep growing in the present. But that path may well leave us dependent on a scientific long shot to save the climate and ourselves. It's called geoengineering, and if nothing else, it is most certainly the stuff of gods.

It might be the German accent, but Klaus Lackner has a way of speaking that lends an air of authority to his statements, even when what he's suggesting seems to be science fiction. "If you asked me fifteen years ago," Lackner told me from his lab in Tempe, Arizona, "I would have said we need to figure out how to stabilize what we're doing to the atmosphere by reducing carbon emissions. Now I'm telling you we're way past that. We have to change carbon levels directly."

Lackner is the director of the Center for Negative Carbon Emissions at Arizona State University and an academic leader in the most important field you've never heard of: carbon capture. Lackner is working to build machines capable of capturing and storing carbon dioxide in the air, a process called carbon sequestration. While most climate policy focuses on cutting future carbon emissions by replacing fossil fuel energy consumption with zero-carbon renewables or even nuclear power, Lackner aims to reduce current levels of carbon dioxide directly by sucking the gas out of the air. If it can be done—and if it can be done economically—it would be nothing less than a technological miracle. And as Lackner himself says, we're at the point where we need miracles.

Emissions of greenhouse gases lead to warming because over time they add to the carbon concentration in the atmosphere. During humanity's preindustrial history—when the climate was like Little Red Riding Hood's last bowl of porridge, not too cold and not too warm—carbon levels were around 280 parts per million (ppm). By 2013 they had passed 400 ppm and will only continue to rise. Even if future emissions are vastly reduced, the time lag of man-made climate change means that carbon concentrations will continue to grow for a while, and the climate will continue to warm. But if Lackner's invention works, we could bring carbon levels down, perhaps closer to that original 280 ppm—even if it proves politically and technologically difficult to reduce carbon emissions from energy consumption.

This would be geoengineering in action—using technology to manually fine-tune the climate, the way we might adjust the picture quality on a television. And in some form it will be necessary. Of the more than one thousand scenarios for future climate change and climate action the IPCC has considered, only 116 actually see us keeping warming below the 3.6-degrees red line—and of those 116, all but eight require carbon removal, or what's also called negative carbon emissions.[99] That's in part because we've already baked so much future warming into the climate with the carbon we've already emitted, and in part because the fossil fuel habit is so hard to break, especially for those parts of the developing world that depend on rapid economic growth and the energy use that accompanies it. The only way to square that fact with the equally pressing need to keep warming below 3.6 degrees is to bake in a technology that doesn't yet exist commercially.

In 2011, a team of experts reported that pulling CO_2 from the air would cost $600 a ton,[100] which would make the bill for capturing the 37 billion tons of CO_2 emitted in 2017[101]—one year's worth—a cool $22 trillion. But progress is being made—in June 2018 a team of scientists from Harvard and the start-up Carbon Engineering published research indicating they might be able to bring that price of capturing a ton of CO_2 down to between $94 and $232.[102] That would mean it might cost between $1 and $2.50 to capture the CO_2 generated by burning a gallon of gasoline, less than the amount of fuel taxes British drivers currently pay.[103] Lackner believes that if he could get 100 million of his carbon capture machines running, he could reduce carbon levels by 100 ppm, taking us out of the danger zone.[104]

If that price keeps going down—a big if—we might be able to save ourselves. And effective and cheap carbon sequestration would have the added effect of sweeping away many of the moral and political conflicts around climate change. If emitting the carbon that causes climate change is a crime, then we are all criminals. But if carbon dioxide is just another form of waste that can be disposed of safely, then we wouldn't feel any worse for emitting carbon than we would for producing our garbage bag full of household trash. Treating carbon emissions as waste to be removed defuses the psychological dissonance that can hinder climate policy—the guilty gap between all that we know about climate change and the little that we actually do about it.

"I would argue by making carbon emissions a moral issue, by saying that the only way to solve the problem is by donning a hair shirt, you actually invite people to resist you," said Lackner. "They just stop listening to you."

Let's hope that carbon capture becomes a reality—although know that Vaclav Smil himself has estimated that sequestering just one-fifth of global carbon emissions would require building infrastructure *twice* the size of the worldwide oil industry. And that's assuming that carbon capture ever becomes a feasible product—many would-be world-changing technologies have expired in the valley of death between the lab and the market. What's more likely is that we'll need to do geoengineering the old-fashioned way: by shrouding the Earth and cutting off the sun.

Supervolcanoes and major asteroid impacts—and for that matter, nuclear war—share the same killer app: rapid climate cooling. Massive amounts of particulates are blown into the atmosphere, where they shroud the planet and block sunlight from reaching the Earth's surface. Temperatures plunge overnight; photosynthesis grinds to a halt; life dies. But the same chemical process that causes drastic climate cooling could theoretically be controlled to offset the warming created by greenhouse gas emissions. It's called solar radiation management (SRM), and it may be the only realistic, economic strategy we have to avert the worst, most potentially catastrophic effects of climate change.

Unlike carbon capture, we know how SRM will work because we've seen it happen in nature as recently as 1991, when the volcano Mount Pinatubo erupted in the Philippines. The resulting cloud of sulfate aerosols temporarily reduced global temperatures by nearly 1 degree.[105] A possible solar radiation management plan already exists, in fact. Gulfstream business jets with customized military engines and the equipment to disperse fine droplets of sulfuric acid would fly around twelve miles above the Earth. The sulfur would combine with water vapor to form sulfate aerosols, which would then be dispersed around the world by wind patterns. Together the aerosols would reflect about 1 percent of incoming sunlight, enough to at least partially offset the effects of man-made global warming.[1]

And it would do so cheaply—according to the calculations of David Keith, a professor at Harvard University and a leader in geoengineering research, by 2040 it would take about eleven jets delivering a quarter million tons of sulfuric acid to offset the warming we can expect from rising levels of CO_2, all at the cost of about $700 million a year.

Like any deal that seems too good to be true, there's a catch. All the other negative effects of increased CO_2 levels—ocean acidification, for example—would continue unabated. Solar geoengineering is imprecise—if scientists have their climate models wrong, we could overshoot or undershoot our targets on warming. Seeding clouds with sulfuric acid causes acid rain, which is toxic to plants and animals. We would have to keep geoengineering going indefinitely. Assuming the world kept emitting carbon, if we stopped, all that delayed warming would hit us in a sudden, catastrophic wave, what's known as a "termination shock." Most of all, solar geoengineering would force us to take responsibility for the planet's climate in a way we may simply not be ready for. Even Keith, who has planned to launch one of the first field trials of solar geoengineering, is on record saying that he hopes we never have to use it.

Yet I believe we will, for the simple reason that the strategy fits who we are. We are not a species that plans deeply into the future. We are not a species that is eager to put limits on ourselves. We are a species that prefers to stay one step ahead of the disasters of our own making, that is willing to do just enough to keep going. And we are a species that likes to keep going. Solar geoengineering won't answer the question of what we owe the future, or prove that we've somehow matured. But it will provide an insurance policy against the worst, most catastrophic effects of climate change, that fat-tail risk that could bring extinction in its wake. It will prove we're just smart enough, even if that means we might yet prove too smart for our own good.

5

DISEASE
Twenty-First-Century Plague

The existential threats I've covered in this book are either historical or theoretical. I haven't lived through an asteroid strike or a supereruption or a full-scale nuclear war—nor has any other human being—and the full shape of risk to come from new technologies like synthetic biology or artificial intelligence is unknowable. But there is one existential threat that I've encountered face-to-face: infectious disease. And that encounter is how I came to a moment, sitting on the sofa in a borrowed studio apartment in Hong Kong, when the thought occurred to me: I might die. We all might die.

It was March 2003, and I was in my second year working as a reporter for *Time Asia* magazine, in my second year out of college. The world was new to me—I had barely traveled outside the United States before I boarded a plane in July 2001 that brought me, two stops and some twenty-four hours later, to this densely populated, vertiginous city of 6.7 million people crammed on the southeastern corner of China. I was lonely and homesick at first, wrung out by the crowds that clung as closely as the heat and the humidity. I spent my first month living in a hostel largely populated by migrant workers from mainland China, and I went to sleep listening to the screams of monkeys in the zoo across the street. But Hong Kong has a way of getting under your skin, and by the spring of 2003 I'd begun to find my place there. I'd even managed to carve out an identity as a journalist, writing about science.

In the spring of 2003, however, there was one story and one story alone dominating the news: the impending invasion of Iraq. Or at least that was the case for other journalists. *Time Asia* was the regional edition of *Time* magazine, and while the map said that Iraq was part of Asia, what mattered to us in Hong Kong were the boundaries drawn by *Time*'s top editors back in New York, which mandated that *Time Asia*'s remit ran only as far west as Afghanistan. That spring the pages of our magazine—usually a mix of original articles about Asia that we produced and pieces pulled from the U.S. *Time*—were full of stories about the preparation for fighting in the Middle East. That left those of us in Hong Kong with not a lot to do. And that may be why we were so eager to pounce on reports of an unexplained respiratory illness that was spreading in the southern Chinese province of Guangdong, which borders Hong Kong.

I'd been in Hong Kong for less than two years, but I already knew what unchecked infectious disease could do to this crowded city, where the densest neighborhoods stuffed more than 35,000 people into a square mile. In a park down the street from my apartment there was a plaque[1] memorializing the last great outbreak of bubonic plague in Hong Kong in 1894, one that killed more than 100,000 people[2]—greater than 10 percent of the city's population at the time.[3] Not far from the park was what Hong Kongers call a "wet market," an open-air store that sold live chickens to Cantonese shoppers who liked their poultry so fresh it clucked. Those markets had been at the center of a frightening brush with the H5N1 avian influenza virus in 1997.

Up to that point, true to its name, H5N1 bird flu was thought to be a danger to bird populations—not human ones. But the virus mutated—as influenza often does—and in April that year Hong Kong became the first city to record human infections of the disease. Eighteen people fell sick, and six ultimately died.

Those numbers were tiny—that same year 252 people in Hong Kong would die from tuberculosis—but the outbreak alarmed international health officials, not just because of what was happening, but where it was happening. The last global influenza pandemic in 1968 had originated in Hong Kong, and it had begun when a mutated flu virus in birds spread to human beings who had no immunological defense against the new strain. More than one million people around the world would ultimately die. The most likely place an influenza pandemic would emerge was where lots of human beings were in close contact with lots of birds—as might be the case, for instance, in a Hong Kong wet market, where flu viruses could mix and match between species. The initial human H5N1 cases might have meant the start of a new flu pandemic, which is why the Hong Kong government took drastic measures to stem the outbreak, killing all 1.2 million live chickens in the territory.[4] It seemed to work—there were no more human cases after the chicken cull. But health officials and flu experts feared that H5N1 would return one day.

So when news of a mysterious respiratory disease in Guangdong began circulating in February 2003, many assumed the bird flu was back. But this was something else. Soon there were cases in Hong Kong with similar symptoms—fevers, chills, shortness of breath that became so severe many patients were put on artificial respirators. News broke that dozens of doctors, nurses, and students had fallen ill at Hong Kong's Prince of Wales Hospital. New cases popped up in Vietnam, Singapore, Germany, and Canada. The World Health Organization (WHO) declared the disease a "worldwide health threat" and gave it a name: severe acute respiratory syndrome, or SARS.[5]

But even the possibility of a global outbreak of an unknown new disease wasn't enough to displace the looming Iraq invasion, and that Saturday night, as we at *Time Asia* closed our weekly issue, there was only room for me to dash off a short piece summarizing the news. My story began: "International health officials are being confronted by everyone's worst nightmare: a highly contagious, potentially fatal disease of unknown genetic makeup and for which there is currently no antidote or vaccine."[6] Looking back, I can see I had no clue what those words really meant—not yet. If anything I assumed that SARS, like H5N1 in 1997, would flare up and then disappear.

The next morning I boarded a flight for the United States to attend the wedding of a close friend in Tucson, Arizona. As I left the Tucson airport a local TV news crew stopped me to ask a few questions— not about the disease I had left behind but about my opinions on the Iraq invasion. (For the record, I was against it.) A couple of days later, sleepless and a bit drunk after the wedding, I searched the TV news obsessively, but it was all Iraq, all the time. I fell asleep to the sound of Colin Powell giving yet another press conference.

Viruses don't care about the news cycle, however, and by the time I landed back in Hong Kong after eight days away, SARS was everywhere. In an airport drugstore I saw the front page of the *South China Morning Post* announcing the news that the chief executive of the Hong Kong Hospital Authority had fallen ill with a suspected case of SARS.[7] On the high-speed airport train back to the city I found myself observing what would become the defining image of SARS: human faces covered by cotton surgical masks. Only the eyes were visible, wary with fear.

A week after I returned, the government announced a quarantine—preventative isolation—of part of the Amoy Gardens apartment complex, one of Hong Kong's many sprawling housing estates, after news broke that hundreds of residents in the building had fallen ill with SARS.[8] This was the moment when the fear became real. Until then SARS cases had mostly been confined to hospitals, or to the occasional unlucky traveler. But the Amoy Gardens outbreak seemed to be proof that SARS could spread where we lived. Scientists still didn't know what caused the disease, and there was no effective treatment regimen yet. (Across the border in Guangdong, speculators were bidding up the cost of vinegar, which was rumored, falsely, to prevent the disease.[9]) If hundreds of people in one apartment building could contract it, what would stop it from burning through one of the most crowded cities in the world? What would keep us safe?

The WHO placed an unprecedented travel advisory on Hong Kong and Guangdong, warning people to avoid unnecessary visits to the region—the first time the international health agency had ever taken such a step.[10] The city's gleaming airport, which usually handled hundreds of thousands of passengers a day,[11] emptied. Spouses and children of expatriate workers in the city's financial sector were sent home, or to resort islands like Phuket in Thailand to wait out the outbreak. Five-star hotels went empty and restaurants and bars were deserted. Even the Rolling Stones—whose well-tested immune systems had surely weathered worse—canceled a planned concert in the city. A few *Time Asia* staffers were chosen to work from home, to ensure the magazine could be put out in case someone in our office got sick and the entire building had to be put into quarantine. We wore masks while working, but the cotton surgical ones grew wet from our breath after a few minutes, making them permeable to germs, while the molded N-95 masks proved too hot and cumbersome to wear for any length of time—and supplies of them were running low in Hong Kong anyway. Every evening around 6 p.m. the Hong Kong government would announce how many new SARS cases had been confirmed that day. At the height of the outbreak fifty people a day were being diagnosed with the disease.[12]

As for me, I was staying in a roach-infested apartment that I was borrowing from another reporter who had left to cover the war in Iraq. I was trapped in a foreign city eight thousand miles away from my family and friends, and if the epidemic worsened, if the disease started spreading without check through Hong Kong, I might die. We all could die.

And yet that moment is when I discovered what I was meant to do, a discovery that set me on a multiyear journey to writing this book.

Not long after the Amoy Gardens outbreak began, scientists at the University of Hong Kong (HKU) announced that they had identified the likely cause of SARS: a never-before-seen coronavirus, the same viral family that leads to pneumonia and the common cold. The day after the press conference I visited the lab of the HKU researchers at Queen Mary Hospital. I peered through an electron microscope at the culprit. It looked like a blue circular splotch studded with the halo, or corona, of viral protein spikes that give the coronavirus its name. It was a killer 40 millionths of an inch long.[13]

The discovery of the SARS coronavirus marked a turning point in the battle against the disease. SARS could kill—10 percent of the people infected by the virus would eventually perish, a death rate that was high for an infectious disease in the age of antibiotics and antiviral drugs.[14] As scary as it was, however, SARS turned out to have its weaknesses. The virus didn't spread easily—most of the early transmissions were in hospitals, largely because doctors initially hooked up patients to respirators that inadvertently spread viral particles through the air like a biological weapon. Once new infection control methods were put into place, hospital transmissions soon halted. The Amoy Gardens outbreak that alarmed us so much turned out to be a one-off—a single infected resident had diarrhea, an occasional symptom of SARS, and plumbing problems in the apartments resulted in viral particles spreading throughout the building.[15] No other wide-scale public outbreaks occurred. On May 23, the WHO finally lifted the travel advisory on Hong Kong, and the following day there were zero new cases in the city for the first time since the beginning of the outbreak. By July 5, SARS—which had infected more than 8,000 people worldwide and killed over 770—was declared contained. By November, even the Rolling Stones had come back to Hong Kong.

SARS was a twenty-first-century plague. It was a zoonotic disease, meaning it jumped from animals to humans—a pathway it has in common with most emerging diseases, including HIV.[16] After years of work, scientists in Hong Kong would eventually conclude that the virus originated in horseshoe bats in southern China. The bats in turn spread the disease to masked palm civets, catlike mammals native to Southeast Asia. Humans entered the picture because civets, along with a menagerie of other wild animals, were kept in the crowded live markets of Guangdong to be sold for food. These markets are unsurpassed viral factories, where wild animals and human beings are in close quarters, close enough for a new pathogen like SARS to jump from one species to the next. Perhaps it spread through a cut or a scratch inflicted on an unwary Guangdong shopper who wanted a freshly killed civet as an ingredient for the exotic "dragon-tiger-phoenix soup," a wildlife dish popular at the time among wealthy Chinese in the region.[17]

The SARS outbreak was the product of a series of accidents and errors, one after the other. First the virus needed to emerge in bats, then spread to an animal more likely to come in contact with people—the civet—and then jump from that intermediate host to a human being. But that wasn't all. Had this series of viral accidents occurred fifty years before, when mainland China was largely isolated from the rest of the world, SARS might never have gone global. By 2003, however, China had opened up, and thousands of people moved between Guangdong and Hong Kong each day.

One of them was a doctor in Guangdong named Liu Jianlun. Liu was already infected with SARS when he traveled to Hong Kong in February 2003 for a wedding. Even then, most SARS patients only infected those who had been in very close contact with them, like nurses in a hospital or family members taking care of sick relatives at home. But for reasons that doctors still don't understand, a handful of SARS patients were "super-spreaders" capable of transmitting the disease to unusually high numbers of people.[18] Liu was one. He would infect sixteen people at the Metropole Hotel, and they in turn traveled home by plane to unwittingly seed outbreaks in Canada, Singapore, Taiwan, and Vietnam.

The last necessary element was government malpractice. SARS spread in Guangdong for months before the Chinese government finally began to admit what was happening, giving the disease time to leak across the border to Hong Kong—which hosts hundreds of international flights each day—and from there to the rest of the world. The secrecy continued even after the disease reached Beijing in March 2003, until a brave doctor named Jiang Yanyong reached out to my *Time* colleagues in the Chinese capital. Jiang told them that the government was lying to the world about the extent of SARS,[19] even going so far as to shuffle dozens of suspected SARS patients between hospitals in an effort to prevent WHO officials from seeing them.[20]

SARS demonstrated the difference transparency makes to an outbreak response, when even one sick patient overlooked or hidden can keep an outbreak going. Beijing's lies made a bad crisis worse. SARS also showed that the connections of a globalized world make us more vulnerable to diseases that can now be spread via the nearly 12 million people who on average travel by air each day.[21] The world was primed for the disease.

Yet those same connections also helped stop the outbreak before it could do even more damage. A network of international labs sharing data over the internet was able to identify the virus that caused SARS barely more than a month after the WHO issued its first global alert.[22] Compare that to the two long years that passed between the first signs of AIDS and the discovery of HIV[23] in the early 1980s. Once it became clear in March just how dangerous SARS was, the WHO showed admirable spine, placing travel advisories on affected countries despite protests by governments that the moves would cost billions in lost revenue—predictions that turned out to be accurate. Aside from Hong Kong, mainland China, Canada, and Taiwan, most of the outbreaks in the more than thirty affected countries were limited to a handful of cases, thanks to rapid isolation of sick patients and preventative quarantine of those who had come into contact with the infected. Though hundreds ultimately died, SARS could have been far more devastating.[24]

We were lucky. There was no vaccine and no antiviral drug for SARS—and there still isn't—but doctors were able to effectively treat the diseases using the kind of basic supportive methods employed for any severe respiratory illness. Most of the infected recovered. SARS also turned out to be an inefficient transmitter. Diseases are defined by their basic reproductive number, or R0—the number of people one sick person will infect on average. Measles, one of the most contagious diseases on Earth, has an R0 of 12 to 18. SARS, after infection control methods were put in place, had an R0 of 0.4,[25] which is why the outbreaks were stopped so swiftly.

Despite all that, SARS brought East Asia to a near standstill for weeks. It cost the global economy as much as $80 billion,[26] which if it were a natural disaster would have made it one of the most expensive in human history. It was the most alarming infectious disease to emerge since AIDS. And yet in the end, my solitary fears in that Hong Kong apartment were wrong. SARS simply lacked the legs to be a truly existential threat.

This is the paradox of infectious disease as an existential risk. Nothing has killed more human beings through history than the viruses, bacteria, and parasites that cause disease. Not natural disasters like earthquakes or volcanoes. Not even war. By one estimate half of the human beings who ever lived were killed by one disease, malaria,[27] which still knocks off nearly half a million people per year.[28] Epidemics have been mass killers on a scale we can't begin to imagine today. The plague of Justinian struck in the sixth century and killed as many as 50 million people, perhaps half the global population at the time,[29] while the Black Death of the fourteenth century—likely caused by the same pathogen—may have killed up to 200 million people. Smallpox may have killed as many as 300 million people in the twentieth century alone,[30] even though an effective vaccine—the world's first—had been available since 1796.[31] At least 25 million, and perhaps far more, died in the 1918 influenza pandemic[32]—numbers that dwarf the death toll of World War I, which was being fought at the same time. The 1918 flu virus infected one in every three people on the planet.[33] HIV, a pandemic that is still with us and which still lacks a vaccine, has killed 35 million people and infected 77 million, with more added every day.[34]

If these numbers surprise, it's because epidemics are rarely discussed in history classes, and seem to fade more quickly from memory than the catastrophes of war or weather. There are few memorials to the victims of disease. The historian Alfred Crosby was the author of *America's Forgotten Pandemic*, one of the great books on the 1918 flu.[35] But Crosby was only prompted to begin researching the pandemic when he stumbled on the forgotten fact that American life expectancy had suddenly dropped from 51 years in 1917 to 39 years in 1918, before rebounding the following year.[36] That plummet in 1918 was because of the flu.

Pathogens make such effective mass murderers because they are self-replicating. Most natural disasters are constrained by area, save the very rare planetary catastrophes we covered earlier, which kill globally by changing the climate. An earthquake that strikes in China can't directly hurt you in the United States. Each bullet that kills in a war must be fired and must find its target. But when a virus—like SARS or the flu—infects a host, that host becomes a cellular factory to manufacture more viruses. (Bacteria, meanwhile, are capable of replicating on their own in the right environment.) The symptoms created by an infectious pathogen—the sneezing, the coughing, the bleeding—put it in a position to spread to the next host, and the next. And because human beings move—while interacting with other human beings in every manner from a handshake to sexual inter-course—they move microbes with them. No wonder that militaries have long tried to harness disease as a tool of war. No wonder that until recently far more soldiers died of disease than died in combat.[37] A pathogen is a perfectly economical weapon, turning its victims into its delivery system.

Yet despite epidemic after epidemic, despite mass killers like smallpox and the 1918 flu, at no point has disease threatened humans with extinction. Even the Black Death, likely the most concentrated epidemic of all time, now appears as little more than a minor down-turn in what has otherwise been a bull market for long-term human population growth. That's true for animals as well. The International Union for Conservation of Nature reports that of the 833 plant and animal extinctions that have been documented since 1500, less than 4 percent can be attributed to infectious disease. Those species that were eradicated by disease tended to be small in number and geographically isolated—very much unlike human beings, who are both numerous and have spread to every corner of the world.[38]

With the exception of HIV—which can now be managed as a chronic condition with antiviral drugs—every major epidemic mentioned above took place before the dawn of modern medicine, before the development of antibiotics and widespread vaccines. Smallpox was even fully eradicated from the wild in 1980[39]—the only known samples of the virus are kept at highly secure government facilities in Atlanta and Koltsovo, Russia.[40] Plague is now so rare that when it breaks out in countries like Madagascar, it makes global news—yet fewer than 600 deaths from the disease were reported between 2010 and 2015. Studies have shown that most of the fatalities from the 1918 flu were actually due to secondary bacterial infections that today could be controlled by antibiotics,[41] which were introduced less than a century ago. Influenza pandemics remain the great fear of infectious disease experts, but the most recent one in 2009 killed only about 284,000 people worldwide.[42] That was fewer than the number of people who die from seasonal flu in an ordinary year.[43]

Modern science has defanged most infectious diseases, at least outside the developing world—and great progress has been made there in recent years—but basic evolution also plays a role in limiting the catastrophic potential of natural disease. Every pathogen faces a trade-off. In general, the more rapidly it kills, the harder it is to spread widely, because an extremely virulent disease would run out of victims and hit an epidemiological dead end. Pathogens that are highly transmissible, like influenza, rarely kill, even absent the countermeasures of modern medicine. The 1918 flu had a fatality rate of about 2.5 percent.[44] That's tremendously high by the standards of the flu, but it still meant that more than 97 out of every 100 patients survived. Even a virus like HIV—which kills slowly and shows no symptoms for years, permitting the infected plenty of time to spread the disease—is hindered because transmission requires direct contact with blood or with bodily fluids. The self-replication that makes infectious disease such an effective weapon also prevents it from becoming a true existential threat. What viruses and bacteria want—if packets of genes and single-celled organisms can be said to want anything—is to survive and to replicate. They can't do that if they kill all humans.

The Nobel Prize–winning virologist Sir Frank Macfarlane Burnet could be forgiven for noting in 1962 that "to write about infectious disease is almost to write of something that has passed into history."[45] Sickness is with us always, and likely always will be. Infectious disease lives with us intimately, in the way a hypothetical supereruption or asteroid strike or nuclear bomb can't. But even now in rich countries we're far more likely to die of noncommunicable diseases like heart attacks or cancer or Alzheimer's, rather than from the pathogens that reaped our ancestors. The decline of infectious disease is the best evidence that life on this planet truly is getting better.

Yet there is no guarantee these trends will continue. The number of new infectious diseases like SARS and HIV has increased by nearly fourfold over the past century,[46] while since 1980 alone the number of outbreaks per year has more than tripled.[47] Over the past fifty years we've more than doubled the number of people on the planet,[48] which means more human beings to get infected and in turn to infect others, especially in densely populated cities. We have more livestock now than in the last ten thousand years of domestication to 1960 combined.[49] As SARS demonstrated, our interconnected global economy, with its long supply chains, is uniquely vulnerable to the global disruption that can be wrought by infectious diseases, even those that kill in relatively small numbers. That same interconnection—the ability to get to nearly any spot in the world in twenty hours or less, and pack a virus along with our carry-on luggage—allows new diseases to emerge that might have died out in the past. Antibiotics have saved hundreds of millions of lives since the serendipitous discovery of penicillin in 1928, but bacterial resistance to these drugs is growing by the year, a development doctors believe is one of the greatest threats to global public health. Thirty-three thousand people die each year from antibiotic-resistant infections in Europe alone, according to a 2018 study.[50] The "antibiotic apocalypse," as England's chief medical officer, Sally Davies, called it,[51] puts us in danger of returning to a time when even run-of-the-mill infections could kill.

For all the advances we've made against infectious disease, our very growth has made us more vulnerable, not less, to microbes that evolve 40 million times faster than humans do.[52] A World Bank study estimated that a severe influenza pandemic along the lines of the 1918 flu could cost our now much richer and more connected global economy $4 trillion, nearly the entire GDP of Japan,[53] and some experts believe it could kill hundreds of millions of people.[54] The WHO, which performed so well under the stress of SARS, has botched more recent outbreaks so badly that experts have called for the entire organization to be overhauled. Climate change is expanding the range of disease-carrying animals and insects like the *Aedes aegypti* mosquitoes that transmit the Zika virus. Even human psychology is at fault—the spread of vaccine skepticism has been accompanied by the resurrection of long-conquered diseases like measles, leading the WHO in 2019 to name the antivaccination movement as one of the world's top ten public-health threats.[55]

Dr. Peter Piot is the director of the London School of Hygiene & Tropical Medicine and the man who in 1976 helped discover the Ebola virus. He and his team traveled from village to remote village in what is now the Democratic Republic of the Congo, tracking the first known outbreak of Ebola at great personal risk. He has witnessed firsthand the worst that biology can throw at us—and he is worried about what's to come. "We face the globalization of risk in infectious disease today," Piot told me in 2017. "In the future that risk will only go up. That is a fact of life."

And the best way to understand that fact is to revisit what happened—and what was only narrowly prevented—when Ebola broke out in the globalized world of 2014.

———————————

Scientists thought they knew Ebola. Like SARS, the virus probably has its reservoir in bats—though researchers have yet to pinpoint the exact origin species—but it would occasionally jump to infect human beings in isolated rural communities in central Africa. When it did so it would kill, and terribly—burning fevers, shortness of breath, vomiting, diarrhea, and even external bleeding, sometimes from the whites of the eyes. But Ebola would kill so quickly, and in such remote territory, that outbreaks would soon burn through available victims. International medical teams in their hazmat suits would show up to contain the virus. Ebola was the stuff of medical nightmares, less a virus than a real-life bogeyman used for scare stories like the 1995 film *Outbreak*. But was it a global health threat? No.

That began to change in December 2013, when a two-year-old boy in the village of Meliandou in the West African nation of Guinea became ill with Ebola. The location was the first surprise—Guinea is as far away from Ebola's usual turf in Central Africa as Las Vegas is from New York City. The next surprise came when the disease kept spreading through West Africa, breaking out of the villages where Ebola was usually found and into the Guinean and Liberian capitals of Conakry and Monrovia, each home to more than one million people. Ebola in a city, with countless human bodies to feast on—this was the sum of all fears. And as the months passed in 2014, this is what was happening.

Tracing the origin of the outbreak months later, scientists wondered if the Ebola virus had changed, if it had mutated to become more transmissible. But it wasn't the virus that had changed so much—it was Africa. Development and infrastructure improvement, including thousands of miles of road built by Chinese investment,[56] had cleared the thick forests that had once kept both the virus and its victims isolated. Deforestation flushed out the animals that carried Ebola, making contact with humans—and the chance of a new Ebola infection, like the one that sickened that first boy in Meliandou—all the more likely. (One study found that deforestation is linked to 31 percent of outbreaks such as Ebola and Zika.[57]) The roads built to carry trucks and logging crews also made it easier for rural villagers to move to Africa's growing cities, bringing the emerging pathogens of the wild with them.

I saw this for myself on a reporting trip to Cameroon in Central Africa in July 2011. Those who live in this region, like other rural parts of Africa, have long depended on hunting the occasional wild animal—porcupines, cat-sized antelopes called dik-diks, even monkeys—to supplement their diets with protein. The product of those hunts are called bush meat, and the appetite for it has increased as even the poorest Africans have grown richer and new roads allow hunters to penetrate deeper into the forest. Driving away from the Cameroonian capital of Yaoundé one hot July day, I counted stand after roadside stand, each selling fresh bush meat.

The act of hunting and slaughtering a live animal is a bloody one, for both predator and prey, and viruses can easily pass between them. In one small settlement I listened as a Cameroonian health official warned villagers about the health dangers posed by bush meat. Then I walked into a hut where a woman and her son were butchering a fresh porcupine. They skinned the animal, then boiled it to strip off the quills. Once the flesh was pink and raw, the woman began tearing into its belly with a machine, pulling out the yellow, glistening viscera. Blood began to flow as she quartered the quivering carcass, kneading the meat with her hands. This was an encounter with another species more microbiologically intimate than sex, with all the attendant medical risks.[58] Multiply that interaction a thousandfold each day—any of which could permit a dangerous virus to jump from animals to humans—and it becomes clear that a major Ebola outbreak was inevitable.

By the end of July 2014, the virus had established itself in the capital cities of Guinea, Liberia, and Sierra Leone. More than 1,200 people had fallen ill, making this by far the biggest Ebola outbreak on record. There was no treatment for Ebola beyond supportive care focused on maintaining hydration and oxygen levels. Doctors tried their best to isolate sick patients, preventing them from spreading the disease, until the outbreak could be contained. But that strategy works only in the kind of sparsely populated territories where Ebola had previously struck. It was impossible to control the outbreak in cities using the same tactics. There weren't enough isolation units, not enough doctors or nurses. Local medical staff were often the first to get sick, even as they found themselves ostracized by their communities for working with Ebola patients. Cultural practices threw unexpected curveballs—it was months into the outbreak before researchers realized that traditional burial habits, including the practice of family members washing the bodies of the deceased, gave the virus access to a buffet of potential hosts.[59] None of this was foreseen, and the result was a health catastrophe that eventually took the lives of more than ten thousand people. And it came so close to being much, much worse.

Dr. Tom Frieden was the director of the Centers for Disease Control and Prevention (CDC) under President Barack Obama for 2,783 days[60]—and the one that scared him the most came on July 23,

2014.[61] That was the day Frieden received the news that a man had died of Ebola in the Nigerian megacity of Lagos. As alarming as the outbreak had been up until then, this case represented a threat of an entirely different magnitude. Lagos is the biggest city in Africa, with a population around the same size as those of Guinea, Liberia, and Sierra Leone combined. While those nations were relatively isolated from the rest of the world, Lagos is a major international air travel hub. "That was the moment of maximum terror," Frieden told me. "If it got completely out of control in Lagos, it could spread through Nigeria and the rest of Africa. It could have continued for months to years. It could still be going on."

The Nigerian government's initial response was sluggish, but things changed when Frieden dispatched CDC staff to Lagos to work alongside a cadre of Nigerians with experience in the international campaign to eradicate polio. Health officials traced hundreds of possible Ebola contacts and were able to confirm and isolate nineteen further cases[62] connected to the original death. They halted the chain of infection, preventing Ebola from gaining a foothold in Africa's most populous country. But today all Frieden can think about is how lucky we were—and how we may not always be so fortunate. "We were not more than a couple of days away from Ebola not being readily controlled," he said. "That shows how close we are to a possible disaster."

Even so, Frieden and the CDC couldn't prevent Ebola from reaching the United States. In mid-September, a Liberian named Thomas Eric Duncan contracted the disease in his home country before flying to Dallas to visit family. He would die in October at Texas Health Presbyterian Hospital, but not before infecting two nurses, Nina Pham and Amber Vinson. It was the first time Ebola had spread on American soil, and Pham and Vinson were both put in specially prepared isolation wards.

The homegrown cases made Ebola suddenly real to Americans, sparking a public panic that was further sensationalized by too many in the media. I had a close encounter with the hysteria myself. One night that month I appeared on MSNBC's *The Last Word with Lawrence O'Donnell* for what was supposed to be a short segment to discuss *Time*'s Ebola coverage, only for the show to pivot and spend an hour watching Pham's ambulance move in real time from an airplane to a government hospital in Bethesda, Maryland, mile by slow mile. All the while I kept searching for something, anything interesting to say on live TV about a situation none of us fully understood. It was like one of those nightmares where you show up for an exam without your pants.

Fortunately Pham and Vinson both recovered, and the domestic Ebola outbreak ended with them, save for a few isolated cases imported from Africa. But had it continued, would even our $2.9 trillion U.S. health care system[63] have been sufficient to prevent a potential wipeout?

At the start of the Ebola outbreak there were only three biocontainment units in the United States that could treat patients such as Pham and Vinson: the University of Nebraska Medical Center, the National Institutes of Health in Bethesda, and Emory University Hospital in Atlanta. As of 2018 there are now fifty such facilities across the United States, but that still adds up to only 144 beds.[64] To respond to just ten cases of Ebola in 2014, the American government spent $119 million on screening and quarantine alone. In an outbreak of any greater size—say, of a new disease that is more contagious than Ebola—that kind of response would not scale.

Those new biocontainment facilities were paid for with the billions of dollars Congress allocated to disease response in the wake of the Ebola outbreak. But that funding ends in 2020, which means financial support could fall off a cliff. The budget of the Public Health

Emergency Preparedness Program, which helps state and local health departments surveil for infectious disease and train epidemiologists, has dropped by more than a third from its $940 million peak in 2002. Local health departments have cut more than 55,000 jobs. These are the frontline workers in any epidemic; laying them off in mass numbers is like firing your infantry before an invasion. The CDC's Global Health Security Agenda—an international program that works to foil diseases overseas before they reach the United States, just as the CDC did with Ebola in Nigeria—is being downsized dramatically, and in 2018 the CDC announced that its work in 39 out of 49 countries would be scaled back or shut down completely.[65] In May 2018, the head of global health security at the White House's National Security Council abruptly resigned after his office was essentially eliminated.[66]

An epidemic of infectious disease can test a national leader like nothing else. The Obama administration had to respond to three major outbreaks: the H1N1 flu pandemic in 2009, Ebola in 2014, and Zika in 2016. In each case President Obama worked closely with respected experts like Frieden at CDC, Dr. Anthony Fauci at the NIH, and Kathleen Sebelius at the Department of Health and Human Services (HHS) to determine how the government would eradicate the disease while reassuring a panicked public. There were decisions that could have been made better or faster, as there always are, but on the whole, being the nation's epidemiologist-in-chief seemed to fit Obama's temperament, his ability to filter out the noise around him and focus on the goals that mattered. Obama's predecessor George W. Bush acquitted himself well during the SARS outbreak and personally spearheaded billions in funding to fight HIV/AIDS around the world.[67]

Donald Trump does not inspire the same degree of confidence. During the 2014 Ebola outbreak—before he became a candidate for president—he called for preventing American health care workers who had been infected with the virus from returning to the country.[68] He warned without evidence on Twitter that "Ebola is much easier to transmit than the CDC and government representatives are admitting."[69] As president, Trump oversees a government that has been hollowed out of health expertise. His first CDC director, Dr. Brenda Fitzgerald, was forced to resign over her investments in tobacco,[70] while his first HHS secretary, Dr. Tom Price, had to resign over taking charter flights at taxpayer expense.[71] It was more than a year and a half into his term before Trump finally appointed a White House science adviser. Obama's first secretary of energy was Dr. Steven Chu, who is a Nobel Prize–winning physicist. Trump's secretary of energy is former Texas governor Rick Perry, who wears glasses.

All of the above are reasons why Trump lacks the talent and the temperament to lead the United States through an outbreak, just as he has proven himself unfit to control the nuclear codes, just as he has shown himself to be a major obstacle on climate change efforts. But there's something else about Trump that would make him dangerous in the face of a new disease. Public health is built on a foundation of trust with the public—trust about the often uncertain facts around an outbreak, and trust that doctors and the government know what they are doing. Whatever you might think of Donald Trump as president, it is impossible to ignore that he has acted again and again to erode trust in his own government—fighting with his own intelligence officials over Russian election hacking or Saudi Arabian assassinations, censoring the science produced by his own agencies.[72] The consequences of those spasms of interference have been bad enough already. If the president conducts a Twitter war against government doctors during an outbreak, people will die.

It's discomfiting to imagine Trump or someone like him in charge during the next SARS or Ebola outbreak, but even a far more capable leader would struggle to contain an outbreak of a new and contagious disease. Take vaccines, one of the major reasons why you're much less likely to die from infectious disease than your great-great-grandparents were. The middle of the twentieth century was a golden age for vaccines, as scientific heroes such as Jonas Salk developed protections against life-threatening diseases like polio and measles. Yet while the worldwide pharmaceutical market is worth more than $1 trillion today, the market for vaccines makes up only 2 to 3 percent of it.[73] Given that it can take ten years of testing and billions of dollars to develop a vaccine[74]—and given that attempts have an estimated 94 percent chance of failure[75]—drug companies have shied away from the business.

Ebola is a case in point. An early-stage vaccine existed for years before the West Africa outbreak, but with little commercial incentive to develop something that would prevent a rare disease that at the time only menaced the developing world, work languished. Global health authorities fast-tracked candidates during the 2014 outbreak but the disease was ultimately contained[76] before a vaccine was close to ready. After the crisis ended, some of the companies involved in preparing the vaccine complained that they had spent millions on a product that no one now wanted to buy. "Vaccines are a major market challenge," said Trevor Mundel, the president of the global health division at the Gates Foundation. "There's just no incentive for any company to make pandemic vaccine to store on shelves."

No disease better illustrates that conundrum than influenza. There is a healthy market for seasonal flu vaccine, and it does prevent tens of thousands of hospitalizations and millions of illnesses each year.[77] But the flu vaccine is far from perfect—the 2017–18 flu shot was only 36 percent effective, even worse than it is during an average year.[78] Because flu viruses mutate constantly, a new vaccine has to be developed and manufactured every year to match the latest strain. Months pass from the development of a candidate strain in a lab to the moment the vaccine is loaded in a syringe in your doctor's office. Scientists developing the vaccine are forced to predict which flu strain will be circulating nine months in the future, when the vaccine will finally be ready for use. Some years—like the 2017–18 flu season, which saw record-breaking hospitalization rates in the United States—they choose poorly.[79]

A mismatched vaccine is an annoyance when the result is a nasty case of seasonal flu. But it would be deadly during a severe influenza pandemic. Such pandemics are the single event that infectious disease experts fear the most—flu, said Dr. Frieden, is the "big one."[80] Humans have little to no immune protection against these new flu strains, so when they emerge they spread rapidly around the world. Hundreds of millions of people get sick, and if the new strain is especially virulent, tens of millions could die. That happened during the 1918 flu pandemic, still the standard for what a global disease disaster could do to a mostly modern world. Three other flu pandemics have occurred since 1918, and scientists know that a new pandemic is inevitable.

Flu vaccines didn't exist in 1918, but they did in 2009, when a new strain in Mexico jumped from pigs to people. Even though the government and the pharmaceutical industry tried to fast-track a new vaccine, however, the first doses weren't available for 26 weeks, and it would have taken 48 weeks to produce enough to vaccinate every American.[81] By then the pandemic would have already run its course.[82] The virus moves swifter than we do.

One way to catch up would be to develop what is known as a "universal" flu vaccine. While seasonal flu vaccines focus on genetic parts of the influenza virus that are easy to target but which mutate constantly—which is why the vaccines quickly fall out of date— would-be universal vaccines aim for sections of the virus that remain stable from strain to strain. If successful, they could provide years or even a lifetime of protection in a single shot—including from future pandemic strains. The work is a scientific challenge—at a meeting on universal flu vaccines in 2018, one expert said that the field was essentially in the same place now as it had been in the 1960s.[83] But we're spending only $160 million a year on universal flu vaccine research, compared to the more than $1 billion annually that goes to HIV vaccine work.[84] I would never argue that we need to spend less money developing a vaccine for a disease like AIDS that has already killed tens of millions of people, but the damage a severe influenza pandemic could do to this planet is far worse. Yet we don't take it seriously enough to properly fund research into the one tool that could retire the risk of a flu pandemic altogether.

Since we're unlikely to have a vaccine available to protect us from the next pandemic, we'll need to get better at containing them before they get out of control. The example of Ebola, though, demonstrates how porous our global response system is. The WHO responded far too slowly to Ebola. It wasn't until August 8, 2014, when there had already been more than 1,700 recorded cases, that the agency declared Ebola in West Africa a public health emergency of international concern—its highest alert.[85] The agency has been largely unable to stop a new Ebola outbreak that began in the Democratic Republic of Congo in 2018 and had surpassed 1,000 cases by March 2019, making it the second-biggest outbreak of the disease on record.[86] But the WHO is asked to do far more than its resources allow. The agency responsible for the health of the entire world has an annual budget of just $2.2 billion—equivalent to that of a single large American hospital.

By the WHO's global standards, the United States has a strong health system—which doesn't mean it's ready for the stresses of a major pandemic. Unlike other natural disasters, which tend to be confined to a single location or region, disease can strike almost everywhere at once. Hospitals that on normal days have enough beds and mechanical ventilators to function would be overwhelmed by needy patients, as would other nearby hospitals, leaving nowhere to turn for help. A global economy that depends on just-in-time delivery for basic parts and long, often international supply chains would be fatally disrupted by a prolonged outbreak. That includes the medical industry—many of the drugs that keep American patients alive are manufactured outside the United States. If police and government officials fall sick, public safety might be the next casualty. "Even though it is an 'act of nature,' a pandemic is much closer to war," writes the public health expert Michael Osterholm in his 2017 book, *Deadliest Enemy.* "As in war, in a pandemic, there is greater and greater destruction day by day, with no opportunity for recovery."[87]

Infectious disease is where the natural and the man-made meet and multiply. Pathogens have always been a fact of human life— leprosy, a chronic infection caused by the bacterium *Mycobacterium leprae,* has been sickening human beings since at least the dawn of civilization[88]—but our infrastructure, our decisions, our climate, and our leaders all influence the course a sickness will take. We've built a world that amplifies the opportunity for a new virus to leap from an animal to human, and from there to any other spot on the globe. That's why the rate of new diseases and new outbreaks is growing. A globalized, interconnected planet of more than seven billion people is a feast for viruses. In a 2015 TED Talk, Bill Gates—who has dedicated his post-Microsoft career to disrupting infectious disease as he once disrupted the software industry—said something that struck me. "When I was a kid the disaster we worried about most was a nuclear war." But today, he continued, "if anything kills over 10 million people in the next few decades, it's most likely to be a highly infectious virus, rather than a war. Not missiles, but microbes."[89]

As frightening as a pandemic is, we know how to solve disease in a way we clearly don't for climate change, or won't for nuclear war. While the existential threats from emerging technologies like artificial intelligence present us with question marks, the fight against infectious disease has the benefit of history—and over the course of history, we're still winning, as I learned on a rainy Boston morning in 2018 when I visited Dr. Marc Lipsitch at his office at Harvard's T. H. Chan School of Public Health.

Lipsitch is one of the most influential epidemiologists in the United States, and one who takes seriously the possibility that disease pandemics might constitute a true global catastrophic risk—which is why I was there to see him. But that morning Lipsitch showed me something I wasn't expecting: a chart that graphed infectious disease mortality in the United States over the course of the twentieth century. He includes the slide in his epidemiology courses, and what it shows is a drastic decline, from around 800 deaths from infectious disease per 100,000 people in 1900 to about 60 deaths per 100,000 by the closing years of the century. There was a brief spike in 1918— that would be the flu—and a slight and temporary upturn during the worst of the AIDS epidemic in the 1980s. But, Lipsitch told me, "death rates from infectious disease dropped by nearly one percent a year, about 0.8 percent per year, all the way through the century."

At first I assumed the graph represented the success of vaccines and antibiotics. It does, but Lipsitch pointed out to me that the decline in death rates began years before the introduction of vaccines to diseases like rabies, typhoid, or yellow fever, and decades before the first antibiotics came into use. Even as American cities were bursting with new arrivals and potential disease hosts in the early twentieth century—New York City's population had already reached 3.4 million by 1900, and would grow by an additional 2 million over the next two decades alone—rates of death from infectious disease had already begun falling.[90] So what happened?

One theory is that the introduction of water filtration and disinfection in cities eliminated the threat from waterborne diseases like typhoid and cholera that had long winnowed urban populations. Another is that people simply became healthier—better fed, more robust, richer—which made them more likely to survive infectious diseases of all kinds. And assuming those trends continue, we should be even more resilient in the decades to come. "I think these are really, really big risks," Lipsitch told me. "But I don't think infectious diseases are really existential risks."

Even the 1918 flu virus, spreading in the days before vaccinations or antibiotics, grew less lethal as the pandemic wore on, which is exactly what evolutionary biologists would have expected. Paul Ewald, the director of the program in evolutionary medicine at the University of Louisville, told me that "particularly nasty" variants of the flu virus arose in the crowded and dirty trenches and army bases of World War I Europe. The horrific and unusual conditions there—a sudden and temporary reversal of the increasingly hygienic twentieth century—meant that even a pathogen that killed in a day, as the 1918 flu sometimes did, could continue spreading unabated, just as a wildfire burns easily in a hot and dry forest. "But the flip side is that as the flu spread around the world it quickly evolved to become milder," said Ewald. "And those milder strains won out in an evolutionary contest."

Ewald argues that evolution favors diseases that can spread easily, and under modern conditions—especially once cleaner water and general hygiene have been factored in—that means milder ones. Unless it happens to emerge in an environment as extreme and rare as a World War I trench, a flu virus that kills in a day is limited in its ability to spread, and so loses out to the milder version that only makes you wish you were dead. "That's why we've never seen that level of virulence again in flu," said Ewald. "If a new virus were somehow both highly lethal and transmissible, the only way it could maintain both of those qualities is if transmission could somehow be easily feasible from sick people. That would rarely happen in nature."

Thanks to new advances in biotechnology, however, nature is coming under the control of humanity. And that changes the game entirely.

BIOTECHNOLOGY
Engineering a Killer

In the darkened ballroom of the Mandarin Oriental Hotel in Washington, D.C., some of the finest minds in government are debating how to stop the end of the world. They're here to take part in a daylong tabletop exercise put on by the Johns Hopkins Center for Health Security, an academic nonprofit focused on biosecurity. The participants—which include former Senate majority leader Tom Daschle and Dr. Julie Gerberding, who headed the CDC during SARS—are playacting the role of presidential advisers convened to respond to a fictional outbreak of a new virus. The scenario on the table has been meticulously crafted by infectious disease experts, and details are doled out to the participants via reports from a Hopkins staffer who plays the role of the national security adviser and through fictional cable news segments that are shown on a TV in the ballroom. It might sound like a very Washington game of pretend, but such tabletop exercises offer officials an invaluable opportunity to test out answers to unprecedented crises—like the global disease pandemic about to strike.

The outbreak begins when groups of people in Germany and Venezuela begin to fall ill with a disease that has no known cause. That sets off alarm bells that grow louder as the disease erupts in new countries, including the United States. Just as they did with SARS, scientists soon manage to identify the mysterious virus causing the outbreak. They name it Parainfluenza Clade X to indicate that it is a member of an unknown branch, or clade, of the parainfluenza family. The news is surprising—parainfluenzas usually cause nothing worse than the common cold, yet this new Clade X spreads as efficiently as the flu and initially kills more than 10 percent of its victims. It is impervious to existing antivirals and vaccine. There is no effective treatment.

Up to this point the simulation has consciously echoed past disease outbreaks like SARS or the 2009 H1N1 flu pandemic, although with what appears to be a much more deadly virus. But the exercise takes a turn when the participants learn that Clade X did not emerge naturally from the wild like SARS, the result of a viral encounter between an unlucky animal and an unlucky human. Clade X, scientists discovered, was created in a lab by members of an environmental extremist group with the goal of immediately and drastically reducing what they view as human overpopulation. With the use of cutting-edge genetic engineering tools, a run-of-the-mill parainfluenza strain was spiked with the neurological virulence genes of a Nipah virus, a real-life pathogen that emerged in Southeast Asia in the late 1990s and can kill as many as three-quarters of its victims during its rare outbreaks.

As the exercise unfolds the participants at the Mandarin Oriental debate whether the United States should close its borders to slow the spread of the disease, and where to concentrate increasingly scarce health resources. But their decisions ultimately make little difference. While we saw in the last chapter that nature demands a virus choose either contagiousness or virulence, the makers of Clade X have used biotechnology to override evolution, ensuring that their creation can retain both the transmissibility of a parainfluenza virus and the deadliness of Nipah. It is the perfect bioweapon: a virus that spreads like the common cold and kills like Ebola. The world is defenseless.

By the end of the Hopkins exercise, twenty months into the fictional pandemic, 150 million people worldwide—2 percent of the global population—have died from Clade X. The global economy has collapsed under the strain, with the Dow Jones average down 90 percent, U.S. GDP down 50 percent, and unemployment at 20 percent. Washington is barely functioning—the president and the vice president are both ill, and one-third of Congress is dead or incapacitated. Former Missouri senator Jim Talent, who is playing the secretary of defense, puts it starkly as the simulation concludes and the lights come up at the Mandarin Oriental. "America," he tells the audience, "was just wiped out."[1]

The Clade X exercise, which took place in May 2018, was the latest in a series of pandemic war games put on by the Center for Health Security. The scenarios are always worst case, which is the point. One earlier exercise, called "Dark Winter" and staged by Hopkins in 2001, simulated a smallpox bioterror attack on the United States. The timing—just a few months before the 9/11 attack—was eerily prescient, as if the organizers had foreseen how the threat of terrorism, including bioterrorism, would come to consume the U.S. government and public in the years to come.

At their best these exercises provide a way to road-test how we might react to health threats that loom just over the horizon. In 2001 that meant the possibility that terror groups like al-Qaeda might get their hands on a sample of smallpox virus and release it into the world as an infectious weapon, seeking to sicken and kill as many people as possible. Such conventional bioterror remains a real threat—al-Qaeda

and more recently ISIS have both sought to find and weaponize existing viruses like smallpox.

The Clade X exercise, however, shows that the threat from
biology is evolving like a virus. Miraculous new biotechnology tools
have been developed over the past few years, including in synthetic
biology—the broad name for the science of rewriting the genes
of living things—and the gene-editing technique CRISPR, which
enables biologists to find and replace bits of DNA in a cell almost
as easily as they might cut and paste letters in a Microsoft Word
document. These tools promise life-changing medical advances,
but one day soon they might also allow an ambitious terror group—
or even a single alienated microbiologist—to tweak the genes in
existing pathogens and create something worse than nature ever
could. Something like Clade X. "Clade X was a fictional pathogen,
but it is based on scientific principles," Dr. Tom Inglesby, the director
of the Center for Health Security, told me after the exercise. "These
kinds of things are absolutely plausible."

I've lived through disease outbreaks, and in the previous chapter
I showed just how unprepared we are to face a widespread pandemic
of flu or another new pathogen like SARS. But a deliberate outbreak
caused by an engineered pathogen would be far worse. We would face
the same agonizing decisions that must be made during a natural
pandemic: whether to ban travel from affected regions, how to keep
overburdened hospitals working as the rolls of the sick grew, how to
accelerate the development and distribution of vaccines and drugs.
To that dire list add the terror that would spread once it became
clear that the death and disease in our midst was not the random
work of nature, but a deliberate act of malice. We're scared of disease
outbreaks and we're scared of terrorism—put them together and you
have a formula for chaos.

As deadly and as disruptive as a conventional bioterror incident would be, an attack that employed existing pathogens could only spread so far, limited by the same laws of evolution that circumscribe natural disease outbreaks. But a virus engineered in a lab to break those laws could spread faster and kill quicker than anything that would emerge out of nature. It can be designed to evade medical countermeasures, frustrating doctors' attempts to diagnose cases and treat patients. If health officials manage to stamp out the outbreak, it could be reintroduced into the public again and again. It could, with the right mix of genetic traits, even wipe us off the planet, making engineered viruses a genuine existential threat.

And such an attack may not even be that difficult to carry out. Thanks to advances in biotechnology that have rapidly reduced the skill level and funding needed to perform gene editing and engineering, what might have once required the work of an army of virologists employed by a nation-state could soon be done by a handful of talented and trained individuals. Or maybe just one.

When Melinda Gates was asked at the South by Southwest conference in 2018 to identify what she saw as the biggest threat facing the world over the next decade, she didn't hesitate: "A bioterrorism event. Definitely."[2]

She's far from alone. In 2016, President Obama's director of national intelligence James Clapper identified CRISPR as a "weapon of mass destruction," a category usually reserved for known nightmares like nuclear bombs and chemical weapons. A 2018 report from the National Academies of Sciences concluded that biotechnology had rewritten what was possible in creating new weapons, while also increasing the range of people capable of carrying out such attacks.[3] That's a fatal combination, one that plausibly threatens the future of humanity like nothing else.

"The existential threat that would be most available for someone, if they felt like doing something, would be a bioweapon," said Eric Klien, founder of the Lifeboat Foundation, a nonprofit dedicated to helping humanity survive existential risks. "It would not be hard for a small group of people, maybe even just two or three people, to kill a hundred million people using a bioweapon. There are probably a million people currently on the planet who would have the technical knowledge to pull this off. It's actually surprising that it hasn't happened yet."

Our best hope against the threat of bioengineered pathogens may be the same tools that can lead to their creation. Cheap genetic sequencing is enabling scientists to diagnose diseases of unknown origin in a matter of days, shrinking the vulnerable window of time when a new outbreak can spread unnoticed. Genetic engineering could speed the laborious process of creating and manufacturing vaccines, so that even an engineered supervirus could quickly be matched by an effective countermeasure. In their wildest dreams, some scientists believe that we might even be able to genetically design human beings who would be biologically impervious to viral infections, taking the ancient threat of disease—natural or man-made—off the table.

That's what makes biotechnology so scary and so exhilarating. It is a dual-use technology, capable of being wielded for both benign and malevolent ends. Just as we saw with the drive to build a nuclear bomb, the discoveries being made by geneticists who only want to help the world could be used to destroy it. The question we face is this: is it possible to harness the gifts of biotechnology without opening the door to a real-life Clade X?

———————

As long as wars have been fought, armies have sought to turn disease into a weapon. In one of the first recorded examples of biowarfare, the Athenian leader Solon poisoned the water supply of the city of Kirrha with a noxious plant in 600 BC. Alexander the Great is believed to have catapulted the bodies of dead men into cities under siege, a tactic later adopted by warriors in the Middle Ages. By one account the Black Death may have been sparked in Europe when invading Mongols hurled the corpses of plague victims over the walls of the Black Sea port of Caffa. During the French and Indian War in colonial America, the British general Sir Jeffery Amherst—whose name was given to the town in Massachusetts and the college later founded there—urged one of his commanders to spread smallpox among the indigenous tribes besieging his fort.[4]

In each of these examples, military leaders, without knowing anything about the existence of germs, used disease much as modern-day terrorists would: to kill their enemies and cripple the morale of survivors. It was at best a crude weapon and barely controllable, one that risked backfiring and infecting the attacker. But as medicine advanced and doctors began to understand how disease spread, it became clear that the same knowledge that could be used to fight infectious disease could also temper it into a more perfect weapon. Medicine itself became a dual-use dilemma, and remains one today.

Every major combatant in World War II—including the United States—ran some type of biological weapons program. The Japanese military had the most extensive one, and they made terrifying use of it in China, repeatedly targeting civilians with bombs filled with plague-infested fleas. Much of the work was carried out by Unit 731, officially under the army's Epidemic Prevention and Water Purification Department. (It would not be the last time an offensive bioweapons program masqueraded as a benign medical project.) Unit 731 carried out horrific tests on human subjects, including vivisecting live people, without anesthesia, who had been deliberately exposed to diseases, all in a twisted effort to perfect biological weapons.[5] After the war the sadistic commander of Unit 731, General Shiro Ishii, traded his research data to the American military in exchange for clemency, and was allowed to live peacefully until his death from cancer in 1959.

It was during the Cold War that biological weapons research reached its peak, however. The United States carried out years of research at Fort Detrick in Maryland. Some of that work was defensive, but much of it involved weapons experimentation, including in secret field trials carried out in major American cities. In one 1950 experiment in San Francisco, a U.S. Navy ship sprayed a cloud of microbes to test how a biological weapons attack might spread through the city. The germs were supposed to be noninfectious, but they later turned out to have caused urinary tract infections in several unlucky San Franciscans.[6]

Yet the U.S. program paled next to the work done by the Soviet Union, which built the largest biological weapons factory in history at Vozrozhdeniya, an island in the inland Aral Sea. By the end of the Cold War more than sixty thousand people in the Soviet Union were involved in the research, testing, and manufacturing of biological weapons. The Soviets produced thousands of tons of deadly pathogens, including anthrax, plague, and smallpox, easily enough to end all human life on Earth.[7]

The United States unilaterally renounced its offense biological warfare program in 1969, a few years before signing on to the Biological Weapons Convention (BWC), an international agreement that officially banned the development, stockpile, and use of germ weapons. The Soviet Union signed the BWC but secretly continued work on biological weapons, convinced the United States was doing so as well. The Soviets, though, soon discovered that germs made for disobedient soldiers. In 1979 the accidental release of anthrax spores at a military complex in Yekaterinburg reportedly killed one hundred workers and townspeople. It was one of many recorded calamities in Soviet bioweapons research. Another may well have been a minor flu pandemic that broke out in 1977, which some researchers have since traced to the accidental release of an old flu virus from a Soviet military lab.[8] It wasn't until the 1990s that a now-independent Russia finally admitted the existence of the Soviet Union's decades-long offensive bioweapons research.

To this day, the BWC represents the only time that the world has agreed to ban an entire class of weaponry. As president Richard Nixon put it when announcing in 1969 that the United States would abandon offensive bioweapons work: "Mankind already carries in its own hands too many of the seeds of its own destruction."[9] The negotiation of the pact was made easier because no country ever determined out how to wield bioweapons reliably. Germ weapons are made out of life, after all, and life is fussy. While a bullet or a missile will go where it is fired, once in the field germs will infect whomever they can. By one count more than a thousand Japanese soldiers fell victim to their country's own germ weapons during World War II. That was why biological weapons, despite being stockpiled by every major power, were so rarely employed on the field of battle. One paper from Oxford's Future of Humanity Institute counts just eighteen uses between 1915 and 2000, nearly all of them during World War I and II.[10]

Another reason that states largely abandoned their programs is that biological weapons are inherently destabilizing. Unlike nuclear arms—which require the kind of rare expertise and expensive materials that effectively limits the weapons to powerful governments—it is futile to physically restrict access to most dangerous pathogens and biotechnologies. There are more microbiologists than nuclear engineers in the United States today;[11] add in the much greater number of general biologists and doctors, and you have far more people in

America who have some experience with the mechanics of disease transmission than with the study of nuclear chain reactions.

More important, the line between what might be considered legitimate biomedical research and work that could turn germs into weapons is a blurry one. A virologist employed in a biological weapons program would mostly use the same tools and techniques as their counterpart in a vaccine research program. That contradiction—the same skills that can heal the body can also be used to harm it—is at the heart of medicine, with its healing blade. It's why new physicians pledge to "first do no harm." It is not the tools that make the difference between a weapon and a salve. It is the intention.

But intentions are difficult to police, which is why even with an international treaty in force banning germ weapons, even though hardened soldiers are horrified by the thought of plague bombs on the battlefield, the risk of biowarfare will never disappear. It is inherent in the medical arts, a dark twin. Biological research "is picked up," Tom Inglesby told me. "It's published. It moves quickly around the world." The only way to fully ensure that germs could never be used to hurt would be to ban medical research altogether, a policy that would surely inflict far more harm than it could possibly avert. As the authors of the landmark Fink Report put it in their 2004 paper for the National Research Council, "The contrast between what is a legitimate, perhaps compelling subject for research and what might justifiably be prohibited or tightly controlled cannot be made a priori, stated in categorical terms, nor confirmed by remote observation."[12]

This is the dilemma of dual-use technologies, and it is key to understanding the existential risk of biotechnology, and almost every other man-made existential risk as well, including artificial intelligence. In both fields it has become increasingly difficult to draw a distinction between research that benefits humankind and work that could lead to our extinction. It was initially true of nuclear weapons as well—few of the physicists involved in fundamental nuclear science research in the 1920s and '30s could have foreseen that the last stop would be Hiroshima (though some, like Leo Szilard, were able to do just that, and did their best to stop the bomb).[13] It's true in a sense of climate change, where the same energy sources that underwrote the Industrial Revolution and the great material boom of the twentieth century are turning our atmosphere into an oven. The dual-use dilemma is really the dilemma of science, how research once begun can lead to any destination, including many that its authors could never have imagined. Science is a method, not a guarantee. It is not ironic that it was the inventor of dynamite, Alfred Nobel, who established the greatest prizes in international science. It's fitting.

To their credit, most of those at the cutting edge of biotechnology are at least conscious of this dilemma—far more so than the physicists who paved the way for a nuclear bomb. Emily Leproust is the CEO of Twist Bioscience, a start-up that manufactures synthetic DNA. Her customers—a mix of academic labs and biotech companies—order custom strands of DNA, like for a specific strain of yeast. Twist then synthesizes the genes, packs them up, and mails them out. If synthetic biology is a new gold rush, then Twist is the company selling the picks and shovels.

I met Leproust at Twist's offices in San Francisco's new biotech hub in Mission Bay, just across the street from where the Golden State Warriors' new arena is rising from the ground. She is dizzyingly tall, with a black bob cut. Born in Tours, France, Leproust received her PhD in organic chemistry at the University of Houston, where she also learned most of her English—"though I didn't pick up the accent," she added in her deeply pitched voice. Leproust is an energetic evangelist for the power of synthetic biology, which she believes will transform medicine, energy, and material science. A few months after we met, in the fall of 2018, her company pulled off a successful initial public offering that added to the hundreds of millions in financing Twist has raised since it was founded in 2013. "We believe we can improve the human condition and the sustainability of the entire world," she told me.

But Leproust also knows that there are significant risks posed by tools that allow scientists to rewrite the operating code of life. "Every invention is a coin, with a positive side and a negative side," she said. "With dynamite, you can build tunnels but you can also kill people. With iPhones, I can FaceTime with my mom, but some people use them to detonate bombs. With every invention there is a good use and a bad use."

Leproust is right—most inventions, if you add creativity and subtract morals, can be used for good or ill. But synthetic biology isn't an invention, not in the way the telephone or the nuclear bomb is an invention. Better to think of it as a technological platform. What makes synthetic biology so revolutionary, and perhaps so dangerous, is not what it can do, but what it can make doable. And one of the most frightening possibilities of how synthetic biology might change what's achievable involves a virus that haunts the nightmares of biosecurity experts: smallpox.

———————

No virus in nature makes for a better bioterror weapon than smallpox. Smallpox can kill as many as 30 percent of the people it infects. It can be transmitted by airborne droplet, and it is highly contagious— on average every smallpox victim will infect three to five unvaccinated people.[14] And since the virus was eradicated from the wild decades ago, immunizations have largely been halted, which means that much of the world—and virtually all of the United States—would be vulnerable to infection. In the Center for Health Security "Dark

Winter" tabletop exercise from 2001, the final death toll in the United States from a smallpox attack reached 1 million people.[15]

Yet as biological weapons go, smallpox can be and is controllable. Eradicated from nature, the only two known samples of the virus are kept at highly secure government facilities. That makes the smallpox virus akin to a nuclear bomb in that it would have to be stolen by terrorists in order to be weaponized—and there are far more nuclear bombs in the world than there are viable samples of smallpox. That creates a double barrier to wielding smallpox as a weapon of terror: first a sample would need to be stolen, and then terrorists would need to figure out how to weaponize it.

The comparison to nuclear terrorism is worth examining. It's possible, of course, that terrorists could make their own nuclear bomb, eliminating the need to steal one. But they would have to find the right radioactive material and recruit people who knew how to build and use a bomb—and it is very hard to do both. That we have yet to experience the nightmare of a nuclear warhead going off in a major city may well have less to do with the success of security services than the fortunate fact that such plots are almost impossible to carry out.

But what if that changed? What if building your own nuclear bomb became only as difficult as, say, programming a computer virus is today? That would fundamentally alter the rules of nuclear terrorism, so much so that it might only be a matter of time before a bomb exploded in an unlucky city. And while we're fortunate that nuclear bombs haven't become any easier to build or use, that is precisely what is happening in the field of synthetic biology.

Scientists have been able to practice for decades what we might think of as basic genetic engineering—knocking out a gene or moving one between species. More recently they have learned to rapidly decode and sequence genes, which makes the book of life readable. But that was just the beginning. Now researchers can edit genomes and even write entirely original DNA. That gives scientists growing control over the basic code that drives all life on Earth, from the most basic bacterium to, well, us. This is the science of synthetic biology. "Genetic engineering was like replacing a red lightbulb with a green lightbulb," said James Collins, a biological engineer at the Massachusetts Institute of Technology (MIT) and one of the early pioneers of the field. "Synthetic biology is introducing novel circuitry that can control how the bulbs turn off and on."

Every bit of living matter operates on the same genetic code, formed in part by the nucleotide bases of DNA: cytosine (C), guanine (G), adenine (A), thymine (T). This is the programming language of life, and it hasn't changed much since Earth's primordial beginnings.[16] Just as the English language can be used to write both "Baa, Baa Black Sheep" and *Ulysses*, so DNA in all its combinations can write the genome of a 0.00008-inch long *E. coli* bacterium[17] and an 80-foot-long blue whale.[18] "The same DNA in humans is the same DNA in every organism on the planet," said Jason Kelly, the CEO of Ginkgo Bioworks, a synthetic biology start-up based in Boston. "This is the fundamental insight of synthetic biology."

The language of DNA may have first been written billions of years ago, but we only learned to read it in recent years. Sequencing DNA—determining the precise order of the C, G, A, and T— was first performed in the 1970s.[19] For years it was laborious and expensive. It took more than a decade and about $2.7 billion for the public and privately funded scientists behind the Human Genome Project to complete their mission: the first full, sequenced draft of the genes that encode a human being.[20] But thanks in part to technological advances driven by that effort—the Apollo Project of the life sciences—the price of sequencing DNA has plummeted. It now costs less than $1,000 to sequence a person's full genome,[21] and it can be done in a couple of days.

But reading a genome is just the beginning. As it has become cheaper and easier to sequence genetic data, the same trends are playing out in the writing of genes, albeit more slowly. This is the synthesis in synthetic biology, the ability to author a genome—or maybe just edit it a little.

What does that mean for smallpox? Before the synthetic biology revolution, a virus was a thing. Not quite living, not quite dead, but it existed only in the real world, whether in the wild in its human hosts or as archived samples in a lab. But a virus is just genetic data, a certain series of DNA or RNA, much as this book is a collection of letters arranged just so. Like any collection of data, the genetic code of a smallpox virus can be copied and shared. But the method matters. A printed book can be copied and shared by hand, as viruses can be grown in a lab by experienced technicians. Just as it's faster and easier, however, to share a digital copy of a book than a printed one, it's faster and easier to share the digital data that makes up a virus. And a biologist with the right and not terribly rare set of skills and tools could take that genetic data and synthesize a sample of their very own smallpox virus. What can be digitized cannot easily be controlled— just ask the record companies that tried to prevent the sharing of digital music after Napster. And now life itself can be pirated.

One response, already in place, is to make it illegal to download the genetic blueprint of certain agents like smallpox or Ebola or SARS. DNA synthesis companies like Twist Bioscience work with the U.S. government to check customer orders for anything suspicious— and that means both the orders and the customer. "We have a strict protocol so that every sequence that comes in is screened," Twist's Leproust told me. "For instance, if it's a sequence for a flu virus and it's from a company that is developing a diagnostic test for flu, that's great. But if someone orders the Ebola sequence to be shipped to a P.O. box in North Korea, we would not do it."

There's a lot of daylight, however, between sending flu virus data to a medical diagnostics company and shipping Ebola blueprints to Kim Jong Un. In 2017 a team of researchers led by David Evans from the University of Alberta stitched together fragments of mail-order DNA to re-create an extinct relative of smallpox called horsepox. The entire experiment cost $100,000 and took about six months.[22] Horsepox itself isn't dangerous to human beings, and Evans—an internationally recognized expert in pox viruses—said he performed the experiment to help create a better vaccine for smallpox. But his work triggered a firestorm of criticism from scientists who fretted that the publication of Evans's research in the open-access science journal *PLOS One*[23] had shown terror groups how to synthesize a smallpox virus of their own.

What Evans and his team did wasn't illegal, in part because it was done with private money, not public funds. Evans has said that he discussed the work with federal agencies in Canada—though doing so wasn't required—and his university's lawyers reviewed his paper for legal issues.[24] Even as Evans was performing his experiments, experts at the WHO were hammering out the rules around synthesizing potentially dangerous viruses. Yet Evans, without really asking anyone's permission, went ahead and did the work on his own, presenting the synthesized horsepox virus to the world as a fait accompli—and then published his methods for all to see. "Have I increased the risk by showing how to do this?" Evans told *Science* in 2017.[25] "I don't know. Maybe yes. But the reality is that the risk was always there."

The horsepox case demonstrates two concepts that are key to understanding the existential risk posed by biotechnology, and all existential risks derived from emerging technologies. The first is "information hazards," a term coined by Nick Bostrom.[26] Information hazards are risks that arise from the spread of information— especially new discoveries—that might directly cause harm or enable someone else to cause harm. They are the unwanted children of science. If the genetic sequence of Ebola were put online for anyone to download, that act would represent an information hazard. Information hazards can also include more general ideas such as employing deep learning to build more effective artificial intelligences. The discoveries need not be immediately weaponizable, and they may appear to have benign consequences, at least at first. The same fundamental work in atomic physics that eventually made the Trinity test possible first gave the world invaluable insights into how matter itself was composed, information that seemed largely harmless to most scientists at the time. The point is that we should be aware that there is a hazard to putting many kinds of information out in the public sphere—even though that is exactly what scientists are trained to do. And that's what makes information hazards so pernicious.

Every weapon, from the first sharpened stone tool to the latest killer drone, began as a discovery. But where once information was handed down from person to person in an analog chain, information is now digitized data. That makes it infinitely easier to spread, and infinitely more difficult to control. "Information wants to be free"[27] goes the old line, coined by our friend Stewart Brand. That's usually meant as a political posture, or sometimes just a belief that we should be able to download music and videos without paying. But what it really describes is an inescapable fact of the digital age, which is also the age of existential risk. Information wants to be free in the way that water wants to flow downhill—and it's just as hard to stop.

Evans didn't think that his experiment made the world more dangerous. (When he said that "the risk was always there," he likely meant that he believes the information hazard of piecing together a smallpox virus existed whether or not he did the actual work.) Based on the critical reaction of the virology community to his horsepox paper, however, Evans's colleagues didn't agree. But they didn't—and couldn't—stop him.

That brings us to another term, also coined by Bostrom: "the unilateralist's curse."[28] Imagine a community of biologists who each have the ability to carry out an experiment that might accidentally show a terrorist how to create a powerful biological weapon. It only takes one person to decide—unilaterally—to carry out that experiment, and thus create an information hazard that exposes everyone to the potential harm of the bioweapon. It doesn't matter if 99 out of 100 biologists decide not to perform the experiment. If one goes forward, the information hazard is born.

The curse is the asymmetry. Since any configuration other than all 100 scientists deciding not to do the experiment means that someone will carry it out, there is a bias toward information hazard. The greater the number of people who have to individually make the decision to not perform the experiment, the more likely it is that someone will go ahead and do it.[29] (Ben Franklin has a useful quote here: "Three can keep a secret, if two of them are dead.") Everything about modern science as an institution—the relentless drive for prestige, the rivalry between major scientific publications for landmark papers, the cutthroat competition between scientists to publish new discoveries first—puts more weight on that bias. Perhaps this is what Robert Oppenheimer meant when he said the following, after the Trinity test: "the deep things in science are not found because they are useful; they are found because it was possible to find them."[30] No scientist was ever awarded tenure for not publishing something.

In November 2018 the world witnessed the unilateralist's curse in action when the Chinese biophysicist He Jiankui shocked the scientific community by announcing that he had created the first babies genetically edited with CRISPR. That a scientist *could* edit a human embryo using CRISPR—and bring those babies to term, as He did—wasn't in doubt. But the mainstream opinion among the mandarins of gene editing was that it *shouldn't* be done—at least not until there was much clearer evidence that such editing wouldn't cause unwanted side effects, and until the public was ready to accept such a fundamental change to what it means to be human. But He showed just how ineffective scientific opinion is in the face of a determined unilateralist. It didn't matter that 99 out of 100 scientists might have refused to gene-edit an embryo. He was the hundredth—and so the work was done. It's impossible to say yet what the ultimate consequences will be, and in the aftermath of He's announcement the gene-editing community mostly reacted in revulsion, calling for a moratorium on similar work.[31] But science rarely moves backward, especially now that data can so easily be shared. Information, after all, wants to be free—and that includes information hazards.

We should fear the possibility that advances in biotechnology will be purposefully weaponized. But we should also be worried about mistakes. It won't matter if the end of the world is intentional or accidental, the product of terror or error. The end is the end.

In 2014 *USA Today* obtained government reports tallying up more than 1,100 laboratory mistakes between 2008 and 2012 involving hazardous biomaterials.[32] More than half of these incidents were serious enough that laboratory workers received medical evaluation or treatment for potential infection. The same year, *USA Today* reported that up to seventy-five scientists at the CDC might have been exposed to live anthrax bacteria after potentially infectious samples were sent to labs that lacked the safety equipment to handle them.[33] In another incident, live samples of the smallpox virus were discovered in a storage room at the NIH. Even after the SARS outbreak had been contained there were incidents in Singapore, Taipei, and Beijing where laboratory workers were accidentally infected by the virus.[34] Altogether between 2004 and 2010 there were more than 700 incidents of the loss or release of "select agents and toxins" from U.S. labs, and in 11 instances lab workers contracted bacterial or fungal infections.[35]

The occasional infection and even death among lab technicians is an occupational hazard of working with virulent pathogens, and it can happen even at laboratories that take the highest precautions. But potentially far more dangerous to the public is the possibility that a lab would willingly create and experiment on an artificially enhanced pathogen. In 2010 and 2011 the respective labs of Yoshihiro Kawaoka at the University of Wisconsin–Madison and Ron Fouchier of Erasmus Medical Center in the Netherlands separately announced that they had succeeded in making the deadly H5N1 avian flu virus more transmissible through genetic engineering. Since it first spilled over from poultry to human beings in Hong Kong in 1997, H5N1 has infected and killed hundreds of people in sporadic outbreaks, mostly in Asia.[36] The virus has a roughly 60 percent fatality rate among confirmed cases. On reporting trips to Indonesia in the mid-2000s—where more people have died from H5N1 than in any other country—I witnessed firsthand the damage the virus could do, and the fear it engendered. But the world was fortunate—H5N1 still almost never spreads from person to person; nearly every infection is due to close contact with infected poultry.

Flu experts, though, worried that H5N1 might mutate—perhaps by swapping genes with a human flu virus in a process called reassortment—and gain the ability to transmit easily from person to person, triggering what could be a disastrous pandemic. That was always possible—but on the other hand, by 2010 H5N1 had been circulating for nearly fifteen years without ever touching off a pandemic. Perhaps, like a thief trying to pick a lock, it hadn't yet come across the right combination—but would do so eventually. Or perhaps something about the nature of virus meant that those changes would never happen, and that an H5N1 pandemic was an impossibility. All scientists could do was wait and see.

But biotechnology offered a new strategy. Kawaoka introduced mutations in the hemagglutinin gene of an H5N1 virus—the *H* in H5N1—and combined it with seven genes from the highly transmissible but not very deadly 2009 H1N1 flu virus. Fouchier and his team took an existing H5N1 virus collected in Indonesia and used reverse genetics to introduce mutations that previous research had shown made H5N1 strains more effective in infecting human beings. Both researchers were trying to do in the laboratory what epidemiologists feared might happen in the wild—a bird H5N1 virus mutating in a way that made it more transmissible to human beings—and then they stepped back and recorded what happened. In both cases, the modified H5N1 flu viruses were able to spread between ferrets in the lab. (Ferrets have long been used as test subjects in flu work because they seem to be infected by influenza in the same manner as humans, so a flu viruse that spreads among ferrets would likely spread between people.) Such work is called "gain of function" research—and it's both a powerful new tool to understand infectious disease and a potential source of existential risk.

The results were useful, indicating that H5N1 did indeed have pandemic potential, but in performing the experiments, Kawaoka and Fouchier engineered altered influenza viruses that potentially possessed the worst of both worlds: the virulence of avian flu and the transmissibility of human flu. In the aftermath of their work the National Science Advisory Board for Biosecurity—established in the wake of the 2001 anthrax attacks—for the first time ever asked scientific journals to hold back on publishing the full details of an experiment, lest potential terrorists use the information as a blueprint for a bioweapon. After both Fouchier and Kawaoka revised their work and volunteered further details about their experiments, the two papers were eventually published in *Science* and *Nature*, respectively, but the scientific community more broadly was split between the Oppen-heimeresque attitude that information should always be open, and fear that what we learned could be misused. In 2014 the U.S. Department of Health and Human Services put a moratorium on such gain-of-function research while regulators tried to sort out the situation.

Harvard's Marc Lipsitch, whom we met in the last chapter, believed the experiments should never have been done. "Is the science so compelling and so important to do that it justifies this kind of risk?" he told me. "The answer is no." Kawaoka and Fouchier—and other respected scientists—obviously disagreed. But Lipsitch and Tom Inglesby of Johns Hopkins pushed further, collaborating on a study in 2014 estimating the chances that a hybrid flu could accidentally infect a lab worker, and from there, spread to the rest of the world.[37] Based off past biosafety statistics, they found that each year of working with the hybrid flu carried 0.01 percent to 0.1 percent chance of triggering a pandemic. While it's impossible to know what the fatality rate would be in a hybrid flu pandemic, let's assume that the modified H5N1, like its wild cousin, would kill three out of every five people it sickened. If about a third of the global population were infected by the new, far more transmissible virus—not unreasonable, since no one would have immunity—the result could be a death toll as high as 1.4 *billion* people, even more than in the fictional Clade X scenario.[38]

Then Inglesby and Lipsitch did something more. They took that 0.01 to 0.1 percent annual chance of escape and multiplied it by the potential death toll. As we saw in earlier chapters, this is commonly done in risk analysis to try to get a sense for how many deaths we might expect per year from an unusual event, such as an asteroid strike or a supereruption. Inglesby and Lipsitch found that even with the extremely low probability that the hybrid virus would escape the lab, the consequences of that rare outcome would be so awful that we could expect between 2,000 to 1.4 million fatalities per year. This doesn't mean that thousands of people each year would die from the hybrid flu research. Just as in the case of a major asteroid strike, either it happens and billions die or—far more likely—it doesn't and no harm is done. But what Lipsitch and Inglesby showed is that the worst-case scenario is so terrible that on a yearly level it could exceed the number of people who die from heart disease in the United States.[39]

"There are really big risks," said Lipsitch. "And do you want to risk that really, really, really low-probability but terrible event?"

In 2017 the NIH lifted the moratorium on gain-of-function research, putting in place new regulations around the work and restricting it to a handful of labs with the highest levels of biocontainment.[40] Any experimenter who wants to boost the virulence of an already dangerous pathogen must prove that the benefits will outweigh the risks and that there is no safer way than gain-of-function methods to answer the questions posed by their study. The proposed experiments must also be reviewed by an independent expert panel. The process, NIH director Francis Collins said when the rules were announced in December 2017, "will help to facilitate the safe, secure, and responsible conduct of this type of research."[41] And in early 2019, *Science* revealed that both Kawaoka and Fouchier had been given the go-ahead to resume their gain-of-function research on flu after a government review. "We are glad the United States government weighed the risks and benefits . . . and developed new oversight mechanisms," Kawaoka told *Science*. "We know that it carry risks. We also believe it is important work to protect human health."[42]

The new rules around such work represent a step forward, although critics were unhappy that the details of the government review weren't made public, which Lipsitch and Inglesby wrote in the *Washington Post* "could put health and lives at risk."[43] But all we have to do is look to the example of He Jiankui and his gene-edited babies to know that there are limits to the scientific community's ability to regulate itself. When science was mostly the province of a handful of countries, it might have been possible for top researchers to effectively control the spread of knowledge, but those days are long past. In 2010 China produced 117,000 PhDs in all fields, more than twice as many as the United States[44] and ten times more than it graduated in 1999.[45] That number has only continued to grow. Even before He, Chinese scientists were the first to CRISPR monkeys and nonviable embryos, and the first to put CRISPR'd cells in a live human adult.[46] It's not that Chinese scientists are less ethical than their Western peers, but they have demonstrated a willingness to push the envelope on biotechnology. The unilateralist's curse is a multiplier, after all—the more scientists who are capable of carrying out potentially dangerous research, the more who will actually go ahead and do it.

Too little attention is paid to the possibility that science is inadvertently creating low-probability but high-consequence existential risks—especially in biotechnology. The default remains: do the science to the best of your ability, and let the chips fall where they may. Scientists have historically been quicker to see the danger of overregulation than underregulation.

But not every scientist.

With a towering frame, a long white beard, and a deep voice that begins at the bottom of his shoes, George Church could audition to play the role of God. And playing God is exactly what Church has been accused of from time to time. The Harvard geneticist is one of the foremost figures in the science of reading, writing, and editing DNA—and he's certainly the boldest. Church has worked with the Long Now Foundation to bring the woolly mammoth back from the grave of extinction by editing the genetic traits of mammoths into the genomes of their close relatives, the Asian elephant. He was among the first to use CRISPR on mammalian cells and has explored the possibility of synthesizing—meaning writing afresh—an entire human genome, all three billion DNA base pairs. If the synthetic biology revolution is going to change how we live, how we work, and even how we die—or don't—Church will get much of the credit. Or the blame.

Church is unusual among the scientists in his field for another reason. While most researchers soft-pedal how new tools like CRISPR might change medicine or society—partly out of fear of public overreaction—Church is more than willing to talk through the full consequences of his revolutionary work, up to and including the possible end of the world. In addition to his labs at Harvard and in the southern Chinese city of Shenzhen, Church has helped found countless biotech start-ups that seek to commercialize his discoveries. (As a result, a conflict-of-interest slide that Church provides at the start of his talks—listing all the various companies he has a financial interest in—is as convoluted as a map of the Tokyo subway.) One of his most exciting experiments involves using CRISPR to edit the genome of pigs so that they could be used to grow organs for direct transplant into human beings—what's known as xenotransplantation—without any concerns over immune rejection.[47] He has a start-up for that discovery, too.

But as eager as Church is to explore what biotechnology can do, he is also well aware that we may not be ready for the power we're beginning to wield. That's why he is on the board of the Centre for

the Study of Existential Risk at Cambridge University and the Future
of Life Institute in Boston. Church takes a position that is unusual
among many scientists, even in the life sciences, with its institutional
review boards and independent ethics committees. He asks to be
regulated, almost as Odysseus asked to be lashed to the mast of his
ship as he approached the Sirens with their deadly songs.

"In 2004, I was worried about the proliferation of DNA synthesis, being able to make anything," Church told me when I met him
at a synthetic biology conference in Boston in 2018. "I was worried
again when I started coding digital information into DNA, that it
would suddenly create a gigantic market for DNA . . . when we start
making an Internet of things and any person can build any option.
I worry that people could make DNA synthesizers that don't follow
the rules and regulations. I worry that once people learn to put pig
organs into humans they'll get viruses from the pigs, and those
will evolve inside an immunosuppressed patient, creating something worse. I'm worried that any cellular-based therapy could cause
cancer, even if the goal is to cure cancer. These are subsets. I could
go on and on."

Church knows that if his field, which is changing the stuff of
life, is seen to be out of control, the backlash from the public could
be disastrous. "You want equitable distribution so that everyone can
have access to the positive aspects of this technology," he told me in
an earlier interview. "You don't want people creating killer viruses,
and that will require some regulation. I am a fan of the regulatory
agencies. Without them we would be at risk for much bigger setbacks.
When a lot of people die—that's when you get the real setbacks."

Biotechnology has already experienced that kind of fatal
setback. In 1999, eighteen-year-old Jesse Gelsinger died during a
gene therapy trial, a tragedy that halted progress in the then-prom-

ising field of correcting genetic disorders through the replacement of defective genes with transplanted healthy ones.[48] In the aftermath of He Jiankui's CRISPR'd babies, scientists worried that politicians might demand a total ban on the use of gene editing in embryos, cutting off what could be an effective technique to treat incurable genetic disorders.[49] But this is the question we must answer, for biotechnology and for other man-made existential risks caused by emerging technologies: how do we regulate a rapidly advancing science that has consequences—positive and negative—we can't yet fully understand?

The truth is that it's almost impossible to successfully thread the needle between allowing new technologies to develop without exposing society, and perhaps the entire human race, to unknown levels of risk, in part because we can rarely know what those risks will be before it's too late. When Facebook first emerged in the mid-2000s, social media was just a small part of the internet, the economy, and our lives. The companies providing it weren't very powerful. It would have been easy for the government at the time to regulate social media—but without a crystal ball, how would Washington possibly have known what to regulate and why? Now, more than a decade later Facebook is worth hundreds of billions of dollars, and social media has been employed to hack a presidential election. Facebook, Twitter, and their peers have changed our habits, culture, and even our minds in ways we're just beginning to comprehend. Yet the government is struggling to figure out how to effectively regulate a global industry in a manner that these now very rich and influential companies can't sidestep or subvert.

This is what is known as the Collingridge dilemma. David Collingridge was an academic at the University of Aston in Britain when he published a book in 1980 called *The Social Control of*

Technology. Collingridge maintained that we can either regulate a technology in the nascent stages, when both its potential risks and its potential benefits are still unknown, or we can wait until it has more fully matured, by which time it may have spread so widely that effective regulation becomes impossible.[50] Either way we may lose.

Facebook isn't going to end the world—probably—but biotechnology could. Synthetic biology holds enormous potential—cures for disease, engineered organs, cell lines resistant to all viruses, biofuels that could replace oil, meat made without killing animals. Holding back the development of this field would cost lives, and may even make it more difficult to combat existential risks like climate change. The potential downside of synthetic biology, however, could be absolute—and if we wait until after an engineered pathogen is released into the public, it will be too late.

On the same trip where I met with George Church, I stopped by the South Boston offices of Ginkgo Bioworks. Ginkgo is the first company in synthetic biology to become a "unicorn," meaning a private valuation of more than $1 billion. Ginkgo bills itself as the "organism company," and its bioengineers design and build custom microbes. Originally its microbes were mostly used to synthesize flavors or fragrances derived from plants. In one case Ginkgo partnered with a French perfume company to create a rose fragrance by extracting the genes from real roses, injecting them into yeast, and then engineering the microbe's biosynthetic pathways to produce the smell of a rose—which, it turns out, smells just as sweet when emitted from a genetically-modified yeast. The company is also working on extracting DNA molecules from preserved plant specimens to synthesize the fragrances of flowers that have gone extinct, like a hibiscus from Maui that vanished from the wild around 1911.[51]

Fragrances were just the beginning, however. In 2017 Ginkgo partnered with the German life sciences giant Bayer on a joint venture that will engineer microbes capable of providing nitrogen directly to plants, reducing the need for artificial chemical fertilizer.[52] Already Ginkgo estimates that it is responsible for at least one-third of all gene-length DNA synthesis.[53] (Twist Bioscience is Ginkgo's biggest supplier.) The facility I visited in South Boston is less a lab than a biological factory, pumping out synthetic organisms the way a Ford factory pumps out F-150 pickups. Christina Agapakis, Ginkgo's creative director, showed me around the factory floor, as casually dressed biology PhDs busily pipetted and assayed. But what struck me—beyond the *Jurassic Park* T-shirts a few of them were wearing, which hit a little too close to home—was how few of them there were. Much of the work at the Ginkgo factory—or "foundry," as they call it—was now automated, a shift that has already taken place in conventional manufacturing.

The synthetic biology revolution isn't just about what scientists can do, but how they can do it. The growing automation at a biotech company like Ginkgo is an example of a trend called deskilling. With each passing year, the scientific expertise needed to pull off a specific experiment in synthetic biology falls. What might have recently required the hard work of a postdoctoral student can now be done by undergraduates, and might soon show up in an ambitious high school student's science fair project. Synthetic biology is getting easier and it's getting faster and it's getting cheaper, which means more and more people can do it. And so the information hazard accumulates.

Deskilling has helped Ginkgo design and produce more than 10,000 genes per month.[54] To give a sense of how much that is, Jason Kelly, Ginkgo's cofounder and CEO, told me he designed all of 50,000 base pairs of DNA throughout the entirety of his time as a grad student at MIT in the mid-2000s. All of his work then was done by hand. But now synthetic biology is shifting from an artisanal practice carried out in the laboratory by highly trained experts to a true industry. Everything is sped up, allowing researchers to design an organism, build it, test it, analyze the test, and then start the whole cycle over and over again. "Things that would have made a great PhD thesis a few years back can now be done by people in two weeks here," Agapakis told me.

The personal history of another one of Ginkgo's founders, Tom Knight, illustrates just how far synthetic biology has come, and where it is poised to go. Knight was a computer engineering prodigy at MIT in the 1960s. Back then, Knight did his programming on bulky computers that required users to manually enter instructions from deck after deck of punched cards. Today programming is just a matter of typing code into a computer, but for decades biological programming moved no faster than Knight's punch-card processing, limiting what could be done. "In any sort of engineering discipline, the critical thing that controls what you can ultimately do is the speed at which you can try things out," said Knight.

During Knight's first act as a computer engineer in the 1960s, '70s, and '80s—before he switched to the nascent field of synthetic biology— he benefited from the computing revolution described by Moore's law. Laid down by Gordon Moore, the cofounder of the chip company Intel, Moore's law predicted that computing power would double roughly every eighteen months. Supercomputers, laptops, iPhones—they were all possible because Moore's law turned out to be correct. And in recent years the same trend has taken place in the reading and writing of DNA, which has fallen in price and increased in speed. This has made possible the industrialization of synthetic biology. Synthetic biologists may never be able to program as seamlessly as their counterparts in tech—biology is made up of bits of life, however tiny, whereas computer code is just code—but they're getting better and faster all the time.

Students today are practically ordered to learn computer coding, but in the future everyone might be an amateur synthetic biologist, programming the stuff of life. If digital apps on our smartphones have transformed how we live and work, imagine what bioengineering apps might do. "Designing genomes will be a personal thing, a new art form as creative as painting or sculpture," as the physicist and writer Freeman Dyson envisioned in a 2007 article for the *New York Review of Books*. "Few of the new creations will be masterpieces, but a great many will bring joy to their creators and variety to our fauna and flora. The final step . . . will be biotech games, designed like computer games for children down to kindergarten age, but played with real eggs and seeds rather than with images on a screen."[55]

That's the optimistic version. But deskilling something as powerful as synthetic biology will inevitably amplify existential risk. The artificial intelligence (AI) risk scholar Eliezer Yudkowsky has a smart take on the effects of deskilling that he has called "Moore's law of mad science": "Every 18 months, the minimum IQ necessary to destroy the world drops by one point."[56] It's not quite as precise as the original Moore's law—nor was it meant to be—but Yudkowsky is on to something. Consider computer programming. The relative ease of coding has created the multitrillion-dollar tech industry as we know it, but it has also empowered thousands of people to create malware for crime, for espionage, and sometimes just for kicks—to make computer viruses, in other words, viruses that cost the global economy more than $600 billion in 2017.[57]

Now imagine if it became almost as easy to program biology as it is to program computers, just as Dyson envisioned. Creativity would be unleashed, but so would our darker impulses. Far fewer people would likely use synthetic biology to program and release a killer virus into the world than the number who create malware for crime, in part because there are far more thieves among the population than there are murderers. But murderers do exist. And this technology could empower them in a way that threatens us all.

———————

In 2013, Cody Wilson—a crypto-anarchist, law student, and one of the fifteen most dangerous people in the world according to *Wired* magazine—posted plans on the internet for a 3-D printable gun. Called "the Liberator," the gun was a single-shot pistol made mostly of plastic. If you had access to a 3-D printer, and a few extra ingredients, you could download and print your own untraceable gun, with no regulation whatsoever from the government. Shortly after the blueprints were put online the State Department ordered them removed, citing a possible violation of firearm export rules. Wilson sued, and in June 2018 the State Department—now under the control of the pro-gun rights Trump administration—decided to settle his case, winning him the right to put the plans back online while also recouping nearly $40,000 in legal fees from the U.S. government. Gun-rights advocates celebrated the settlement as "the end of gun control," in the words of Fox News columnist John Lott Jr.[58]

On August 1, 2018, a federal judge issued a restraining order temporarily halting the release of the blueprints, pending lawsuits by several states, and while Wilson insisted he could still sell the plans online, in September he left Defense Distributed after a separate arrest on sexual assault charges.[59] But the Liberator is an early sign of how technologies like 3-D printing and synthetic biology can radically lower the bar of expertise and empower individual rogue actors while eroding the ability of the government to protect us from ourselves.

If everyone eventually gains the power to potentially end the world, and governments are largely helpless to stop them, then the continued existence of the world depends on the collective action of all of us—all 7.7 billion and counting—to actively choose not to destroy it. The Stanford political scientist James Fearon developed a thought experiment he outlined in a 2003 talk, back when the global population was closer to 5 billion. He imagined a time when each person had the ability to destroy the world by pushing a button on their cell phone. "How long do you think the world would last if five billion individuals each had the capacity to blow the whole thing up?" he asked. "No one could plausibly defend an answer of anything more than a second. Expected life span would hardly be longer if only one million people had these cell phones, and even if there were 10,000 you'd have to think that an eventual global holocaust would be pretty likely. 10,000 is only two-millionth of five billion."

Even the advanced biotechnology of the future will require more expertise than pushing a button on a cell phone—but perhaps not that much more, and certainly far less than is required to use current weapons of mass destruction. "What we now have [with biotechnology] is the mirror opposite of nuclear weapons, which require great infrastructure, access to controlled materials and knowledge that you can't simply go and look up online," Gabriella Blum, a professor at Harvard Law School and the coauthor of the book *The Future of Violence*, told me. "So we ask ourselves: what's people's propensity to inflict harm, and in what ways?"

It doesn't even have to be deliberate harm. Because viruses and bacteria can self-replicate, even an accident could be just as catastrophic as a deliberate attack. (After all, every natural disease outbreak is, in a sense, an accident, and those accidents have taken the lives of billions of humans.) This is why biotechnology ultimately poses the single greatest existential risk humans will face in the years to come, as the science continues to mature. Biotechnology takes our ingenuity, our thirst for discovery—and turns it against us. It leaves us only as strong as our weakest, maddest link. It gives us promise and it gives us power, the most dangerous gifts of all.

7

ARTIFICIAL INTELLIGENCE

Summoning the Demon

I received my first computer when I was eight years old, a Christmas gift my family gave to itself in 1986. It was an Apple IIe, a plump beige desktop with a cathode-ray tube monitor and an external drive for floppy disks. I remember the way the screen would buzz when it was turned on, giving off the slightest electric charge if you put a fingertip to the glass. I remember how the letters and numbers would sink into the keyboard when pressed, back when computer keyboards were still meant to evoke typewriters. I remember the sounds the computer would make when it was working, which I learned to diagnose like a doctor with a stethoscope pressed to a patient's chest. When everything was operating cleanly—not often—the computer would hum contentedly. But when something went wrong, it would issue an angry wheeze from inside its plastic chassis, as if protesting what it was being forced to do by its inferior human users.

Our family had owned pieces of technology before the Apple: TVs and VCRs, telephones and microwaves, alarm clocks and remote controls, even an old Atari 2600 video game console. But what set

the Apple IIe apart—though I didn't know it at the time—was that
it was a general-purpose machine. Every other electronic product
we owned had a single function—the microwave heated up food;
the VCR recorded television; the Atari played Pong. But the Apple
was designed to do anything that you could program it to do: operate a payroll program, do graphic design, create spreadsheets, run a
word processor, or, as was mostly the case at our house, play games.
Computers like the Apple IIe were limited only by quality of their
programs and the power of their hardware.

That power has grown exponentially in the years since. The
iPhone 7 I carry around with me is already more than three years
old, but it has over 30,000 times more RAM than my old Apple IIe.[1]
Computing power has increased by more than a trillionfold since the
mid-century days of room-sized mainframes, and that dizzying rate
of improvement—captured in Moore's law, which I referenced in the
previous chapter—is still continuing.[2] And not only are computers
more powerful, they're now ubiquitous. Computers are found inside
our desktops and our laptops and our smartphones, but also our
watches and our TVs and our speakers and our clocks and our cars
and our baby monitors and our lightbulbs and our scales and our
appliances and our alarm systems and our vacuum cleaners. The
development of the Web and mobile data transmission means that
those computers can now talk to each other, and the creation of the
cloud—the offloading of processing and storage to remote server
farms accessed via the internet—means that computing is much less
limited by the power of an individual device. The venture capitalist Marc Andreessen—founder of Netscape and an early investor in
Facebook and Twitter—has an apt description for the universaliza-
tion of computerization. "Software," Andreessen wrote in 2011, "is
eating the world."[3]

It's nearly eaten this book as well. Without the computing revolution, it would be impossible for scientists at NASA to track and potentially deflect incoming asteroids, just as it would be impossible for volcanologists to create a global monitoring system for supereruptions. The first nuclear bomb may have been developed without the help of modern computers, but the Gadget's far more powerful successors—and the global nuclear war they threatened—wouldn't have happened without advanced computing. The electricity that feeds the software that is eating the world is contributing to climate change, but powerful computing also helps manage energy efficiency and speeds the development of cleaner and renewable sources of power. The computing revolution made cheap genetic sequencing and now genetic synthesis possible, which is a valuable tool for the battle against infectious disease—and an extinction-level threat thanks to the new tools of biotechnology.

Until recently, computing was powerful because it made us powerful—for better and for worse. Computers were general-purpose tools, like my Apple IIe, and they were our tools, controlled by us, working for us. They amplified human intelligence, which is the same thing as amplifying human power. It may feel as if our computers make us dumber, and perhaps they do, the way that riding in a car instead of running can make you weaker over time. Yet just as a human driving an automobile can easily outpace the fastest runner who ever lived, I know I'm much, much smarter with an iPhone in my hand, and all it can do, than I would be if left to my own non-Apple devices.

But what happens if the intelligence augmented by explosive computing power is not human, but artificial? What happens if we lose control of the machines that undergird every corner of the world as we know it? What happens if our tools develop minds of their own—minds that are incalculably superior to ours?

What may happen to us is what happens when any piece of technology is rendered obsolete. We'll be junked.

———————————

There's no easy definition for artificial intelligence, or AI. Scientists can't agree on what constitutes "true AI" versus what might simply be a very effective and fast computer program. But here's a shot: intelligence is the ability to perceive one's environment accurately and take actions that maximize the probability of achieving given objectives. It doesn't mean being smart, in a sense of having a great store of knowledge, or the ability to do complex mathematics. My toddler son doesn't know that one plus one equals two, and as of this writing his vocabulary is largely limited to excitedly shouting "Dog!" every time he sees anything that is vaguely furry and walks on four legs. (I would not put money on him in the annual ImageNet Large Scale Visual Recognition Challenge, the World Cup of computer vision.) But when he toddles into our kitchen and figures out how to reach up to the counter and pull down a cookie, he's perceiving and manipulating his environment to achieve his goal—even if his goal in this case boils down to sugar. That's the spark of intelligence, a quality that only organic life—and humans most of all—has so far demonstrated.

Computers can already process information far faster than we can. They can remember much more, and they can remember it without decay or delay, without fatigue, without errors. That's not new. But in recent years the computing revolution has become a revolution in artificial intelligence. AIs can trounce us in games like chess and Go that were long considered reliable markers of intelligence. They can instantly recognize images, with few errors. They can play the stock market better than your broker. They can carry on conversations via text almost as well as a person can. They can look at a human face and tell a lie from a truth. They can do much of what you can do—and they can do it better.

Most of all, AIs are learning to learn. My old Apple IIe was a general-purpose machine in that it could run any variety of programs, but it could only do what the program directed it to do—and those programs were written by human beings. But AIs ultimately aim to be general-purpose *learning* machines, taking in data by the terabyte, analyzing it, and drawing conclusions that it can use to achieve its objectives, whether that means winning at StarCraft II or writing hit electronic dance tracks—both of which are possible today.[4] This is what humans do, but because an AI can draw on far more data than a human brain could ever hold, and process that data far faster than a human brain could ever think, it has the potential to learn more quickly and more thoroughly than humans ever could. Right now that learning is largely limited to narrow subjects, but if that ability broadens, artificial intelligence may become worthy of the name. If AI can do that, it will cease to merely be a tool of the bipedal primates that currently rule this planet. It will become our equal, ever so briefly. And then quickly—because an AI is nothing if not quick—it will become our superior. We're intelligent—*Homo sapiens*, after all, means "wise man."[5] But an AI could become superintelligent.

We did not rise to the top of the food chain because we're stronger or faster than other animals. We made it here because we are smarter. Take that primacy away and we may find ourselves at the mercy of a superintelligent AI in the same sense as endangered gorillas are at the mercy of us. And just as we've driven countless species to extinction not out of enmity or even intention, but because we decided we needed the space and the resources they were taking up, so a superintelligent AI might nudge us out of existence simply because our very presence gets in the way of the AI achieving its goals. We would be no more able to resist it than the far stronger gorilla has been able to resist us.

You're reading this book, so you've probably heard the warnings. Tesla and SpaceX founder Elon Musk has cited AI as "the biggest risk we face as a civilization,"[6] and calls developing general AI "summoning the demon."[7] The late Stephen Hawking said that the "development of full artificial intelligence could spell the end of the human race."[8] Well before authentic AI was even a possibility, we entertained ourselves with scare stories about intelligent machines rising up and overthrowing their human creators: *The Terminator, The Matrix, Battlestar Galactica, Westworld.* Existential risk exists as an academic subject largely because of worries about artificial intelligence. All of the major centers on existential risk—the Future of Humanity Institute (FHI), the Future of Life Institute (FLI), the Centre for the Study of Existential Risk (CSER)—put AI at the center of their work. CSER, for example, was born during a shared cab ride when Skype co-creator Jaan Tallinn told the Cambridge philosopher Huw Price that he thought his chance of dying in an AI-related accident was as great as death from heart disease or cancer.[9] Tallinn is far from the only one who believes this.

AI is the ultimate existential risk, because our destruction would come at the hands of a creation that would represent the summation of human intelligence. But AI is also the ultimate source of what some call "existential hope," the flip side of existential risk.[10] Our vulnerability to existential threats, natural or man-made, largely comes down to a matter of intelligence. We may not be smart enough to figure out how to deflect a massive asteroid, and we don't yet know how to stop a supereruption. We know how to prevent nuclear war, but we aren't wise enough to ensure that those missiles will never be fired. We aren't intelligent enough yet to develop clean and ultra-cheap sources of energy that could eliminate the threat of climate change while guaranteeing that every person on this planet could enjoy the life that they deserve. We're not smart enough to eradicate the threat of infectious disease, or to design biological defenses that could neutralize any engineered pathogens. We're not smart enough to outsmart death—of ourselves, or of our species.

But if AI becomes what its most fervent evangelists believe it could be—not merely artificial intelligence, but superintelligence— then nothing is impossible. We could colonize the stars, live forever by uploading our consciousness into a virtual heaven, eliminate all the pain and ills that are part of being human. Instead of an existential catastrophe, we could create what is called existential "eucatastrophe"—a sudden explosion of value.[11] The only obstacle is intelligence—an obstacle put in place by our own biology and evolution. But our silicon creations, which have no such limits, just might pull it off—and they could bring us along.

No wonder that a Silicon Valley luminary as bright as Google CEO Sundar Pichai has said that AI will be more important than "electricity or fire."[12] AI experts are so in demand that they can earn salaries as high as $500,000 right out of school.[13] Militaries—led by the United States and China—are spending billions on AI-driven autonomous weapons that could change the nature of warfare as fundamentally as nuclear bombs once did. Every tech company now thinks of itself as an AI company—Facebook and Uber have hoovered up some of the best AI talent from universities, and in 2018 Google rebranded its entire research division as simply Google AI.[14] Whether you're building a social network or creating drugs or designing an autonomous car, research in tech increasingly is research in AI—and everything else is mere engineering.

Those companies know that the rewards of winning the race to true AI may well be infinite. And make no mistake—it is a race. The corporations or countries that develop the best AI will be in a position to dominate the rest of the world, which is why until recently little thought was given to research that could ensure that AI is developed safely, to minimize existential risk and maximize existential hope. It's as if we find ourselves in the early 1940s and we're racing toward a nuclear bomb. And like the scientists who gathered around the New Mexico desert in the predawn morning of July 16, 1945, we don't know for sure what our invention might unleash, up to and including the end of the world.

Oppenheimer, Fermi, and the rest could only wait to see what Trinity would bring. But we can try to actively shape how AI develops. This is why existential risk experts are so obsessed with AI—more than any other threat the human race faces, this is where we can make a difference. We can hope to turn catastrophe to eucatastrophe. This is a race as well, a race to develop the tools to control AI before AI spins out of control. The difference could be the difference between the life and the death of the future.

This is AI—the cause of and solution to all existential risk. As Hawking wrote in his final book, published after his death: "The advent of super-intelligent AI would be either the best or the worst thing ever to happen to humanity."[15]

Unless, of course, superintelligent AI is a fantasy. That's something else that sets AI apart from other existential risks. We know we are warming the climate. We know we can make pathogens more virulent. And we know a nuclear button exists. But superintelligent AI might not be possible for centuries. It might not be possible at all. Which should either give us comfort or cause concern. Because one constant over the multi-decade history of AI is that whatever prediction we make about artificial intelligence is almost certainly going to be wrong.

In 1956 a group of researchers in the field of neural nets and the study of intelligence gathered at Dartmouth University. They convened for what has to be an all-time ambitious summer project, after proposing the following plan to their funders at the Rockefeller Foundation: "An attempt will be made to find how to make machines use language, form abstractions and concepts, solve kinds of problems now reserved for humans, and improve themselves. We think that a significant advance can be made in one or more of these problems if a carefully selected group of scientists work on it together for a summer."[16] (And what did you do on your summer vacation?)

As it turned out, the Dartmouth Summer Project did not exactly find how to make machines use language or improve themselves— two challenges that still bedevil AI researchers. But it did help kick off the first boom in AI, as scientists rushed to create programs that pushed the boundaries of machine capability.

There were programs that could prove logic theorems, programs that could solve college-level calculus problems, even programs that could crack terrible puns. One of the most famous was Eliza, a natural-language processing computer program that mimicked the response of a nondirective Rogerian psychologist. Created by Joseph Weizenbaum at the MIT Artificial Intelligence Laboratory between 1964 and 1966, Eliza allowed users to carry on a conversation via keyboard with a facsimile of the most facile kind of psychologist, the sort who turns every statement a patient makes back into a question. ("I'm having problems with my mother." "How does that make you feel?")

Weizenbaum actually designed Eliza to demonstrate how superficial communication was between humans and machines at the time—though that didn't stop users, including his own secretary, from sharing their secrets with the program—but the creation of an

AI that could converse seamlessly with a human being has been the holy grail of AI since before the field properly existed.

In 1950 the English mathematician and computer scientist Alan Turing invented what he called the "Imitation Game," now known as the Turing Test. In the game a human judge carries on a conversation via text with two entities—one a fellow human, the other a computer program. The judge doesn't know whether they are communicating with a machine or a person, and if they can't tell the difference between the two from the responses, the machine is said to have passed the Turing Test. To this day the test remains a popular marker of the progress of AI research—there's an annual contest, the Loebner Prize, that judges chatbots against the Turing Test—though it has been criticized for being limited and anthropocentric. If you've ever wanted to strangle a customer service chatbot, you can thank Alan Turing—and also for breaking the Nazis' Enigma code and helping to win World War II.

Turing predicted that by the year 2000 a computer with 128 MB of memory—a colossal amount by the standards of 1950—would have a 70 percent chance winning his Imitation Game. He was wrong on both counts. Computers had far larger memories by 2000 than 128 MB—which today would barely be enough to store a digital copy of Pink Floyd's double album *The Wall*—but still weren't passing his test. More recently chatbots have performed better on the Turing Test, although it's impossible to know whether that's because the bots are becoming more like humans, or because texting-addled, screen-addicted humans are becoming more like bots.[17]

Turing always knew his test was more a game than it was an exacting measurement of intelligence, but he did believe that the creation of authentic AI was possible. And he predicted that when that happened, it would forever alter our place on this planet. In 1951 Turing wrote: "Once the machine thinking method had started, it would not take long to outstrip our feeble powers. There would be no question of the machines dying, and they would be able to converse with each other to sharpen their wits. At some stage therefore we should have to expect the machines to take control."[18]

Turing wouldn't live to see the first AI boom. He died in 1954, persecuted by British authorities because of his homosexuality.[19] But Turing's visions of ultraintelligent machines endured. In 1965 his former colleague I. J. Good furthered Turing's line of thought, outlining a concept now called the "intelligence explosion." Good wrote:

> Let an ultraintelligent machine be defined as a machine that can far surpass all the intellectual activities of any man however clever. Since the design of machines is one of these intellectual activities, an ultraintelligent machine could design even better machines; there would then unquestionably be an "intelligence explosion," and the intelligence of man would be left far behind. Thus the first ultraintelligent machine is the last invention that man need ever make, provided that the machine is docile enough to tell us how to keep it under control.[20]

This paragraph contains everything you need to know about AI as an existential risk—and as an existential hope. A machine is invented that is more intelligent than human beings, its creators. Just as human beings strive to improve, to make ourselves smarter and better, so will this machine. The difference is that the machine can do this by upgrading its software and hardware directly, by writing better code and building better versions of itself, while humans are stuck with 73-cubic-inch brains[21] that can currently be upgraded only by the very slow process of evolution. The AI would be engaging in what is called recursive self-improvement—the process of improving one's ability to make self-improvements—and it would lead to smarter and smarter machines, at a speed faster than human thought, with no foreseeable end point. That is an intelligence explosion. And as Good wrote, if we can keep the machine under control, we'll never need to make anything else, because our ever-improving AI could do it far better than we mere humans ever could.

Good began his original article with these words: "The *survival* of man depends on the early construction of an ultraintelligent machine" (emphasis added). Good was writing just a few months after the Cuban Missile Crisis, when the extinction of man by his own hand seemed not unlikely and handing off control to an ultraintelligent machine might have seemed prudent. There's your existential hope. But it's noteworthy that when Good wrote an unpublished memoir in 1998—as James Barrat discovered in his book *Our Final Invention*—he added that perhaps the line should now read: "The *extinction* of man depends on the early construction of an ultraintelligent machine" (emphasis added).[22] And there's your existential risk.

Many AI researchers believed at the time that such an ultraintelligent machine might be just around the corner. But as the 1960s gave way to the '70s, AI research hit dead end after dead end. The money dried up, leading to the first "AI winter"—a period of shriveled funding and interest. The field experienced a renaissance in the 1980s, only for the bubble to pop again—a second "AI winter." Nearly all technological fields are subject to hype cycles followed by disappointment, as early research inevitably fails to fulfill its promise,[23] but AI has been especially marked by extreme booms and busts.

This is due in part to the difficulty of measuring just what AI actually is. Researchers ruefully note that as soon as an AI achieves something that had been considered a mark of real intelligence, that achievement is suddenly downgraded precisely *because* a machine can now do it.[24] Take chess—no less than Turing thought that only a truly intelligent AI could defeat the world's top chess players.[25] Yet while IBM's Deep Blue beat world champion Garry Kasparov more than twenty years ago, we're still waiting for the robopocalypse. As the AI researcher Rodney Brooks said in 2002: "Every time we figure out a piece of it, it stops being magical. We say, 'Oh, that's just a computation.'"[26]

That has changed somewhat in recent years, however, thanks to advances in computation that really do seem indistinguishable from magic. Take machine learning, which is when algorithms (a set of rules followed by a computer) soak in data from the world, analyze the information, and learn from it. The recommendation engine that drives Netflix is one example. The streaming service's machine-learning algorithm compiles the data you generate by watching movies and TV shows, weighs it, and then spits out recommendations for, say, Goofy Dance Musicals or Latin American Forbidden-Love Movies. (Both are actual Netflix categories.[27]) The more often you watch—and the streaming service's average American customer spends ten hours a week on Netflix[28]—the more data you generate, and the smarter the Netflix recommendation engine becomes for you.

The key here is data. A machine-learning algorithm depends on data to become smarter—lots and lots and lots of data, terabytes and petabytes of data. Until recently, data was scarce. It was either locked away in media like books that couldn't easily be scanned by a computer, or it simply went unrecorded. But the internet—and especially the mobile internet created by smartphones—has changed all that. As we now know, sometimes to our chagrin, very little that we do as individuals, as companies, and as countries goes unrecorded by the internet. As of 2018, 2.5 quintillion bytes of data were being produced *every day*, as much data as you could store on 3.6 billion old CD-ROM disks.[29] And the amount of data being produced is constantly increasing—by some counts we generate more data in a single year now than we did over the cumulative history of human civilization.[30] What oil was to the twentieth century, data is to the twenty-first century, the one resource that makes the world go— which is why tech companies like Facebook and Google are willing to go to any lengths to get their hands on it.

Raw machine learning has its limits, however. For one thing, the data an algorithm takes in often needs to be labeled by human beings. The Netflix algorithm isn't able to micro-categorize all the movies and TV shows in the service's catalog on its own—that required the labor of human beings Netflix paid to watch each and every one of their offerings, as Alexis Madrigal reported in a 2014 story for *The Atlantic*.[31] If a machine-learning AI makes a mistake, it usually needs to be corrected by a human engineer. Machine learning can produce remarkable results, especially if the algorithm can draw from a well-stocked pool of properly labeled data. But it can't be said to produce true intelligence.

Deep learning—a subset of machine learning—is something else. Fully explaining deep learning would take more space than we have (and, probably, more IQ points than this author possesses), but it involves filtering data through webs of mathematics called artificial neural networks that progressively detect features, and eventually produces an output—the identification of an image, perhaps, or a chess move in a game-playing program. Over time—and with a wealth of data—the AI is able to learn and improve. The difference is that a deep-learning neural network is not shaped by human programmers so much as by the data itself. That autodidactic quality allows an AI to improve largely on its own, incredibly fast and in ways that can be startlingly unpredictable. Deep learning results in artificial intelligence that appears actually intelligent, and it powers some of the most remarkable results in the field.

In 2016, the AI start-up DeepMind, now owned by Google, shocked the world when its AlphaGo program beat the South Korean master Lee Sedol in a series of the ancient board game of Go. Go has rules that are simpler than chess but is far more complex in practice, with more possible positions in a game than there are atoms in the universe. Unlike chess, Go can't be brute-forced by a machine's superior memory and computation speed, which is what made AlphaGo's victory so stunning. (IBM's Deep Blue was able to defeat Garry Kasparov because it could quickly simulate all potential moves and choose the best one, something that not even a human chess grandmaster can do.) But AlphaGo—which had been trained using deep-learning methods—also demonstrated what appeared to be true creativity. In the second game against Lee the program pulled off a move so unexpected that the human master had to leave the room for fifteen minutes to compose himself. AlphaGo went on to win four out of the five games in the series.

AlphaGo's victory represented what the computer scientist Stuart Russell terms a "holy shit" moment for AI, but it's far from the only one.[32] While the original AlphaGo was programmed with millions of existing Go games—meaning its success was built on the foundation of human experience—in 2017 DeepMind produced a program that began with the rules of Go and nothing more. AlphaGo Zero improved by playing itself over and over and over again millions of times—a process called reinforcement learning—until after just three days of training it proved capable of beating the original, human-trained AlphaGo, 100 to 0.[33] For good measure, DeepMind created AlphaZero, a generalized version of the program that taught itself the rules of chess and shogi (Japanese chess) in less than a day of real time and easily beat the best existing computer programs in both games.[34] The accomplishments were a textbook example of the power and the speed of recursive self-improvement and deep learning. "AIs based on reinforcement learning can perform much better than those that rely on human expertise," computer scientist Satinder Singh wrote in *Nature*.[35]

That same year a program called Libratus, designed by a team from Carnegie Mellon University in Pittsburgh, trounced four professional poker players.[36] Poker was another game that was thought to be beyond the capabilities of current AIs, not necessarily because of the level of intelligence it demands but because unless they're counting cards—meaning cheating—poker players have to compete with incomplete information of the playing field. That's closer to the way the real world works, as is the occasional bluff, which should advantage team human. As recently as 2015 the world's best human poker players beat Claudico, at the time the top AI player.[37] Yet in 2017 Libratus cleaned house. And in 2019, another DeepMind AI, called AlphaStar, beat top-level humans in StarCraft II, a highly complex computer strategy game that requires players to make multiple decisions at multiple time frames all at once, while dealing with imperfect information.[38]

It's not just fun and games. Medical algorithms can detect disease, and police algorithms can predict crime. Personal digital assistants like Siri and Alexa are integrating themselves into our daily lives, getting smarter the more we use them (and the more they listen to us). Autonomous cars—which demand precise image recognition and decision making from AIs—are edging closer to reality. Actual robots out in the real world are navigating terrain and mastering physical challenges with a fluidity that would have seemed preposterous a few years ago. In a 2017 survey, hundreds of AI experts predicted that machines would be better than humans at translation by 2024, writing high school essays by 2026, driving a truck by 2027, working in retail by 2031, writing a bestselling book by 2049 (uh-oh), and surgery by 2053. The experts gave a 50 percent chance that machines would outperform humans in all tasks within 45 years, and that all human jobs would be automated within the next 120 years. Sooner or later we all may be Lee Sedol.[39]

Every great technological leap has been accompanied by social and economic disruption. The original Luddites were nineteenth-century British weavers and textile workers who revolted over the introduction of automated looms that threatened their artisanal livelihoods. Mechanization reduced the number of farmers, sending agricultural workers streaming into the cities. The first cars caused so much panic in the countryside that anti-automobile societies sprang up to resist them; one group in Pennsylvania even proposed a law requiring that cars traveling through the country at night send up a rocket every mile and stop for ten minutes to let the road clear.[40] Some people were initially wary of telephones, fearing they could transmit electric shocks through their wiring.[41]

Sooner or later, however, we adjust to new technology, and then we take it for granted. The short-term job loss that causes so much fear is usually eased by the productivity growth enabled by labor-saving advances. We end up richer and better off—for the most part.

AI, though, could be different. Existential risk experts may obsess over superintelligence, the possibility that AI will become immensely smarter than us, but AI doesn't have to become superintelligent— assuming that's even possible—to create more upheaval than any other technological advance that has preceded it. AI could exacerbate unemployment, worsen inequality, poison the electoral process, and even make it much easier for governments to kill their own citizens. "You've got job loss," Andrew Maynard, the director of the Risk Innovation Lab at Arizona State University, told me. "You've got privacy.

You've got loss of autonomy with AI systems. I think there are a lot of much clearer issues than superintelligence that are going to impact people quite seriously, and that we have to do something about."

Despite one of the longest economic expansions in U.S. history, real wages for the working class have largely remained stagnant—and the spread of AI may be playing a role. One 2018 survey found that wages slipped in job areas where automation and AI was taking hold[42], and a survey by the McKinsey Global Institute estimated that 800 million jobs worldwide could be lost to automation by 2030.[43] Past experience suggests that broad job loss should be temporary—and indeed the McKinsey report found that only 6 percent of all jobs were at risk of total automation—as investment shifts from declining sectors to rising ones. If AI is as revolutionary as its most ardent advocates claim, however, then the past is no longer a reliable guide to the future, and we may be in for flat wages followed by widespread job loss. AI may do work, but it isn't labor—it's owned and controlled by capital. As AI becomes a more integral part of our economy, its gains seem likely to come at the expense of workers, and accrue to owners, intensifying inequality.

If little is done to share the wealth generated by AI, the results could be socially explosive. The San Francisco Bay area, home to tech companies like Google and Facebook, can claim to be the world center of artificial intelligence research, a region that sets the terms of the future. It's also a place of striking income inequality, where some of the richest people in the world live side by side with the most destitute. A 2018 report by the Brookings Institution ranked San Francisco as the sixth-most unequal city in the United States. The raw dollar gap between the poorest and the richest is so high there largely because the very rich are richer than almost anywhere else.[44] And that's because in the tech economy, rewards overwhelmingly flow to owners versus a broad class of workers. In 2016 Apple,

Alphabet (the parent company of Google), and Facebook all made well above $1 million in revenue per worker. Their total combined revenues of $336 billion were actually much less than what the top three Detroit car companies of Chrysler, General Motors, and Ford made in 1996, adjusted for inflation. But the tech companies—in addition to producing much less physical stuff—employ far fewer workers. That means less of the value of these companies—some of it generated by AI—has to be paid out to labor and more can be kept for owners and investors.[45]

It may not be machine overlords we need to fear, but the human overlords who own the machines.[46] There's a reason why some of Silicon Valley's most voracious capitalists have begun putting their weight behind plans to give people universal basic income (UBI)—a way for the average person to share, at least a little bit, in an increasingly capital- and tech-driven economy. (There's also a self-serving reason—without UBI, consumers may no longer have the money to keep businesses operating.) Christine Peterson of the Foresight Institute—a Silicon Valley–based think tank that examines emerging technologies—even has a novel idea for something called Inheritance Day, which would give every human alive an equal share of an unclaimed portion of the universe, with the assumption that these shares will eventually become valuable as we spread into space. Think of it as homesteading for the AI age.[47] It's extreme—but then so is the thought of an economy utterly hollowed out by machines.

The danger from near-term AI isn't only economic. The same machine-learning capabilities that can make AIs better at recommending movies or playing games can also enable them to become unparalleled killers. In 2017 Stuart Russell and the Future of Life Institute produced a striking video dramatizing a near-future scenario where swarms of tiny mechanical drones use AI and facial recognition technology to target and kill autonomously—so-called

slaughterbots.[48] The video is terrifying, and almost certainly a
glimpse of the near future. We already have remote-piloted drone
aircraft that are capable of delivering lethal payloads, but autonomous weapons would outsource the final decision to pull the trigger
to an AI. In the future drones might be programmed to assassinate a
particular target, like a politician, or even carry out ethnic cleansing
by hunting a specific racial group. The development of such weapons
would represent the third great revolution in warfare, after gunpowder and nuclear arms. They threaten to reduce the restraints on war
and make it far easier to kill at will—for states at first, but eventually
for criminal gangs, terrorists, and even lone individuals.

So it should be worrying that it's not just businesses pouring
money into AI research—it's militaries, too. In 2018 the Pentagon
launched a $1.7 billion Joint Artificial Intelligence Center, while the
Defense Advanced Research Projects Agency (DARPA)—the military think tank that helped develop the internet—announced its own
$2 billion AI campaign the same year.[49] One cutting-edge military
program, Project Maven, was designed to aid computer systems in
identifying targets from aerial footage. This would be very useful if
you wanted to, say, develop a drone that could pick out targets and
fire on them autonomously. The Defense Department had partnered
with Google on Project Maven, but when news of that collaboration
became public, the company faced a backlash from its own work-
force. Thousands of Google employees signed an open letter protesting the company's involvement with the project, and in May 2018
Google announced that it would not renew the contract.[50] A few days
later Google's Sundar Pichai released an open letter promising that
the company's AI projects will not include weapons or technologies
that cause or are likely to cause overall harm.[51]

That was a positive sign for the future of AI, and in keeping with
the sensibilities of the majority of artificial intelligence researchers.

So was an open letter sent in 2017 to the United Nations and signed by tech luminaries like Elon Musk and Demis Hassabis, the cofounder of DeepMind, that called for a ban on lethal autonomous weapons— killer robots, in other words.[52] But with a $700 billion budget, the U.S. military has deep pockets, and if companies like Google eschew working on AI programs that could be used for defense—or offense— you can bet other firms will be more than willing. Both Microsoft and Amazon have made it clear that they will continue to work with the Department of Defense on tech and AI, with Amazon CEO Jeff Bezos saying in October 2018 that "if big tech companies are going to turn their back on the Department of Defense, then this country is going to be in trouble."[53]

Bezos has a point. Whatever American tech companies choose to do, China has made it clear that it is determined to close its artificial intelligence gap with the United States—and then surpass it. By one estimate China spent $12 billion on AI in 2017, and is poised to spend at least $70 billion by 2020.[54] In 2017 China attracted half of all AI capital in the world,[55] and by the same year the country was producing more highly cited AI research papers than any other nation, including the United States.[56]

Some of that money has been dedicated to building the world's most powerful—and intrusive—facial recognition software. On a 2018 reporting trip to China, I was struck by how ubiquitous facial recognition systems had become there, whether in security lines at airports or at hotel front desks, where guests could use their faces to pay their bills.[57] In cities across China, facial recognition cameras scan crowds, searching for both wanted criminals and those who "undermine stability."[58] China is also using AI to build an unprecedented social credit system that will rank the country's 1.4 billion citizens not just according to their financial trustworthiness, but also their "social integrity."[59]

If Chinese officials use AI to turn China into a technologically Orwellian state—as seems likely—that's a tragedy for China, and potentially a "mortal danger" for more open societies if those tools are exported, as the financier and philanthropist George Soros put it in a speech in 2019.[60] But even more worrying for the world is the growing likelihood that the United States and China will embark on a twenty-first-century arms race, only one involving AI instead of nuclear weapons. Such a race would pose an existential risk in itself—not just because artificial intelligence might threaten humans, but also because of the possibility that sudden leaps in AI capability might upset the balance of power needed to keep peace between nuclear-armed states.

Imagine, for instance, if the United States were to develop sophisticated military AI that could remotely disable China's nuclear forces. Even without using such technology—even without threatening to use it—Beijing may be put in a position where it feels it needs to launch a strike against the United States before Washington could mobilize its new AI weapon. Mutually assured destruction only works if both countries maintain rough technological parity. AI threatens to overthrow that parity. "Weaponized AI is a weapon of mass destruction," said Roman Yampolskiy, a computer scientist at the University of Louisville who has closely studied the risks of AI. "An AI arms race is likely to lead to an existential catastrophe for humanity."

Weapons are just the most obvious ways that AI can break bad, however—and it's not always easy to tell the dangerous uses of AI from the beneficial ones. In February 2018, fourteen institutions— including the Future of Humanity Institute and the Centre for the Study of Existential Risk—collaborated on a hundred-page report outlining malicious uses of AI. The research singled out the way AI can accelerate the process of phishing through automation—instead of one person pretending to be a Nigerian prince dying to give you his money, an AI phishing program can send out endless emails, and use reinforcement learning to refine them until someone bites. AI will soon be able to hack faster and better than humans, and spread propaganda by creating fake images and videos that will soon be indistinguishable from reality. And like biotechnology, AI is a dual-use technology—the same advances can be employed for good or ill—which makes regulation challenging. In the near future—perhaps as soon as 2020—we'll look back on the Russian election hacking efforts in 2016 as a beta test, a Trinity before an AI Hiroshima.[61]

The possibility of an AI-induced accidental nuclear war is nightmarish. But what should truly frighten us is that we don't understand the AI systems that we're using—not really. And I don't mean ordinary people googling "artificial intelligence takeover"—I mean the experts. The very characteristics that make deep learning so powerful also makes it opaque. An artificial neural network can have hundreds of layers, adding up to millions and even billions of parameters, each tuned on the fly by the algorithm as it tries to draw conclusions from the data it takes in. The process is so complex that even the creators of the algorithms often can't say why a particular AI makes a particular decision. Recall AlphaGo's match against Lee Sedol, when the program produced a move on the Go board that utterly shocked the human grandmaster. No one—including AlphaGo's makers at DeepMind—could explain how and why the program decided to do what it did. And of course, neither could AlphaGo. AI doesn't explain—it acts.

But when AI acts, it can make mistakes. A Google algorithm that captions photographs regularly misidentified people of color as gorillas.[62] An Amazon Alexa device offered porn when a child asked to play a song.[63] A Microsoft chatbot called Tay began spouting white supremacist hate speech after less than twenty-four hours of training on Twitter.[64] Worst of all, in March 2018 a self-driving Uber car on a test run in Arizona ran over and killed a woman, the first time an autonomous vehicle was involved in a pedestrian death. The automobile detected the victim six seconds before the crash, but its self-driving algorithm classified her as an unknown object until it was too late to brake.[65]

Some of those working on developing autonomous cars, like Elon Musk at Tesla, argue that early problems in the technology should be balanced against the fact that human drivers cause deadly crashes all the time.[66] 37,133 people in the United States died in motor accidents in 2017, about one victim every fifteen minutes.[67] Musk has a point, but error in an AI algorithm is qualitatively different than human error. A single bad human driver might kill a few people at most when they make a mistake, but an error in a single AI could spread across an entire industry, and cause far greater damage. IBM's Watson for Oncology AI was used by hundreds of hospitals to recommend treatment for patients with cancer. But the algorithm had been trained on a small number of hypothetical cases with little input from human oncologists, and as STAT News reported in 2018, many of the treatments Watson suggested were shown to be flawed, including recommending for a patient with severe bleeding a drug that was so contraindicated for their symptoms that it came with a black box warning.[68]

No technology is foolproof, especially in its early days. But if a bridge collapses or an airplane crashes, experts can usually look over the evidence and draw a clear explanation of what went wrong. Not always so with AI. At a major industry conference in 2017, Google AI researcher Ali Rahimi received a forty-second standing ovation when he likened artificial intelligence not to electricity or to fire—like his boss Sundar Pichai—but to medieval alchemy. Alchemy produced important scientific advances, including the development of metallurgy, but alchemists couldn't explain the scientific basis of *why* what they were doing worked when it worked. AI researchers, Rahimi suggested, are currently closer to alchemists than scientists.[69]

While most AI researchers are against autonomous weapons, the fact remains that there is little regulation in the field. A comparison to biotechnology is illustrative. Biological weapons are banned by international law. In the life sciences there are ethics boards at universities to review experiments and reject ones they find wanting. There is the Department of Bioethics at the National Institutes of Health, to think deeply about the direction and methods of medical research. There is the National Institutes of Health, period. Biologists are trained in lab safety and ethics. That's why when someone in the field goes rogue—like the Chinese scientist who gene-edited human embryos—outrage tends to follow.

AI ethics, by contrast, is in its infancy. There is nothing like the National Institutes of Health around AI, no significant independent boards to oversee experiments. While there has been an international push to ban autonomous weapons, no such law exists—and in fact, a proposed treaty was blocked by countries including the United States and Russia in 2018.[70]

Some positive change is on the way. Brent Hecht is a young computer scientist at Northwestern University who chairs the Future of Computing Academy, a part of the world's largest computer science society. Hecht advocates that peer reviewers in computer science and AI who judge articles submitted to scientific journals should evaluate the social impact that a piece of research might have, as well as its intellectual quality. That might seem like a small thing, but it's unprecedented for the field. "Computer science has been sloppy about how it understands and communicates the impacts of its work, because we haven't been trained to think about these kinds of things," Hecht told *Nature*. "It's like a medical study that says, 'Look, we cured 1,000 people,' but doesn't mention that it caused a new disease in 500 of them."

Large tech companies are beginning to take notice as well. When DeepMind was purchased by Google in 2014, the AI start-up insisted on the creation of an AI ethics board, to ensure that its work should "remain under meaningful human control and be used for socially beneficial purposes."[71] In 2017 the Future of Life Institute convened a major conference at Asilomar Conference Grounds in California, where more than four decades ago biologists met to hash out the ethical issues around the new science of recombinant DNA. The result was the creation of the Asilomar Principles, a road map meant to guide the industry toward beneficial and friendly AI.[72] In 2019 Facebook spent $7.5 million to endow its first AI ethics institute, at the Technical University of Munich in Germany.[73]

At a moment when tens of billions of dollars are flowing to AI research and implementation, however, just a tiny amount is being spent on efforts to keep AI safe. According to figures compiled by Seb Farquhar at the Centre for Effective Altruism, more than fifty organizations had explicit AI safety-related programs in 2016, with spending levels at about $6.6 million. That's a fourfold increase from just a couple of years before, but it hardly compares to the scale of the challenge.[74] And while attention to the possible risks of AI has grown in recent years, the field as a whole is far more focused on breaking new ground than double-checking its work. "You are dealing with creative geniuses who pour their whole mental abilities into a difficult problem," said Christine Peterson of the Foresight Institute. "To say we also want you to think about the social issues and the hardware insecurities is like asking Picasso to think about how his paint is made. It's just not how they're wired."

There may also be a competitive disadvantage to slowing down the pace of AI research. As impressive as AI projects like AlphaZero are, they still represent what is called "narrow AI." Narrow AI programs can be very smart and effective at carrying out specific tasks, but they are unable to transfer what they have learned to another field. In other words, they're unable to generalize it, as even the youngest human being is able to generalize what he or she learns. AlphaZero learned how to play chess in nine hours and could then beat any human who has ever lived,[75] but if the room where it was playing caught on fire, it wouldn't have a clue what to do. Narrow AI remains a tool—a very powerful tool, but a tool nonetheless.

The ultimate goal of AI researchers is to create artificial general intelligence, or AGI. This would be a machine intelligence that could think and reason and generalize as humans do, if not necessarily in a humanlike manner. A true AGI wouldn't need to be fed millions upon millions of bits of carefully labeled data to learn. It wouldn't make basic mistakes in language comprehension—or at least not for long. It would be able to transfer what it learned in one subject to another, drawing connections that enable it to become smarter and smarter. And it would be able to do all this at the accelerated speed of a machine, with an infallible memory that could be backed up in the cloud. To us slow, carbon-based humans, an AGI would seem to have superpowers, and there may be very few limits, if any, to what it could do. In this way, Sundar Pichai would be right—the development of artificial general intelligence really would be more significant than electricity or fire.

The first country or company to develop powerful AGI would be in a position to utterly dominate its competitors, even to the point of taking over the world if it so chooses. This would be a winner-takes-all competition, since the first move the owner of a working AGI might be smart to make would be to sabotage any rival efforts to develop artificial general intelligence. And that should worry us, because the closest analogue we have is the race to a nuclear bomb during World War II. Safety concerns at the Manhattan Project took a backseat to the all-important goal of developing the bomb before the Nazis. That worked out well enough—the Allies won the war, and you'll recall that the atmosphere did not ignite and end life on Earth after Trinity—but the same dynamics could prove disastrous if the groups racing toward AGI are tempted to cut ethical corners to cross the finish line first. And that's because if human-level AGI is actually achievable, it's not likely to stay at human level for long.

Nor is there any guarantee that human beings will remain in control of something that will wildly exceed our own capabilities in, well, everything. As Nick Bostrom wrote in his 2014 book, *Superintelligence*, which introduced the existential threat of AI to the general public: "Before the prospect of an intelligence explosion, we humans are like small children playing with a bomb. . . . We have little idea when the detonation will occur, though if we hold the device to our ear we can hear a faint ticking sound."[76]

———————

There is a classic Hollywood film that perfectly captures the confusion, the betrayal, and the chaos that could follow the creation of a superpowerful artificial intelligence—and despite all the times you may have seen a still of Arnold Schwarzenegger above a story about an AI takeover, it is not *The Terminator*. It's *The Sorcerer's Apprentice*, the famous Mickey Mouse short from the 1940 animated Disney film *Fantasia*.

It begins with Mickey's motivations. Tired of fetching water to fill a cistern in the sorcerer's cave, Mickey (the apprentice) wants the broom (the AI) to do his menial labor for him. He puts on the sorcerer's magic hat, and then, using a spell he barely understands, Mickey brings the broom to life. He instructs it to carry buckets of water from the fountain outside to the cistern. And the broom does so—the program works. The broom is doing exactly what it's been told to do—so well, in fact, that the apprentice can relax in his master's chair, where he drifts off and dreams of the unlimited power that will be his once he becomes a full-fledged sorcerer.

But Mickey is soon awoken when the chair is knocked over by the force of the rising water. Has the spell been broken? Quite the opposite. The broom is still doing exactly what it was programmed to do: fill the cistern with water. And while the cistern looks full and then some, to the broom, as it would to an AI, there's always a tiny probability that the cistern is not 100 percent full. And because all the broom knows is its spell—its software code—it will keep fulfilling those instructions over and over again, in an effort to drive that probability slightly higher. While a human—or a cartoon mouse—knows that instructions aren't meant to be followed to the absolute letter, the broom doesn't have that basic common sense, and neither would an AI.

When Mickey tries to physically stop the broom, it marches over him. It doesn't matter that Mickey is the programmer who instructed the broom to fill the cistern. The instructions weren't to listen to Mickey at all times—they were to ensure that the cistern is filled. Mickey is now not something to be obeyed, but rather an obstacle to be overcome, just as any human getting in the way of a superintelligent AI—even for the best reasons—would become an obstacle to be overcome. When a desperate Mickey chops up the broom, all he does is create an army of brooms, each bent on carrying out his original commands. The broom has adapted, because even though self-preservation isn't part of the broom's instructions, it can't fill the cistern if it's been destroyed. So it fights for its life—just as a superintelligent AI would if we tried to unplug it.

Mickey, now close to drowning in the sea of water that fills the cave, grabs the book of spells and frantically looks for some magic words—some code—that can reprogram his broom. But he's not smart enough, just as we wouldn't be smart enough to reprogram a superintelligent AI that would surely see those attempts as obstacles to fulfilling its goals—and respond accordingly. It's only when the far more knowledgeable sorcerer himself returns that the broom can be disenchanted—reprogrammed—and the apprentice is saved. But as we should know by now, when it comes to man-made existential risks, we're all apprentices. (The *Sorcerer's Apprentice* analogy is taken from the work of Nate Soares at the Machine Intelligence Research Institute.)[77]

Trying to comprehend the nature of an AI superintelligence is akin to trying to comprehend the mind of God. There's an element of the mystical in attempting to define the intentions of a being infinitely more intelligent and therefore infinitely more powerful than you—a being that may or may not even exist. There's even an actual religion, founded by a former Uber and Waymo self-driving engineer, dedicated to the worship of God based on artificial intelligence.[78] But the existential fear of superintelligence isn't just spiritual vaporware, and we have a sense, at least, of how it might come to be.

Researchers like Eliezer Yudkowsky suggest that the development of superintelligence could start with the creation of a "seed AI"—an artificial intelligence program that is able to achieve recursive self-improvement. We've already done that with narrow AI—AlphaGo Zero started out knowing nothing about the game Go, but quickly managed to improve to a human level and then far beyond. If scientists could crack the secret to artificial general intelligence, then they may be able to create a seed AI capable of growing into a superintelligence. If that happens, we may not have much warning before an AI grows to human intelligence—and then well beyond. The result would be a "fast takeoff," a sudden leap in an AI's capabilities that would catch the world—and quite possibly the AI's creators—fatally off guard.

A fast takeoff would be the most dangerous way that a super-intelligent AI could develop, because it would leave us with little to no time to prepare for what could come next. Imagine if the race for a nuclear bomb had resulted not in a single weapon, but thousands upon thousands of nuclear missiles all at once. Would the world have survived if the nuclear arms standoff of 1983 had been transplanted to 1945, when there was no experience with atomic weapons and none of the safety infrastructure that had been built up through decades of the Cold War? I doubt it.

The more time we have to prepare before superintelligent AI is achieved, the more likely we'll survive what follows. But you'll get total disagreement from scientists on how soon this might happen, and even whether it will happen at all. Ray Kurzweil, a computer scientist at Google and a futurist who has long heralded the revolutionary potential of AI, has predicted that computers will achieve human-level intelligence by 2029 and something like superintelligence by 2045.[79] That makes him slightly more optimistic than Masayoshi Son—the CEO of Softbank, a multibillion-dollar Japanese conglomerate that is heavily invested in AI—who says 2047.[80] Even more confident is Vernor Vinge, a former professor at San Diego State University and the man who invented the concept of the technological singularity. He projected that it would come to pass between 2005 and 2030— although in Vinge's defense, he made that prediction way back in 1993.

The point is that no one knows, and there are many in the field of AI who believe we are still far away from developing anything close to a program with human-level general intelligence, and that something like superintelligence would require an actual miracle. Anyone who lives with an Alexa or tries to use Siri knows that consumer AI remains, at best, a work in progress. Andrew Ng, the former head of AI at the Chinese tech company Baidu, once said that "the reason I don't worry about AI turning evil is the same reason I don't worry about overpopulation on Mars."[81] In other words, neither problem is something we'll likely need to deal with anytime soon.

It's easy to think that just because we see AI improving rapidly at a growing number of tasks—and greatly exceeding our own abilities along the way—it won't be long before an AI could be developed that would be at least as smart as a human being. Human beings are intelligent—but not that intelligent, right? After decades of study, however, we still don't understand the basics of the mysterious thing known as human consciousness. We don't know why we're intelligent and self-aware in a way no other life-form appears to be. We don't know if there's something special about the biological structure of our brains, produced by billions of years of evolution, that can't be reproduced in a machine.

If designing an AGI is simply impossible, then we don't have to worry about superintelligence as an existential risk—although that means we won't have AI to help defend us from other existential risks, either. But there's also no reason to assume that intelligence can only be the product of biological evolution. Machines have significant advantages over minds made of meat. Our brains become fatigued after several hours of work, and start to decline after just a few decades of use. The human brain may store as little as one billion bits, which barely qualifies it to act as a smartphone.[82] Computers can process information millions of times faster than human beings, and most important, they do not forget. Ever. If this book had been written by a machine brain, it may or may not have been better—but it would definitely have been finished faster.

Science has a track record of overpromising and underdelivering. As the venture capitalist Peter Thiel said dismissively of Silicon

Valley's proudest twenty-first-century achievements: "We wanted flying cars, instead we got 140 characters." But we should also be wary when scientists claim that something is impossible. The brilliant physicist Ernest Rutherford said in 1933 that nuclear energy was "moonshine"—and less than twenty-four hours later Leo Szilard came up with the idea of the nuclear chain reaction, which laid the groundwork for Trinity.[83] There's another saying in Silicon Valley: we overestimate what can be done in three years and underestimate what can be done in ten.

In a 2013 survey, Nick Bostrom and Vincent C. Müller asked hundreds of AI experts for their optimistic and pessimistic predictions about when AGI would be created. The median optimistic answer was 2022, and the median pessimistic answer was 2075.[84] There's almost an entire human lifetime between those two answers, but if we're talking about an event that could be, as I. J. Good put it, humanity's "final invention," the difference may not matter that much. Whether we're close or whether we're far, we can't simply dismiss a risk of this size, any more than we can dismiss the risk from climate change because scientists aren't perfectly certain how severe it will be. Too often even those most knowledgeable about AI seem to think that if superintelligence won't happen soon, it effectively means that it won't happen ever. But that conclusion is incautious at best. As Stephen Hawking wrote in 2014: "If a superior alien civilization sent us a message saying, 'We'll arrive in a few decades,' would we just reply, 'OK, call us when you get here—we'll leave the lights on'? Probably not—but this is more or less what is happening with AI."[85]

If artificial general intelligence is developed before we're ready, we might be screwed. The first challenge would be control. Those developing the seed AI might try to keep it confined or "boxed," as the term goes, preventing it from being able to connect to the external world—and more important, the digital wealth of the internet—except through a human interlocutor. It's similar to how medical researchers would handle a very, very dangerous virus. But remember—a superintelligent AI would be much, much smarter than you. Not smarter than you like Albert Einstein was smarter than you, but smarter than you in the way that Albert Einstein was smarter than a mouse, as Bostrom has put it.[86] And intelligence isn't just about the mathematical skill and technical ability that AI has demonstrated so far. We can expect a superintelligent AI to be able to instantly decode human society, psychology, and motivation. (Human beings aren't that complicated, after all.) If kept off the internet, it could figure out a way to convince a human being to connect it. Blackmail, flattery, greed, fear—whatever your weak point is, a superintelligent AI would find it and exploit it, instantly. And in a world as networked as ours, once it was online there would be no stopping it.

So let's say it happens—a superintelligent AI is born, and it escapes our feeble attempts to control it. What happens next? Much of what we think about AI comes from fictional narratives, from *Franken-stein* through *Ex Machina*, stories that assign human motives and emotions to something that would be utterly unhuman. An AI wouldn't act to take revenge on its human parents, and it wouldn't be driven to conquer the world—those are human motivations that would have no logical place in a machine. Remember the parable of *The Sorcerer's Apprentice*. An AI, even a superintelligent one, is in its silicon soul an optimization machine, designed to discover the most efficient path to its programmed goals. The reason why we couldn't simply unplug a superintelligent AI is that the machine cannot accomplish its goal—whatever that goal is—if it ceases to operate, which is why it will stop you if you try. If humans get in the way of its goals—if we're perceived to be getting in the way, or even if omnicide offers a slightly more efficient path to achieving its goals—it would try to remove us. And being superintelligent—intelligence, after all, is the ability to apply knowledge toward the achievement of an objective—it would succeed.

If that sounds like it wouldn't be a fair fight, well, welcome to natural selection. Evolution has nothing to do with fairness, and an encounter between human beings and a superintelligent AI is best understood as evolution in action—likely very briefly. The same narratives that informed our picture of an AI takeover tend to feature similar endings, with human beings banding together to defeat the evil machines. But those are stories, and stories need to give protagonists a fighting chance against antagonists. A fist squashing a mosquito isn't a story, and that's what our encounter with all-powerful AI might be like.

Since controlling a superintelligent AI once it has become super-intelligent seems impossible, our best chance is to shape it properly first. Fictional stories about robot uprisings aren't helpful here, but fairy tales are, because they teach us to be very, very, very careful when we're asking all-powerful beings to grant our wishes. Nick Bostrom has a term for what happens when a wish has an ironic outcome because of the imprecise way it is worded: perverse instantiation.[87] King Midas learned this the hard way when he wished to have the power to turn everything he touched to gold. Precisely that happened, but Midas didn't foresee that his wish meant that his food and drink and his beloved daughter would turn to dead and cold gold when he touched them. Midas got precisely what he wished for but not what he wanted.

So it might be with an AI. Perhaps you program the AI to make humans happy, which seems like a worthy goal—until the AI determines that the most optimal way to do so is to implant electrodes that repeatedly stimulate the pleasure centers of your brain. You may be very "happy," but not in the way you presumably wanted. Or perhaps you program the AI to solve a fiendishly complex mathematical problem, which the AI does—but to do so, it needs to turn the entire planet, and every living thing on it, into energy to fuel its computation. Or maybe an AI is put in charge of an AI paper clip factory and is programmed to maximize production—so, with all of its optimization power, it proceeded to transform everything in the universe into paper clips.

That last example is taken from Bostrom's *Superintelligence*— you can even play a computer game where you take charge of a superintelligence and try to turn the universe into paper clips.[88] If it sounds absurd, it's meant to be. We already have some experience with perverse instantiation in far more basic AI. Victoria Krakovna, a researcher at DeepMind, built a list of existing AI programs that followed orders, but not in the way their creators expected or necessarily wanted. In one case an algorithm trained to win a computer boat racing game ended up going in circles instead of heading straight for the finish line. It turned out that the algorithm had learned that it could earn more points for hitting the same gates over and over again, rather than actually winning the race.[89] In another example AI players in the computer game *Elite Dangerous* learned to craft never-before-seen weapons of extraordinary power that enabled the AIs to annihilate human players.[90] That one was a bit on the nose—the gaming blog Kotaku titled its post on the story: "Elite's AI Created Super Weapons and Started Hunting Human Players. Skynet Is Here."

What we might call gaming the system, the AI sees as achieving its goal in the most efficient way possible. The point is we can't expect an AI, even a superintelligent AI, to know what we mean, only what we say—or what we program. "Once AIs get to a certain point of capacity, the result will depend on them regardless of what we do,"

Seth Baum of the Global Catastrophic Risk Institute, an existential risk think tank, told me. "So we either better never give them that capacity, or make sure they are going to do something good with it when that happens."

Pulling that off is the art and science of AI alignment, the key to defusing artificial intelligence as an existential risk. If we're going to take the steps to create a superintelligence, we need to do everything we can to ensure it will see our survival, our human flourishing, as its only goal. All we need to do to program a friendly AI, as Max Tegmark of the Future of Life Institute has put it, is to "capture the meaning of life."[91] No problem.

Here's a present-day illustration of how challenging it will be to get AI alignment right. As companies like Uber and Google develop autonomous cars, they need to train the self-driving AI to respond safely and correctly in almost every potential situation. For example: what should an autonomous car do if the only way it can avoid hitting and killing a pedestrian is to swerve in a way that puts its driver at risk? Should it prioritize the life of the driver, or the life of the pedestrian? What if there are two pedestrians? What if the pedestrian is elderly—meaning they have fewer life-years to potentially lose—and the driver is young? What if the situation is reversed? What's the ethically correct decision—in every situation?

What I'm describing is known as the trolley problem, a classic thought experiment in ethics. There is no perfect answer to the trolley problem, but how you choose to respond—save the driver or save the pedestrian—says a lot about where you stand on moral psychology. But an autonomous car will be programmed by someone who will have to answer that question for the AI. Ethicists have been debating the trolley problem for decades with no clear answer, so how will it be possible for computer scientists to figure it out, let alone the countless other philosophical disputes that would need to be solved to create a superintelligent AI that is also friendly to human beings? And don't forget that those who will be charged with inputting those ethical values are not exactly representative of the broad spectrum of humanity—one survey found that just 12 percent of AI engineers are women.[92] Do you really want the people who brought you social media to program our future machine overlords?

Determining the right human values is hard enough; somehow transmitting those values into an AI might be harder. "Even if we knew the values, how do we communicate those values to a system that doesn't share anything with us?" said Allison Duettmann, an AI safety researcher at the Foresight Institute. "It doesn't share intuition with us, or history or anything. And how do we do it in a way that is safe?"

Yudkowsky, at least, has put forward an idea: "coherent extrapolated volition."[93] In his rather poetic words, it is: "our wish if we knew more, thought faster, were more the people we wished we were, had grown up farther together; where the extrapolation converges rather than diverges, where our wishes cohere rather than interfere, extrapolated as we wish that extrapolated, interpreted as we wish that interpreted."[94] In other words, we would want the AI to understand us as our best friend would understand us—with infinite sympathy, with the ability to see past our mere words to the heart of what we mean. A best friend who could do that, and had the power to make all of our wishes come true.

That's our existential hope. It's why getting AI right, as the Silicon Valley venture capitalist Sam Altman has said, "is the most important problem in the world."[95] Not just because of the catastrophe that could befall us if we get it wrong, but because of all that we could win if we somehow get it right. Asteroids, volcanoes, nuclear war, climate change, disease, biotech—given enough time, one of those existential threats will get us, unless we show far more wisdom than the human race has demonstrated to date. If we can somehow safely crack the AI problem, however, we will not only enjoy the boundless benefit that superintelligence could bring, but we'll prove that we have the wisdom to deserve it. It may be a slim hope. It may even be an impossible one. But what more can we hope for?

ALIENS

Where Is Everybody?

What will become of us? Through the course of this book, I've looked to the past, examining our planet's deep history and the geological and astronomical catastrophes that have already ended life. I've looked to the present, scrutinizing the current threat of nuclear war and the growing danger of climate change. And I've looked to the future, exploring the new risks that will be born from emerging technologies like synthetic biology and artificial intelligence. Yesterday, today, and tomorrow—somewhere among them, we'll find our own fate, and just maybe, discover how to escape it.

One of the toughest challenges presented by existential risk, however, is that such threats are "too big to learn," in the words of the Duke University law professor Jonathan Wiener.[1] Our species has always advanced through trial and error. But by definition we've never experienced a full-scale existential threat—otherwise we wouldn't be here—and if one strikes in the future before we're ready, there won't be anyone left to write the after-action report. We only get one shot at this planet, one attempt to run the great experiment known as human civilization. But you can't learn anything from your own extinction.

Our planet, however, is not the only one in this vast universe. New research suggests that there are thousands, maybe millions, maybe even billions of planets in our own galaxy that could potentially host life.[2] If there is life, there is the potential for intelligence, including civilizations not unlike our own. And that would mean there is not merely one experiment testing what will happen to intelligent life, but countless other trials running out there among the stars. Alien life, facing its own existential risks.

Aliens, of course, could be classified as an existential threat themselves. That's what fiction teaches us, from H. G. Wells's Victorian-era story *The War of the Worlds* to the mid-century thriller *The Body Snatchers* to the 2018 blockbuster film *A Quiet Place*. As different as these narratives are, each emblematic of the cultural fears of their moment, they share common themes. Advanced extraterrestrials come to conquer Earth—whether secretly or in force—and humans band together to repel them. But in any extraterrestrial intelligence that possesses the technological capability to cross the stars for Earth and the hostile intentions to attack it would almost certainly crush us, as quickly and as efficiently as a malevolent superintelligent AI. If such ETs exist and they're angry, then aliens would be an existential threat—albeit one we could do little to counter. If they exist.

But do they? That's the question that really matters for existential risk—not military plans to track UFOs or panicked tales about alien abduction. If intelligent aliens live, if they have ever lived, then they hold out the possibility of another experiment, another attempt on a distant star to survive and thrive in a universe so often hostile to life. If they've managed to outlast the same kind of existential threats that we face, then they offer us a strand of hope that we too can make it, that our destiny isn't death. Perhaps they may even be able to guide us, as many of the scientists who launched the first searches for extraterrestrial intelligence so fervently hoped. "We are at a very dangerous moment in human history," said Carl Sagan in 1978. "We are not certain that we will be able to survive this period of what I like to call technological adolescence. Were we to receive a message from somewhere else, it would show that it's possible to survive this kind of period. And that's a useful bit of information to have."[3]

If they don't exist, however, then we face another question. If we are truly as alone in the universe as we now appear, then perhaps intelligent life is an unlikely accident of cosmic fate, and we really are one in a 100 billion—the number of stars in the Milky Way. If so, our fate is even weightier, for *Homo sapiens* would become the cosmos's sole conscious observers. Or worse, perhaps intelligent life has arisen elsewhere in the galaxy, only we see no evidence because aliens have been destroyed over and over, or have destroyed themselves. Maybe every time the experiment of intelligence has been conducted, it has failed. If that is the case—if the apparent lack of intelligent life in the galaxy is the product of repeated cosmic catastrophes that destroyed alien civilizations, as we ourselves are in danger of being destroyed— then are we too doomed to extinction?

We should be worried. The Search for Extraterrestrial Intelligence, or SETI, has been under way for decades, and a new wave of science has shown that our galaxy may be far more hospitable to life than we first suspected. Yet we have found nothing. Not a scrap of evidence of extraterrestrial intelligence, not even the slightest hint of an alien microbe, the sort of foundational life that has existed on our planet for billions of years. So far, at least, it's all quiet on the interstellar front.

It's a big galaxy and an even bigger universe. Jill Tarter, the doyenne of SETI and the inspiration for Jodie Foster's alien hunter in the film *Contact*, has said that all our attempts to search for alien life amount to plunging a single glass into the world's oceans.[4] But the longer our searches are met with silence, the more we will ask a question that first occurred to the physicist Enrico Fermi decades ago: where is everybody? And the answer—if it ever comes—could be the most important we will ever receive.

In 1961, an eclectic band of scientists and engineers met to talk alien hunting in Green Bank, West Virginia. Green Bank is nestled in the Allegheny Mountains, amid some of the most beautiful land in the United States, but the scientists didn't choose the town for its scenery. A couple of years earlier, the physicists Giuseppe Cocconi and Philip Morrison had published a paper in *Nature* that laid out what would become the fundamental strategy for the search for intelligent life.[5] They knew that unless extraterrestrial life showed up on our doorstep, chances were that our telescopes would never actually catch a visual glimpse of aliens. Cocconi and Morrison argued that we should instead try to listen for the signals that aliens might send, just as an amateur radio enthusiast might scan the radio-frequency spectrum hoping to find someone to talk to. The best tool to search for such signals on a cosmic scale was a radio telescope, an antenna and a receiver that can catch radio waves from the stars. And it just so happened that the National Radio Astronomy Observatory (NRAO) had recently been built in none other than Green Bank.

One of the attendees at the Green Bank meeting was a young astronomer named Frank Drake. The year before Drake had launched the first SETI search, using Green Bank's new radio telescope and employing Cocconi and Morrison's methods. But Drake's biggest contribution to the hunt for alien life came not at the controls of a radio telescope, but at a blackboard. The 1961 meeting—which included Cocconi and Morrison among others, as well as Sagan— aimed to codify the nascent science of SETI. If you're going to hunt something, it helps to have some idea of what's out there, and how common your target is. So Drake created a variable-filled equation estimating just how many alien civilizations might exist in our galaxy. Here it is:

$$N = R^* \times fp \times ne \times fl \times fi \times fc \times L^6$$

N equals the number of "broadcasting civilizations"— extraterrestrial intelligences that have reached the point of being able to transmit radio signals that we might be able to detect on Earth. That's the target for SETI. To find that number, the Drake equation multiples the average rate of star formation, times the fraction of those stars that have planets, times the number of planets per solar system that could support life, times the fraction of habitable planets that *do* support life, times the fraction of life-bearing planets where intelligence arises, times the fraction of those alien civilizations that go on to develop broadcasting technology, times the average broadcasting life span of those civilizations.

It's a lot more succinct in the math.

As theories go, the Drake equation is brilliant. If we knew the value of all of the variables that the equation lays out, we would know how many other civilizations are out there, possibly broadcasting to Earth. The problem is that other than the first variable—the average rate of star formation—everything else in the equation was largely unknown when Drake wrote it, so somewhat educated guesses were used to fill in the rest. The group at the 1961 meeting estimated that there were between 1,000 and 100 *million* civilizations in the galaxy—an optimistic hypothesis even at its more pessimistic end, and one that gave the search for intelligent life plenty to search for. And SETI is nothing if not optimistic.

Hopes that humans would rapidly make contact came to nothing, however. From the start SETI was hampered by skeptical mainstream scientists and by government officials who viewed the search for little green men as a waste of money. SETI never fully overcame the "giggle factor" that initially hobbled the study of other unlikely existential risks like asteroid impacts, and while it enjoyed a vogue through the far-out 1960s and into the 1970s, Congress eventually forced NASA to cut all funding for SETI research in 1993.[7] Public funding has never been restored.

Since then, the SETI Institute has been forced to make do with what philanthropic funding it can scrounge up, grabbing observation time on university-run radio telescopes. In 2007 the Allen Telescope Array—named after its benefactor, Microsoft cofounder Paul Allen—finally went online in Northern California as a radio telescope dedicated to SETI searches. But that project had to be downsized when operating funding couldn't be secured from the National Science Foundation. The number of antennas was limited to 42— which, as fans of *The Hitchhiker's Guide to the Galaxy* series know, is the supposed answer to life, the universe, and everything.[8] (SETI researchers slip more fan service into their work than the directors of a Marvel movie.)

SETI was at risk of petering out, another relic of a time when Americans were more hopeful and more adventurous. But recent scientific discoveries—and new deep-pocketed benefactors—gave fresh life to the search for alien life. The Kepler Space Telescope, launched by NASA in 2009, was designed to locate Earth-like planets orbiting other stars, and it succeeded far beyond what its mission planners could have hoped. By the end of its nine-year life in deep space, Kepler had discovered more than 2,600 exoplanets—planets outside our own solar system—30 of which are less than twice the size of Earth while falling within their star's so-called Goldilocks zone, the orbital region where liquid water could pool on a planet's surface.[9] (Our sun's own Goldilocks zone—not too hot, not too cold—includes Earth, of course, but also Mars.) Astronomers have now found more than 3,500 exoplanets, with more being added to the list all the time.[10]

This has obvious ramifications for SETI. While Drake had to estimate the number of stars with planets, we now know that planets are everywhere, and that it is far more common for a star system to have planets than to lack them. By one estimate, there may be tens of *billions* of potentially habitable planets in our galaxy.[11] That means plenty of possible places where at least basic life as we know it could arise, and eventually develop into broadcasting civilizations. "This has changed the whole perspective that scientists and astronomers have about the possibility of life in the universe," Andrew Siemion, director of the Berkeley SETI Research Center at the University of California, told me. "We now know for certain that all the necessary conditions for life to flourish exist ubiquitously, which has sharpened the questions for SETI and astrobiology."

SETI has undergone a revolution of its own. In July 2015 the Russian-Israeli tech entrepreneur Yuri Milner launched Break-through Listen, a $100 million, ten-year program to accelerate SETI efforts to search for radio signals from extraterrestrial civilizations. Milner's funding, which is channeled through the Berkeley SETI Research Center, is securing thousands of hours of searching time on radio telescopes at Green Bank in West Virginia and the Parkes Observatory in Australia. Breakthrough will scan the nearest 1 million stars, a search radius 700 light-years across.[12] That's about as far as light emanating from Earth carrying evidence of our own technological development should have reached, meaning that any aliens in the territory Breakthrough is searching may already be aware of our existence. (A key point: interstellar distances are far away in space *and* time. Light that reaches us from a place a thousand light-years away was emitted a thousand years ago, and vice versa for any potential aliens observing Earth.)

Breakthrough and the SETI Institute—a Silicon Valley–based nonprofit research group that has long steered the search for alien life—are together conducting the broadest-ever SETI missions, with telescopes now covering frequencies between 1,000 and 15,000 mega-hertz.[13] Breakthrough will also scan the Milky Way's galactic plane—the slice of space where most of the galaxy's mass exists—and scores of the nearest galaxies. And Breakthrough will employ optical SETI searches at the Automated Planet Finder at the Lick Observatory in California, looking for laser signals from distant worlds—the inter-stellar equivalent of the beacon towers that were once used to pass messages along the Great Wall of China. A single day of work at Breakthrough is equal to a full year of any other search that has ever been performed, according to Siemion.[14]

Couple that with rapid growth in the same kind of machine learning that is improving surveillance of near-Earth objects like asteroids, and you can see why SETI experts are confident that contact is just a matter of time. SETI has long defended its failure to find evidence of extraterrestrial intelligence by citing—fairly—how little of the sky humans have yet been able to search. But we're getting better and better at looking. "I bet everybody a cup of Starbucks that we'll find ET within a couple of decades," Seth Shostak, senior astron-omer at the SETI Institute, told me on a visit to his overstuffed office in Mountain View, California. "And when that day comes, it will be the biggest story of all time."

So exciting is the possibility of contacting aliens that there are those who don't want to wait around for aliens to contact us. They want to shout "Hello" into the cosmos, carrying out what is known as active SETI or messaging extraterrestrial intelligence (METI). But that impatience creates an existential risk of its own.

———————

SETI is a passive search. The hope is that there are extraterrestrial intelligences out there trying to message us, and we just need to be ready with our radio telescopes to hear. Listening isn't just a search method, though, but a philosophy. Early SETI advocates like Sagan took it for granted that any alien civilization we might come into contact with would be more advanced than us, likely far more advanced, technologically and even ethically. (The universe, after all, had been around for nearly 14 billion years before human beings showed up.[15]) Given our assumed place as a young species in the cosmic hierarchy—and given all that we might hope to learn from our alien betters—a core SETI belief is that we should listen before we speak. Shouting into the cosmos, Sagan said, was "deeply unwise and immature,"[16] the act of a toddler calling attention to themselves.

Proponents of messaging, however, believe it's a mistake to assume that any technologically mature alien civilization will automatically take the first step to establish contact with us. If we're not signaling, after all, maybe they won't signal first either. It's possible that extraterrestrials are no longer using radio, and that sifting through radio waves while searching for alien life is like trying to find evidence of other people in 2018 by looking for smoke signals. Perhaps they assume our silence is a sign that we just don't want to talk, which means it would be up to us to start the conversation.

In truth SETI has often included the occasional effort to send a message into space. The first known attempt was undertaken by Soviet scientists in 1962 at the Unique Korenberg Telescope Array. Using Morse code, they transmitted the words "MIR," "LENIN," and

"SSSR" to Venus, on the assumption that any advanced life-form would obviously be communist and would get their references.[17] Laugh now, but that's really no different than Carl Sagan assuming alien civilizations would have the same values as Carl Sagan.

The Soviet attempt was interplanetary, but Frank Drake himself was responsible for sending the first deliberate interstellar message, on November 16, 1974. To commemorate the rechristening of the Arecibo Observatory in Puerto Rico, then the largest radio telescope in the world, Drake blasted 168 seconds of two-tone sound toward the star system M13. It was noise to the listeners in Puerto Rico, but any aliens who happened to receive it might have noticed a clear, repetitive structure indicating its origin was non-natural. Also encoded in the message were the numbers 1 to 10, the atomic numbers of several basic elements on Earth, and a graphic of the solar system indicating the planetary origin of the transmission.[18]

Given that M13 is 25,000 light-years away from Earth, it's going to take thousands upon thousands of years before Drake's message ever reaches its destination, and at least another 25,000 years before a reply would ever reach us. Yet just days after the message was transmitted, Martin Ryle, then Britain's Astronomer Royal, sent an angry letter to Drake. It was "very hazardous to reveal our existence and location to the Galaxy," Ryle wrote. "For all we know, any creatures out there might be malevolent—or hungry."[19]

To this day, mainstream SETI has eschewed active messaging in part out of the concern, however remote, that something malevolent or hungry might be on the receiving end. The editorial board of the journal *Nature* has cautioned that "the risk posed by active SETI is real,"[20] and in 2006, when the International Academy of Astronautics convened a committee on SETI but refused to push for a ban on active messaging, two prominent members resigned.[21] Before his death, Stephen Hawking was on record saying he didn't think it was a very good idea to invite extraterrestrials to come calling. "I imagine they might exist in massive ships . . . having used up all the resources from their home planet," he said. "Such advanced aliens would perhaps become nomads, looking to conquer and colonize whatever planets they can reach."[22]

But those fears haven't stopped a breakaway group of space scientists from launching the new group METI, led by the former SETI staffer Douglas Vakoch. Milner's Breakthrough Initiative also has a Breakthrough Message component, with a crowdsourced competition to devise a letter to the stars. The interest in METI is in part a product of the exoplanet revolution. With so many potentially life-supporting planets out there, the thinking goes, why not target them and begin beaming out messages of greetings? If Drake's Arecibo message was like shouting at random in the middle of a forest, METI can direct signals to where there's a chance of life, including some planets that are as few as 100 light-years away. Given the vastness of space, METI advocates believe, anything that increases the chances that we might make contact is worth trying. And they argue that the risk of active messaging is overstated. After all, for decades humans have been leaking radio and TV signals into space that could be picked up by sufficiently advanced extraterrestrial intelligence, so it's not as if we've been keeping quiet.

The deeper debate about METI isn't necessarily the act of messaging, but the content. If there really are ETs out there silently listening, the signal from METI or one of the other active messaging groups may be the first thing they ever hear from Earthlings. That's an enormous responsibility. Why should any single group get to decide how Earth says hello—or whether it says anything at all? "What's just so interesting is the assumption of the right to do this," Kathryn Denning, an anthropologist at York University in Canada who studies the strange tribe of alien hunters, told me. "To make these choices on behalf of all humanity. And a refusal to look carefully at the potential consequences. The idea is that all innovation is good, and let history sort it out."

There's something of the move-fast-and-break-stuff ethos of Silicon Valley in the METI movement, a willingness to disregard risk if risk gets in the way of potential reward. That might be acceptable if the reward is a new search engine or social network. It is considerably less acceptable when innovation brings with it the possibility, however remote, of a world-ending threat, as we've already examined in the cases of biotechnology and artificial intelligence. This is the debate over anthropogenic existential risk, played over again. If an act carries even a minuscule risk of human extinction, shouldn't we err on the side of safety, given the ultimate stakes at play? And if that's the case, perhaps we should pause before sending an unknown alien civilization of unknown technological capability and unknown intentions Google Maps directions to our home planet?

Olle Häggström, a mathematician at Chalmers University of Technology in Sweden and the author of the existential risk book *Here Be Dragons*, thinks so. "There are optimists who say that good things can come out of establishing communications," Häggström told me. "We could learn wonderful things from them. But an extraterrestrial civilization of very advanced technology might be a threat—and they might want to get rid of us before we become a threat to them. There are real evolutionary-style arguments pointing in that direction. Maybe we'd be better off observing exoplanets for ten or twenty years until we're in a better position to assess the risk of communication. The risk is too great."

I personally think Häggström is right. The potential benefits of active messaging don't outweigh the risks that would come with it. To understand why, we should look at what first contact might actually be like.

There is, Seth Shostak of the SETI Institute assured me, a protocol in place in the event of the discovery of alien life. And contrary
to Hollywood, there are no Men in Black ready to spirit away the
evidence to Area 51 before it becomes public. SETI itself, like the
mother science of astronomy from which it sprang, has a communi-
tarian ethos, one where the contributions of amateurs are welcomed
and openness is taken for granted. "There's no policy of secrecy,
because as soon as you find a signal that's even the least bit interesting, people today would be tweeting it or sending emails to their
relatives," said Shostak. "You'd need to tell people anyway because
you'd want someone else at another observatory to verify it. Otherwise you wouldn't really believe yourself. There are too many things
that could go wrong."

Shostak experienced his own close call on June 24, 1997, when
a blip came through from a star called YZ Cet, some 12 light-years
away—practically next door by the vast distances of interstellar
space.[23] That day the SETI Institute was making use of a 140-foot
radio telescope in Green Bank, while Shostak was monitoring the
work back at the institute's offices in California. To spend your scientific career searching for intelligent life is to spend much of it in a
state of suspended disappointment. Every seeming signal indicating that an alien civilization might be out there has turned out to
be a false alarm, often either the misheard background hum of the
cosmos or the electronic chatter of humankind's own satellites. But
the search runs on hope, and in the early hours of that morning on
June 24, Shostak found himself hoping.

It wasn't meant to be. Even as Shostak was fielding a call from a
curious *New York Times* reporter and pondering just how SETI might
announce to the world that aliens in fact exist, further checking

showed that the signal from YZ Cet wasn't from YZ Cet at all. It turned out to be a telemetry signal from the Solar and Heliospheric Observatory, a joint U.S.-European satellite that studies the sun from space. The transmission had bounced around the radio telescope in West Virginia in a way that temporarily mimicked an alien source. Another null result; another false alarm. "A close call that wasn't that close," Shostak told me.

But had that signal been real—and confirmed by another observatory—Shostak would have notified the International Astronomical Union, as well as the United Nations and any other relevant research organizations. The discoverer is supposed to get the right of first public announcement—which would be the press conference to end all press conferences—but the data would be made available to anyone who wants it. (The exception would be the actual coordinates of the signal source, to prevent anyone from simply starting up an interstellar conversation on their own.) Then humanity, or some part of it, would have to decide whether to send a response—and what should be in it. Which would be a very, very interesting debate.

Unless one side of the conversation learns to break the laws of physics and travel faster than light, however, actual communication with a civilization via radio would likely unfold over centuries, if not longer. That's how vast the distances are between the stars. We'd know that someone, or some thing was out there—but what would that knowledge actually mean to human beings?

Sagan believed that the discovery of extraterrestrial intelligence would be a great unifying force. Instead of seeing ourselves as fractious members of separate nation-states or sects or races, we would come together as shared citizens of this pale blue dot called Earth. "Just the knowledge that we're not alone would be philosophically profound," said Jacob Haqq-Misra, a research scientist at the

Blue Marble Space Institute of Science and a member of the METI movement. "When people saw the images of Earth from space, it was relevant to inspiring modern environmentalism, to realize our planet isn't an infinite resource. I like to think the discovery of intelligent life would have a similar effect where now we're all Earthlings, because we know there's an other."

As Kathryn Denning noted, with affection, this cosmic optimism was part of the "pot-smoking hippie side of Sagan." As an anthropologist, though, Denning is more pessimistic about how humans would handle contact. "Our evidence about humanity is that we're profoundly xenophobic and really rather nasty," she said. "We don't like things that we don't understand, which is almost a given in any contact scenario. The likelihood that we would have the upper hand is rather small. So I don't really see how that would bring out the best in anybody."

This is what makes the existence of aliens—and therefore attempts to actively contact them—an existential risk. Humans and aliens are likely to be separated not merely by gulfs of space but of time as well. If extraterrestrial civilizations have rates of development anything like human beings—meaning rapid technological growth following industrialization—a head start for the aliens of just a century could be enough to create a massive military gap, akin to the difference between a World War I–era army and the modern U.S. military. We *Homo sapiens* only developed the ability to send and receive space signals, and therefore be capable of making contact, less than a hundred years ago, but the universe has existed for about 140 million times longer. As the writer Steven Johnson pointed out in a 2017 story for the *New York Times Magazine*, "The odds that our message would reach a society that had been tinkering with radio for a shorter, or even similar period of time would be staggeringly long. Imagine another planet that deviates from our timetable by just a tenth of 1 percent: If they are more advanced than us, then they will have been using radio (and successor technologies) for 14 million years."[24]

The chances of us making contact with an alien species technologically on par with us are incredibly small, less likely than randomly running into a high school acquaintance in a foreign airport. Since any alien civilization less advanced than human beings probably wouldn't yet be capable of sending or receiving signals, it's much more likely that our alien pen pals would have outpaced us. That would be doubly so if the extraterrestrials could somehow traverse interstellar space and visit us in person, a feat that is unimaginably beyond human capability.

In a 1979 book called *Xenology: An Introduction to the Scientific Study of Extraterrestrial Life, Intelligence, and Civilization*, the author Robert Freitas created a scale that measured the imagined power differentials between alien societies and human beings. To Freitas, humanity in all its accomplishments compared to a species that could cross interstellar space might be no more than a single amoeba against the entire United States. "Not only would such a civilization *seem* godlike to us," Freitas wrote, "it would actually *be* God for any practical purpose that can be imagined."[25]

Humans have some experience with what happens when two geographically separated civilizations at differing levels of technological development encounter each other. It is the European invasion of the Western Hemisphere, and it did not go well for the people who were already living here. Recent research estimates that the indigenous population of the Americas in the years before Columbus reached Hispaniola was as high as 60.5 million.[26] Thanks largely to the infectious diseases imported by Europeans—diseases to which the indigenous peoples had no immunity—as well as the invaders' policies of exploitation, enslavement, and massacre, the population of the Americas was reduced by 90 percent in just a century. Civilizations like the Aztecs and the Incas were utterly destroyed. The scale of death was so extreme that tens of millions of hectares of now-untended farms were reclaimed by forests, which in turn sucked so much carbon dioxide from the air that global temperatures actually fell.[27] The legacy of the Americas' first contact with Europeans is one of slaughter, genocide, and cultural extirpation. And that may well be our fate in the event of alien arrival.

In 2015 a number of science and tech luminaries—including Elon Musk, who seems to view aliens with the same instinctive wariness that he does AI—signed a statement opposing METI efforts, unless there was extensive discussion first.[28] And no less an academic star than Jared Diamond—whose landmark book *Guns, Germs, and Steel* argued that geographical differences explained the fatal technological and biological gap between conquering Eurasian civilizations and indigenous Americans—has warned that it would be "suicidal folly" to try to contact aliens. "If there really are any radio civilizations within listening distance of us then for heaven's sake let's turn off our own transmitters and try to escape detection, or we are doomed," Diamond wrote in his 1992 book, *The Third Chimpanzee*.[29]

Surely, though, there's some kind of military plan to handle an alien invasion, one that might bring all of humanity together, finally fighting on the same side? President Ronald Reagan told an audience of high school students in Maryland in 1985 that the arrival of hostile aliens might broker peace between the United States and the Soviet Union. "If suddenly there was a threat to this world from some other species from another planet outside in the universe," he said, "we'd forget all the little local differences that we have between our countries and we would find out once and for all that we really are all human beings here on Earth together."[30]

If there was a threat to this world from an alien species, however, chances are the only thing we'd do together is die. Give the Russians some credit for honesty on this score. In a 2013 press conference at Russia's Titov Main Test and Space Systems Control Centre, a journalist asked Sergey Berezhnoy, the center's deputy chief, whether the Russian military could protect the country from an extraterrestrial invasion. "So far, we are not capable of that," said Berezhnoy. "We are unfortunately not ready to fight extraterrestrial civilizations."[31]

The Pentagon appears no more prepared. The U.S. Air Force had a postwar program to study unidentified flying objects (UFOs), called Project Blue Book, but the military terminated it in 1969 after concluding that the alleged UFOs presented no national security threat, represented no technological development beyond current knowledge, and almost certainly weren't extraterrestrial in origin anyway.[32] More recently, the *New York Times* revealed in 2017 that the Defense Department had spent $22 million on the semi-secret Advanced Aerospace Threat Identification Program, which investigated reports of UFOs.[33] The program had been shut down in 2012, and it's still not clear whether its existence had more to do with a genuine concern about the threat posed by aliens, or former Senate majority leader Harry Reid's personal interest in the subject.[34] (Reid represented Nevada, which after all is home to the mythic Area 51.)[35] In 2018 President Trump proposed creating a new branch of the U.S. military called the Space Force, but its responsibilities would have more to do with using outer space to fight wars on Earth, not defending the planet from extraterrestrial invaders—assuming Trump's vague notion ever becomes a reality.

Even if the military can't save us, is it possible that Earth itself would offer some kind of home-field advantage against encroaching aliens? In Wells's *The War of the Worlds*, the invading Martians are so advanced that they possess "minds that are to our minds as ours are to the beasts that perish," and were armed with heat rays and chemical weapons.[36] Humanity wins the war only because the aliens succumb to Earth diseases against which they have no immunity. But in real life there's no reason to expect that we'd be saved by patriotic germs. As happened in the European invasion of the Americas, it's often the more technologically advanced invaders who carry killer diseases, not the other way around. Smallpox, measles, typhus, and cholera were Europe's fatal gifts to the Americas, and were primarily responsible for the ethnic cleansing of the New World.[37]

Worrying about germs assumes that any arriving aliens would be biological, just like us. But that assumption might be nothing more than organic chauvinism. As we saw in the last chapter, there is at least the possibility that we human beings could eventually develop artificial general intelligence, and that such AI could become super-intelligent, displacing us as the dominant power on the planet. If an alien civilization had centuries or longer to develop than we've had, that would give them even more time to create AI of their own—and for that AI to take over and begin looking to expand to the stars. If nothing else, AI aliens would complicate any attempts at communication. Not only would we need to try to talk to an alien species that had evolved on a separate planet light-years away, with little shared frame of reference, but it might be a species that had already shed its organic roots. If it's difficult enough to imagine what superintelligent AI developed by human beings might be like, just try to conceive of alien AI.

An extraterrestrial intelligence that is artificial might be more likely to actually reach our planet, since it presumably wouldn't be held back by finite biological life spans. Even if aliens were organic, they might well prefer to send robot probes to colonize the galaxy, for the same reason that all of humanity's exploration of the solar system, save our nearby moon, has been done by machines, not astronauts.

Our greatest defense against a hostile extraterrestrial intelligence, whether organic or artificial, is interstellar distance. But aliens could wreak havoc on us simply by sending a signal, without ever visiting Earth. Unless aliens shout "Hi!" over the cosmic loudspeakers, a message from space transmitted by radio would almost certainly need computer analysis, which brings with it a threat any computer user should be aware of: malware. Computers have antivirus software, but in a 2018 paper, Michael Hippke of the Sonneberg Observatory in Germany and John Learned of the University of Hawaii note that there would be no way to be certain that we could decontaminate an alien message before we opened it. Such a message could well contain a virus that might disable global technology—or something even worse. Opening this message would therefore constitute an existential risk in itself—after all, as Hippke and Learned write, "it is cheaper for [aliens] to send a malicious message to eradicate humans compared to sending battleships."[38]

For our species to meet its doom because of poor email hygiene would be tragic and yet somehow fitting, at least if you've ever been part of an email reply-all-pocalypse. But there would be one positive takeaway from contact with extraterrestrial intelligence, whatever happens next. We would know that at least one independently evolved intelligence had managed to survive all the existential threats that the universe could throw at it—asteroids, gamma rays, abrupt climate change—and all that it could do to itself. It would provide a measure of existential hope to balance out the existential risk that suffuses life on Earth in the early twenty-first century.

But what if it turns out that there isn't anyone out there, that we really are alone in the universe? Then that existential hope could crumble, to be replaced by existential dread.

————————

Implicit in the search for extraterrestrial intelligence, whether passive or active, is the assumption that there is someone out there to be

found. Frank Drake's equation estimating the number of broadcasting civilizations in the galaxy has millions of possible solutions, but
the one answer that he and everyone else involved in SETI would
never believe is this: one—meaning our civilization, and no other.
Yet after decades of craning our ears to the heavens, all we have heard
in response is what the scientist and writer David Brin termed "the
Great Silence."[39] Despite predictions like Drake's, despite the sheer
size and age of the galaxy, which would seem to give plenty of opportunity and space for intelligent civilizations like our own to make
their presence detectable, we see nothing and hear nothing. And that
should give us pause, and maybe, just a touch of dread. An empty
universe may not merely be lonely, but actively hostile to intelligent
life—including our own. Or to ask Enrico Fermi's question again:
where is everybody?

For someone whose name was put on the question that has
defined the debate over extraterrestrial intelligence, there's little
evidence that Enrico Fermi thought or even cared much about aliens.
After the Manhattan Project, Fermi continued to consult at Los
Alamos, working on Edward Teller's design for the far more powerful hydrogen bomb. One day in 1950 Fermi was lunching at the lab
with his colleagues Teller, Emil Konopinski, and Herbert York when
the conversation turned to a cartoon in *The New Yorker* showing
little green men carrying trash cans stolen from the street. Fermi
suddenly asked a question: "Where is everybody?" It was a typically
Fermian outburst, a tossed-off query that nonetheless cut to the heart
of the debate. If aliens did exist, then why in a vast universe billions
of years old hadn't we seen them yet? Why hadn't they shown up on
Earth to conquer our planet, or at least steal our trash cans? Where
is everybody?[40]

Exactly what Fermi thought the answer might be isn't clear—
the physicist died a few years later, and his lunch companions all

remembered the conversation differently. Teller and York thought Fermi was talking about whether interstellar travel was even possible—1950, remember, was before Sputnik and the first attempts to search for extraterrestrial signals. If it was indeed impossible to travel faster than light—as Einstein's theory of special relativity holds—then perhaps it wouldn't be so surprising that we hadn't found any evidence of extraterrestrial intelligence. As Teller told the physicist Eric Jones years later, Fermi wondered if "we are living somewhere in the sticks, far removed from the metropolitan area of the galactic center."

To this day the brute fact of Einsteinian speed limits and interstellar distances helps explain why we've never had a confirmed alien visitation, even if intelligence did exist elsewhere in the galaxy. But York remembered Fermi probing deeper, following up his question with a quick calculation on the probability of Earth-like planets, the probability of life arising, the duration of a technological civilization—a Drake equation before Drake. "He concluded on the basis of such calculations that we ought to have been visited long ago and many times over," York told Jones. "As I recall, he went on to conclude that the reason we hadn't been visited might be that interstellar flight is impossible, or, if it is possible, always judged to be not worth the effort, *or technological civilization doesn't last long enough for it to happen*" (emphasis added).

Fermi's lunchtime question came to be known as the Fermi Paradox, although it's not really a paradox, and it was later writers who eventually built it out, well after Fermi's death.[41] There are dozens upon dozens of possible solutions to the Fermi Paradox. One of the most likely is that the sheer amount of space in outer space means that there could be many independently evolved civilizations in the galaxy that would nonetheless never run into each other, a possibility Fermi himself noted. But the problem with that answer is that the galaxy is not just huge, but old. Even if it remains physically impossible to travel close to the speed of light—let alone beyond it like the starship Enterprise on *Star Trek*—an alien civilization could spread out at sub-light speeds and still colonize plenty of the galaxy, if it had enough time.[42]

According to the mathematicians Duncan Forgan and Arwen Nicholson of Edinburgh University, spacecraft traveling at one-tenth the speed of light—fast but not inconceivable—could cross the entire Milky Way galaxy in just 10 million years. That's a long, long time by human standards—more than a thousand times longer than human civilization to date—but less than 0.1 percent of the age of the galaxy.[43] It's doable not just once, but many times over throughout the history of the Milky Way, just as empires on Earth have risen and fallen over and over again. And if those alien empires have been superseded by artificial intelligence, which isn't hampered by the unfortunate organic trait called death, such expansion would presumably be even easier.

So we can judge that galactic expansion is possible, yet we see no evidence of it. There are no visible colonies here on Earth—the most habitable planet in the galaxy, as far as we know—or in our solar system or anywhere we point our telescopes. So again—where is everybody?

It's at this point that attempts to solve the Fermi Paradox start to get creative. One hypothesis is that aliens exist but are deliberately hiding their presence from us. The MIT astronomer John Ball proposed that the first alien civilizations to arise over the history of the galaxy would accrue so much power—because they would inherit all the resources of space—that they would essentially run the cosmos.[44] Perhaps just as we have chosen to create nature preserves and leave some remote tribes of indigenous people uncontacted—because we know from experience that an encounter would destroy their culture—these superpowerful aliens would leave us alone to develop.

Or perhaps aliens are trying to talk to us, but instead of using radio, they're employing a technology that we haven't yet mastered. Imagine a modern human trying to call up a caveman on a cell phone. At least they would share a biology and perhaps some points of common reference, but a truly advanced civilization attempting to communicate with us might be more like a modern human trying to talk to an ant. Would you be interested in putting forth the effort that would be needed to understand what little this tiny insect is capable of communicating? Probably not—you have important human things to do. And so it might be with any sufficiently advanced extraterrestrial intelligence, which is to say, almost any that we might encounter. Instead of the Great Silence, we would have the Great Indifference.

But even if aliens refused to make contact with us—whether out of enlightened policy or simple apathy—it doesn't explain why, if they exist, we would see no evidence of their presence. An alien civilization that expanded to the stars would presumably use tremendous amounts of energy. Between 1775 and 2009, energy consumption in the United States alone increased by more than 450 times as it went from thirteen preindustrial colonies to the world's preeminent economic and political power.[45] Evidence of America's unquenchable thirst for energy can be seen everywhere from the vast pits that have been dug up to extract coal to the measurable increases in global temperatures due to greenhouse gas emissions. Imagine how much more energy would be required to run an interstellar empire, and imagine then how much physical evidence it should leave. The physicist Freeman Dyson, whom we met in chapter 6, even pictured artificial power stations that could be built enclosing a star in order to harness raw stellar energy at the source—so-named "Dyson spheres." While Dyson spheres are thought experiments, not anything ET might actually build, Dyson believed that meeting the energy needs of an advanced technological civilization would require something at this scale, and that we might look for such megastructures as we searched for intelligent life. Yet all our stargazing has found nothing of the sort—indeed, nothing clearly artificial at all.

Perhaps then the aliens are hiding—and not just hiding from us because they don't want to interfere with our development, but because they don't want to be found by anyone or anything. If we're worried about potentially hostile aliens, the aliens might be worried too. Rather than call attention to themselves, extraterrestrial civilizations could learn to stay quiet and out of the way—perhaps because those races that stuck their head out met an early demise.

This is one solution to the Fermi Paradox that should send a shiver down our spines, because it asks us to wonder what might happen to us as our cosmic profile begins to rise. Humanity has been releasing radio and other signals into space for decades, and any aliens looking closely enough could notice the electric lights that burn on our planet's nighttime side. Perhaps alien life-forms are observing us in the dark, and if we begin to demonstrate dangerous qualities—say a tendency toward relentlessly using up all the resources we can find—the cosmic cops will put a stop to us. If this sounds familiar, it's the plot of the 1951 sci-fi classic *The Day the Earth Stood Still*—and the 2008 remake with Keanu Reeves.

It's also possible that aliens are not hiding but hibernating. Anders Sandberg and Stuart Armstrong of the Future of Humanity Institute, and Milan Ćirković of the Astronomical Observatory of Belgrade,[46] suggest that highly advanced alien civilizations might choose to go into a state of dormancy that could last for billions of years, all in the name of energy efficiency. As the universe expands and ages, it cools, and one side effect of cooler temperatures is that computing power becomes more efficient. (Technically the aliens wouldn't be hibernating but aestivating, which is a dormant period that takes place during summer rather than winter.) Wait long enough to awaken—trillions of years—and the aliens might be able to get 10^{30} more computing per unit of power than they could wield now. So aliens exist in this scenario—we just happen to be awake while they are sleeping.

Solving the Fermi Paradox is like fan fiction for SETI devotees: weird, creative, and, because we have few ways of actually falsifying our guesses, potentially endless. But there is one solution that should keep us awake at night, even if it means that we'll never have to worry about an invasion by ET. It's this: aliens existed once, but they're all gone. The Great Silence means absolute silence—and that silence could be our future as well.

Absence of evidence is not evidence of absence. It's a scientific maxim, and one relied on by SETI advocates to explain why we have yet to find proof of alien life. We've so far looked at thousands of star systems out of a couple hundred billion,[47] which amounts to much less than 1 percent of what's out there. "I didn't see any gazelles in my front yard this morning when I backed out of my driveway," Seth Shostak told me. "Maybe gazelles don't exist, or maybe just none of them are in my driveway."

But the longer we search and find nothing, the harder that argument is to make.

Let's go back to the original Drake equation. The second variable is the fraction of stars that have planets. When Drake wrote his equation in 1961, scientists had no evidence of planets outside our solar system. But it turned out that such exoplanets existed—we just didn't yet have the technology needed to see them.[48] The next variable in the Drake equation is the number of planets per star system with an environment suitable for life. That's less certain, but many of the exoplanets we've discovered are in an orbital range that suggests they could support life as we know it.

This puts to bed one solution to the Fermi Paradox: the Rare Earth hypothesis, the idea that planets like Earth capable of supporting life are scarce in the universe. That shouldn't surprise us—since Copernicus displaced the Earth as the center of the solar system, advances in astronomy have shown that our beloved planet is more ordinary than we thought. That means there should be plenty of planetary space for life to arise, however, which makes the Great Silence all the more confounding.

Enter "the Great Filter." First coined by the economist Robin Hanson, the Great Filter posits an explanation for the Fermi Paradox that is vitally important to our future on this planet. Hanson theorizes that there is, essentially, a great filter somewhere along the evolutionary path from the emergence of organic molecules on a life-supporting planet to the development of a civilization capable of leaving a mark on the stars. Whether that Great Filter falls before humanity's current point of development or after it is, as the philosopher Phil Torres puts it, "the ultimate question for existential risk scholars."[49]

Picture the Great Filter this way. Imagine that you're a year away from your fortieth high school reunion. You look around and realize that no one from the class ahead of you—the class that would be having its own fortieth reunion this year—is still alive. What happened? If most of the deaths occurred when the alumni were 25 years old, or 35, or 45, that's tragic—but you can take some comfort in the fact that you've already passed those thresholds. But if most of the deaths occurred a year ago—when those alumni would have been the same age you are now—you should be very worried, as it would mean that you're about to hit the Great Filter of, in my case at least, Central Bucks High School East.

You'll remember that the fourth variable in the Drake equation is the fraction of hospitable planets on which the earliest forms of life develop. This is the first possible spot for the Great Filter. It may be that abiogenesis—the process by which raw organic ingredients evolve from nonliving matter to become what we would recognize as basic microbial life—simply doesn't happen very often. Or, following the chain of the Drake equation, it might be that basic life is common, but the billions of years of survival and evolution and sheer luck required for that life to become something we would term "intelligent" is a bar that isn't often cleared. Remember how many times mass extinctions nearly wiped out life on Earth, and how close humanity came to extinction in the past, well before we could ever send signals to the stars. The story of Earth—from the first instances of life 3.5 billion years ago[50] to space-faring *Homo sapiens* today—could be a serendipitous one, and one unlikely to be repeated even in a galaxy as vast and as old as the Milky Way.

If the Great Filter falls somewhere before humanity's current level of climate changing, nuclear-powered technological development, it would be disappointing for those hoping to find other intelligences to share the universe with—but we could also breathe a sigh of relief. It would mean that there is nothing in the history of the cosmos to suggest that we face a Great Filter of our own in the future. We've already passed the test. We could still be annihilated by any number of existential threats outlined in this book, but our odds of survival would increase.

We don't know yet where the Great Filter may fall, but evidence is beginning to trickle in that basic life, at least, can come into existence elsewhere. In 2018, NASA's Curiosity rover identified a variety of organic molecules on Mars that could form the foundation of life, as well as evidence that contributions of methane—a gas that on Earth is mostly produced by living organisms—cycles seasonally in the Martian atmosphere.[51] Mars isn't the only possibility in the solar system—Saturn's moon Titan has a chemically active and carbon-rich atmosphere, and may even host liquid water beneath its frozen icy surface.[52] Where there is water, there may be life.

And that life could be hardy. A 2017 study by researchers from the Max Planck Institute for Astronomy and McMaster University sity argues that life sprang from meteorites and ponds only a few hundred million years after Earth had cooled enough to allow liquid water to form.[53] Microbial life known as extremophiles can be found in punishing environments ranging from the minus-250-degree temperatures of submarine hydrothermal vents to the alkaline lakes of high-altitude Chile[54] to the punishing undersea pressure of the Marianas Trench.[55] NASA astrobiologists actually undertake field trips to the most extreme environments on Earth to study the kind of life that might be able to survive off planet. "Life may be sacred," Shostak writes, "but it may also be commonplace."[56]

If that turns out to be the case, it will be great news for astrobiologists, but a "bad omen for the future of the human race," in the words of Nick Bostrom.[57] The more proof we have that the development of life isn't all that challenging, the more worried we should be. "If the Great Filter is ahead of us, then we're in trouble," Olle Häggström told me. "And when you discover that basic life is out there, it tells us that the filter is later rather than earlier."

Worst of all would be the discovery of "necrosignatures"— evidence of extinct alien civilizations. Just as archaeologists sift through the dirt in search of vanished human civilizations, astronomers can hunt for evidence of extinct aliens. In 2015 the astronomers Adam Stevens, Duncan Forgan, and Jack O'Malley-James published a guide to searching for necrosignatures that could be observed from Earth. If a species destroyed itself with a biological weapon that killed everyone on the planet, microbes would feast on the corpses, excreting methanethiol and ethane in levels that might be detectable. Evidence of a nuclear holocaust could be discovered in the depletion of a planet's ozone layer, or an increase in the air-glow brightness that would leave the atmosphere emitting a green radiance. If an inhabited planet had actually been destroyed—perhaps by some kind of "death star"—astronomers might be able to detect artificial compounds in the disk of debris.[58]

The astrophysicist Adam Frank of the University of Rochester has pondered the possibility of alien extinction through the lens of an existential threat we're very familiar with here on Earth: climate change. Frank—who writes about his research in his excellent 2018 book, *Light of the Stars*—theorizes that if other intelligent species arose elsewhere in the galaxy, they may have employed energy sources that cause global warming as a side effect, just like us.[59] Climate change in this view is something that universally accompanies technological development—which is bad news, because of the four potential outcomes of Frank's models, three are catastrophic.[60]

In a die-off, population and temperature rise together rapidly, before the numbers peak and then crash, leaving a small remnant—too small, perhaps, to broadcast their presence to the universe. In a collapse without resource change, the aliens keep using climate-change-causing resources—like the coal and oil we still burn on Earth—until their planet passes a tipping point, leading to eventual collapse and potential extinction. The same thing happens in the third outcome even if the population eventually switches to more sustainable resources, but too late to halt the momentum of catastrophic change. Only in the fourth outcome, sustainability, does the alien population switch to lower-impact resources in time, and achieve a survivable equilibrium with their planet. Climate change becomes an all-purpose solution to the Fermi Paradox—one that augurs poorly for our future here on Earth, where the heat is very much on. "What's happening here may have happened millions of times before," Frank told me. "If we don't do anything about it, that's our folly."

We haven't seen any astronomical evidence of alien civilizations that have been destroyed by climate change—but we also haven't seen any evidence of civilizations that have managed to live in sustainable harmony with their environment. Climate change is as good an explanation as any, because it stands to reason that if there is a Great Filter that cuts short the life span of intelligent civilizations, the cause might be a common one, embedded in the way life uses energy and develops technologically, just as past civilizations here on Earth have tended to rise and collapse in common and even predictable ways.

It doesn't have to be climate change, however—almost any of the existential threats we examine in this book could be the culprit. A nuclear chain reaction isn't something that only works on Earth; a version powers our sun, and all stars. Joshua Cooper, a mathematician at the University of South Carolina, has suggested that civilizations may reach the space-faring age around the same time that they gain the ability to manipulate their own genetic code, and that they therefore may be wiped out again and again by sophisticated bioweapons of their own making.[61] Just about the only man-made— or alien-made—existential risk that doesn't fit in with the Great Filter is artificial intelligence. A superintelligent AI would presumably survive the species it had displaced and go on to "make a physical impact on the universe," as Hanson told me—perhaps by turning the galaxy into paper clips.

All of this is speculation, because we still know so little about the universe that surrounds us, and our place in it. So we search on. "The first evidence of extraterrestrial intelligence may come to us from the remains of less prudent civilisations," Stevens, Forgan, and O'Malley-James write in the conclusion of their paper on necrosignatures. "In doing so, such information will bring us not only knowledge, but wisdom."[62]

Knowledge, yes; wisdom, I'm less sure of. SETI pioneers believed that the discovery of extraterrestrial intelligence would kindle hope that we could survive what Sagan termed our "technological adolescence." Perhaps the aliens, like friendly older siblings, would even guide us through that transition. But the discovery of dead civilizations—if coupled with a continued failure to make contact with any living alien intelligences—would tell us that we may instead be fated to destruction. That last variable in the Drake equation— the life span of a species—would be revealed, and we would know that the reason we seem to be alone in the universe is that the alien civilizations we might share this universe with have died off. And perhaps so will we.

9

SURVIVAL
The Day After

Everyone wants to save the world, and almost every time we tell a story about the end of the world, that's precisely what happens. At the last moment our heroes defuse the countdown to nuclear annihilation, or invent the vaccine to stop the killer disease, or triumph over a malevolent army of machines or an invading horde of aliens. And so it has been in our actual brushes with end times. Sometimes heroes like Vasili Arkhipov made impossible decisions that prevented global catastrophe. More often, we've simply been lucky. But every time the world has come close, the apocalypse has been postponed. So far.

When I set out to write this book, I hoped that it would play some part, however small, in encouraging us to do what can be done to prevent existential catastrophes—in other words, to save the world, for my child and for all the children of the future. I believe we can. For most of our existence, humanity's survival was indeed a matter of luck, our good fortune—or our anthropic shadow—the main reason we didn't go extinct like nearly every other species in this planet's history. No longer. Now whether we endure or die will come down chiefly to our own decisions, to whether our wisdom is equal to the task of saving ourselves—including from ourselves.

Even the wisest versions of ourselves, however, will get something wrong eventually. Spool out the tape of the future far enough and a global catastrophe will occur. It might be the nuclear war we've dreaded. It might be a supervolcano that we can't predict. It might be the virus, hatched in nature or the lab, that burns through humanity. The doomsayers are always right, eventually. Sooner or later, winter will come.

But the end isn't the end—or at least, it doesn't have to be. A handful of existential threats—like the sudden rise of a hostile AI superintelligence or an asteroid even larger than the one that wiped out the dinosaurs—would be so massive and strike so quickly that they would leave no one behind. Other global disasters, though, would become existential only if we let them. Survival is the difference between mere catastrophe and extinction—and that distinction is all that matters. Existential risk is a game played with permanent stakes. If we can survive and rebound from disaster, no matter how severe, our story will go on, and that disaster may end up as a historical footnote, just like the Toba supereruption tens of thousands of years ago. But if we can't survive, our story is over—forever.

What energy we manage to marshal on existential risk is largely spent on preventing disasters before they ever occur, however, not preparing for the awful aftermath. That makes a kind of emotional sense—who wants to contemplate the last, slow dying of their world? But it leaves us vulnerable to what the U.S. military calls "right of boom"—the seconds and days and years after a catastrophe.[1] We need not just prevention in the face of the worst to come, but resilience.

Everyone wants to save the world. But it may be that the most important work we can do now is to prepare ourselves for the day when we fail to save the world—and ensure there's a plan to pick up the pieces.

———————

In the first seven days after Donald Trump's election in 2016, 13,401 Americans registered with the immigration authorities in New Zealand.[2] This was not because they all had a burning desire to see the locations of the *Lord of the Rings* trilogy. New Zealand—remote, English-speaking, bucolic, and did I mention remote?—has become the backyard fallout shelter for the superrich, an entire nation reachable by private plane where our plutocrats can hole up in the event of a major catastrophe. New Zealand is for the survivalism of the elite what the Grand Caymans have long been for their bank accounts. Tech titans like Peter Thiel have purchased estates in New Zealand for the express purpose of surviving the apocalypse. Thiel even went so far as to secure New Zealand citizenship after spending just twelve days in the country.[3]

The boom market in New Zealand boltholes is a perfect expression of an age when the rich lay claim to more of everything— including the life jackets. But it's not only the ultrawealthy who have plans for surviving the worst. In 2017, two-thirds of Americans reported stockpiling supplies to survive a natural disaster.[4] More than three million Americans calls themselves doomsday preppers,[5] meaning they are actively preparing to survive not just run-of-the-mill natural disasters but also world-ending calamities. They connect through organizations like the American Preppers Network and read magazines like *Recoil OffGrid*. There are prepper podcasts and prepper blogs and the Self-Reliance Expo, a national conference for preppers. There are tips for doomsday prepping with children and prepper gift ideas for Father's Day, like the Stovetec Rocket Stove, which is a great tool for cooking in the event of the total collapse of the electrical grid—or, alternately, for tailgating.[6]

A survey by the National Geographic Channel found that 40 percent of Americans thought that preparing a bomb shelter or stocking up on supplies was a better investment than funding a 401(k).[7] And if you're part of that 40 percent, there are plenty of places that will take your money. You can buy a luxury bunker from a company called Vivos for $20,000 to $50,000, including in a former army munitions depot in South Dakota that is billed as the "largest survival community on Earth."[8] If you'd rather wait out the apocalypse at home, Atlas Survival Shelters—which has been selling bomb shelters since the Cold War—will build you a personal doomsday habitat. It's a mere half a million for the concrete dome version,[9] though you might not have much left over for the $4,995 Prepster Black Ultra Luxe Emergency Bag. That would be a shame, because you'd miss out on the toiletries from Malin+Goetz, the Marvis toothpaste, and the Mast Brothers chocolate bars.[10] What are the end times, after all, without artisanal candy?

There's real fear at play in these trends, and as I've shown throughout this book, much of that fear is justified. But instead of rolling up our sleeves and tackling the collective challenge of existential risk, we seem to prefer indulging in individualist survivalist measures. Each of us should set aside supplies to endure a natural or man-made disaster that disrupts the flow of water or food for several days. Extra medicines, a go bag, photocopies of important documents—everyone should have them handy. A survival plan that focuses solely on individuals and families, however, won't be sufficient in the face of an existential catastrophe. It might fit the self-reliant American character—or at least our idealization of it—but it is perverse that the same country that leads the world in dooms-day prepperness has proven criminally unwilling to help its most vulnerable citizens after disasters like the 2017 Puerto Rico hurricane, and criminally incapable of addressing long-term threats like climate change.

At least some of the failure of the United States to plan nationally for major catastrophes may be a legacy of grand Cold War–era civil defense plans for nuclear war that overpromised and underdelivered. Between the mid-1950s and the late 1960s, some 7,000 volunteers participated in more than 22,000 man-days living in emergency shelters that ranged from family size to over 1,000 people. The tests led to the development of national standards for underground shelters meant to help Americans survive a nuclear holocaust. (Those standards allotted just 10 square feet of room per person—barely more than three times the space allotted to prisoners in the Nazis' Bergen-Belsen concentration camp, a comparison the U.S. government itself made, for some reason.)[11] But the shelters—or the Eisenhower administration's admonishment to Americans to keep an emergency food stash officials called "Grandma's Pantry"—would have been of little use in the event of a full-scale nuclear overkill, as Americans themselves eventually began to realize.

Once it had become clear to even the most hawkish national security officials that the survivors would envy the dead after a nuclear war between the United States and the Soviet Union, earnest plans for civil defense became a punch line. Those opposed to nuclear weapons thought that exposing the absurdity of civil defense would underscore the madness of nuclear war—and they were right. But today, instead of preparing as a society for nuclear war and other existential threats, most of us simply choose to ignore the risk, while a minority of survivalists and preppers obsess over it, plotting how they and their families can outlast the apocalypse.

It's wrong to assume that national survival and civil defense plans are pointless, but it's just as wrong to believe that an individual prepper could endure a global existential catastrophe on their own, no matter how well prepared—or rich—they are. "Being human is not about individual survival or escape," as the media theorist and writer Douglas Rushkoff put it. "It's a team sport. Whatever future humans have, it will be together."[12] That has to include survival as well.

———————

Three of the existential catastrophes covered in this book—a supervolcano eruption, an asteroid strike, and nuclear war—share a common killing method: rapid climate cooling, mostly through a long-lasting and planet-wide reduction in sunlight. The initial detonations or impact or eruption will kill tens of millions instantly, but the extinction blow would be global starvation. The food chain as we know it begins with sunlight. Blot out the sun, and even the best-prepared survivalist, a master of the wilderness, will starve to death along with everyone else once the existing global food supply runs out, perhaps in as little as a year.[13] "There would be chaos, a total breakdown of cities," David Denkenberger told me. "We'd basically lose industrial civilization."

If we're going to prevent human collapse and even extinction after a sun-blocking catastrophe, we're going to need to feed ourselves without the sun, potentially for years. Fortunately, Denkenberger has some ideas. A civil engineer at Tennessee State University, Denkenberger is the coauthor of the book *Feeding Everyone No Matter What: Managing Food Security After Global Catastrophe.* Like a lot of scholars in the existential risk community, Denkenberger sought to find a neglected research area where his work could make a difference to the future of the human race. He settled on postcatastrophe food supplies—in part because very little research had been done on the subject—but quickly realized that simply storing enough backup food to feed the world would be impossible. To keep all 7-plus billion people fed for the years it would take normal sunlight to return would require so many barrels of food that they could be stacked from the Earth to the moon—forty times over. It would also cost more than $10 trillion,[14] about one-tenth of global GDP,[15] and take food from the mouths of the more than 800 million people who already have too little to eat.

Denkenberger understood that to survive, we'd need to get very creative, both in terms of the food we'd raise and the food we'd be willing to eat. He became inspired when he learned that after the dinosaurs were wiped out, one species thrived in the cold and dark and death that followed: fungi. "Well, maybe when humans go extinct the world will be ruled by fungi again," Denkenberger told me. "And then I thought, 'Why don't we just eat the mushrooms and not go extinct?'"

Those mushrooms are key. Trillions of trees will die in the aftermath of a climate-cooling global catastrophe. Those trees are of little direct use to human beings, since we can't digest cellulose. But mushrooms can, and by raising mushrooms on the dead trees, we can transform the caloric energy trapped in the trees into something humans can actually eat. A log roughly 3 feet long and 4 inches in diameter should produce 2.2 pounds of wet mushrooms over the course of four years. Not a lot, but if we're ruthlessly efficient—and, quite realistically, if the global population has already been thinned by the initial disaster—mushrooms might provide enough food to help us survive.[16] "You can use people to grow mushrooms on trees and harvest the leaves, grind them up and get human food out of it," said Denkenberger. "The ground-up leaves could be made into tea to provide missing nutrients like vitamin C, or fed to ruminant animals like cows. Or rats."

Yes, rats. What cellulose that isn't digested by fungi could be fed to rats, which in turn could be fed to us. Some humans already eat rats, but the current global population of rats would only be enough to provide 0.1 percent of our current food requirements. But the good—or perhaps horrifying—news is that rats grow to sexual maturity in just six weeks, and give birth to seven to nine offspring every seventy days. Rats don't need or even particularly like sunlight, so a blocked sun shouldn't stunt their growth.[17] If everything goes to plan, Denkenberger believes that it would take approximately two years until there were enough rats to provide 100 percent of human food.[18] "Furthermore," he and his coauthor Joshua Pearce write, "other rodents would provide additional food."

And then there are the insects. Eating bugs may seem repulsive to many of us, but more than two billion people globally are entomophagists, meaning they consume insects.[19] In countries like Mexico and Thailand food markets are stocked with commercially raised water beetles and bamboo worms. Insects are a highly sustainable source of protein—it takes far less water to raise a third of a pound of grasshoppers than the 869 gallons needed to produce the same amount of beef. A 3.5-ounce portion of cooked *Usta terpsichore* caterpillars—commonly eaten in Central Africa—contains about 1 ounce of protein, slightly more than you'd get from the same amount of chicken. Water bugs have four times as much iron as beef.[20] And they are abundant—by one estimate the total weight of the world's insect population is 70 times that of the human population.[21] Nature is very good at making insects.

The same qualities that make insects so abundant and so persistent would allow many species to weather even the most extensive, climate-changing existential catastrophes. Beetles can feast on dead wood, and humans can feast on beetles. But would we? I once attended an insect food fair in Richmond, Virginia, for a *Time* magazine story. There I sampled Orthopteran Orzo, a pasta dish with ground cricket serving as the meatballs, as well as the popcorn-sized larvae of deep-fried mealworms. They were both passable, but it might actually take the end of the world for me to eat the tarantula tempura, a battered, fist-sized arachnid on a plate that I can still see in my nightmares.[22]

If I were starving, though, I'd manage. As Denkenberger and Pearce write: "We believe hunger would likely overwhelm humanity's current arbitrary selective tastes given a serious crisis."[23] If that still sounds awful, our diet would be even worse if we failed to prepare for a disaster and food stores ran low before we could ramp up mushroom and rat production. We would be forced to rely on the very bottom rung of the food chain: bacteria. There are species of bacteria that can feed on natural gas, and we could feed on the bacteria once it is run through a process of pasteurization. The advantage is that bacteria can double every twenty minutes—faster even than rats—which means that the microbes could theoretically provide 100 percent of human food needs in just two months.[24] As for the taste, well, it's as appetizing as it sounds, Denkenberger said, although he noted that we already consume supplements like spirulina. "People call [the nutritional supplement] spirulina algae, but it's actually a form of bacteria."

Denkenberger has more ideas, including ideas about what won't work—like trying to create vast indoor farms powered by artificial sunlight. (The conversion rate from fossil fuels to electricity to artificial lights to crops is so low that even if we devoted all the electricity in the world to indoor farms, we would be able to provide only about 5 percent of global food demand.) But more than ideas, Denkenberger has a rationally optimistic vision of how we should respond to what would be the most terrible test our species may ever face. We will survive—we will endure—only if the survivors pull together. "If you want to save everyone, then it certainly has to be large-scale coop-eration," he said.

This notion—that we would unite in a moment when everything falls apart—goes against what we've been told will happen after the collapse, everything that survivalists are preparing to survive. Ask most people to picture the aftermath of a horrifying catastrophe that kills billions and blocks out the sun for years, and you'll get some-thing like Cormac McCarthy's novel *The Road*, a story where civili-zation and government have utterly collapsed, leaving the survivors in a world where suicide seems to be the best option and cannibalism is practiced. (About cannibalism: In 2017 a group of students at the University of Leicester decided to calculate how long the human race would last if humans only had other humans to eat. They found that one person would remain after 1,149 days—though the researchers noted that the lone survivor would be at risk from contracting prion diseases, "which," they wrote, "you get from eating human brains."[25])

For all our fear of what would come after, for all our bleak stories, collapse and conflict aren't givens after a disaster. Sociologists who study postcatastrophe societies report that communities often grow tighter, like the scar tissue that forms following a wound, even as they endure what seems to be unendurable. Human beings help each other, including in those times when it doesn't seem to be in their interest. That's likely how *Homo sapiens* survived its closest brush with extinction—the Toba supereruption—and it's the only way we would survive the next one.

Survival, though, will require the triumph of more than just the human spirit. An existential catastrophe will be like no disaster we have ever faced, an emergency with no end in sight and no safe haven. We will need preparation—one of the most important acts we can undertake now is to fully research and ready plans for alter-native foods like the kind Denkenberger has studied. Most of all, we will need leadership, because without it, a terrible disaster could turn terminal.

Given our current politics, that might be the biggest existential risk of all. We've become paralyzed as a country in the face of what would surely seem like insignificant differences after an existential catastrophe. Washington can't plan a budget for an entire year, let alone devise and implement a strategy to save the world. Global politics are no better. But the survival of humanity is more important than anything else, not just for today but for all our tomorrows. And so we will need to take even more drastic preventative measures, to ensure that the future of the human race doesn't die out with the present.

———————————

The survivalist shelters you can buy on the internet won't save you from an existential catastrophe. The human race as a whole, though, would benefit from the insurance of a refuge. Not a shelter or a bunker but a true refuge that could withstand any catastrophe. It would save the individuals who find protection there, but it wouldn't be for them—not exactly. It would be an investment in the future of the species, to ensure that there will be living human beings left to restart some form of civilization. To keep the future going.

A proper refuge is to a commercial doomsday shelter as an aircraft carrier is to a tugboat. Robin Hanson—the economist we met in the last chapter—has suggested the proactive construction of one or more subterranean refuges, perhaps deep inside a mine shaft and completely independent of the surface, stocked with enough food and supplies to last for years. It would need to be isolated, and it would probably need to be secret, to keep it safe from survivors on the outside desperate for protection. That might seem cruel, but if such refuges came into play, any survivors left outside its doors would be sacrificed. The refuge's purpose would be to maintain a suitable and varied group of human beings capable of repopulating the Earth once the worst had passed.

If all of this—mine shafts, refuges, repopulating the Earth—is giving you a Cold War shiver, don't be surprised. Both the Soviet Union and the United States built elaborate, deeply protected bunkers for civilian and military leaders to survive a nuclear attack. One of them, the Mount Weather bunker in Virginia, is still operational, and it's where America's top civilian and military leaders would be relocated in the event of a nuclear attack or similar global catastrophe. There's also the Cheyenne Mountain Complex in Colorado, where hundreds of military staff protected by 2,500 feet of granite man the nation's missile surveillance system.[26] These are the closest examples the United States has to doomsday refuges, though they're both designed to help the world keep going, not survive its end.

What Hanson entertains is something different, better thought of as a bank, but for human beings. The closest existing analogue is a vault built to hold the world's agricultural resources. The Norwegian government constructed the $9 million Global Seed Vault on the remote island of Svalbard, far north of the Arctic Circle. Also known as the "Doomsday" vault, it serves as a backup to the world's 1,700 national seed banks, each of which holds a collection of native seeds that contain that country's agricultural inheritance. Seeds are the foundation of human civilization, and Svalbard holds more than 930,000 varieties.[27] If a line of seed is lost—and they can be lost through war, through drought, through disease—that plant is lost forever. Seed banks are an insurance against that loss.

In the event of an actual doomsday, of course, there may not be anyone left to make a withdrawal from the Seed Vault. (And the Seed Vault itself is far from indestructible—in 2017 the vault flooded because of unusually warm weather, although no seeds were lost.[28]) But if we could create a similar refuge that would safe-guard a selection of the global population, the human world could be restarted after its end, just as we might use Svalbard to restart global agriculture.

It wouldn't be easy. If the refuges were kept empty, people would have to be moved into them once a disaster was already under way, or even afterward. That might be possible in the event of pandemic, or something that can be predicted in advance, like a major asteroid strike. It's less feasible in the event of a near-instant wipeout like a nuclear war. An effective existential insurance policy would require the refuges to be populated full-time, which would mean a rotat-ing cast of individuals who would agree to serve a tour of duty as humanity's backup.

Simply to avoid inbreeding, a refuge would need a minimum of 50 people, and probably closer to 80. (It's believed that fewer than 70 people managed to colonize Polynesia, which now has a population of around 700,000.)[29] Five hundred would be a better figure, to ensure sufficient genetic and skill variety. The number would also depend on how long the refuge population would need to stay underground. A study on very long-term, multigenerational space missions indicates that 160 people would be needed to keep an isolated population going for two hundred years.[30] Anders Sandberg at the Future of Humanity Institute has said that you'd really need closer to 5,000 people to be assured that the species would endure for another thousand years at least.[31] Either way it would get very crowded down there. Feeling claustrophobic yet?

There's also the question of who should go. There would presum-ably be no shortage of volunteers in the event of an actual catastro-phe, but it might take something close to a draft to convince a rotating group of people to spend months or even years of their lives underground so they could rebuild humanity in the extraordinarily unlikely event of an existential disaster. You might expect we'd want the best and the brightest—scientists, artists, soldiers. But these ultimate refugees would most likely be reentering a world that was no longer modern. No electricity, no infrastructure, nothing that supports what most of us spend our lives doing. The best option might be subsistence farmers and actual hunter-gatherers, human beings who would already be familiar with the basic lifestyle that would now be necessary, once modernity had been swept away.

Unless the U.S. government is much better at keeping secrets than it appears to be, no such doomsday refuges exist, and there are no plans to build any. But humanity may find itself in a place where—as seems to have been the case after Toba—a small population does manage to survive a great catastrophe. Even with the worst our planet might be able to throw at us, total extinction—the death of every man, woman, and child on Earth—isn't a given, because of what has been called the "last few people problem." The same adaptive traits that have enabled *Homo sapiens* to spread to every spot of land on the planet and reach more than 7 billion in numbers also make us very hard to kill off completely.

Given enough time, we might even get back to where we are today. The Hutterite community in North America holds the distinction of having the highest rates of population growth in history, increasing 18-fold in just 70 years.[32] Starting from a single surviving couple, if each woman had a mere 8 children, we could theoretically get back to 7 billion people in less than 600 years—and start this whole thing over again.[33]

Redevelopment would have to take a different path, however, because it took us tens of thousands of years to develop the technologies needed to move from hunting and gathering to settled farming, and thousands of years more to reach the Industrial Revolution. In fact there's no guarantee that industrialization would happen again—

the birth of industry happened only once, in nineteenth-century Britain, before it spread around the world, and if we reran the tape of history it might not happen at all.[34] Survivors would presumably have residual expertise that could be handed down to their descendants, but our first go-round with industrialization was powered by fossil fuels that were easy to take out of the ground. Those supplies of coal and oil are long gone, however, and they won't be replaced anytime soon. The warming created by the earlier burning of those fossil fuels would continue, which means that survivors would also have to contest with gradual climate change for decades, at least until carbon concentrations in the atmosphere began to decline.

In his book *The Knowledge: How to Rebuild Civilization in the Aftermath of a Cataclysm*, the British academic Lewis Dartnell explores how we might be able to transmit the best of human experience and knowledge to a band of postapocalyptic survivors. Dartnell suggests creating a start-up manual for civilization—not a library of our most glorious cultural and scientific achievements, but an actual how-to book for the practical knowledge needed to help a broken society pick up the pieces. Think engineering, farming, and basic medicine, not liberal arts. Though Dartnell believes that keeping it in book form might not be the best idea. "I think you want an apocalypse-proof, unhackable Kindle," he told me. "Paper books have a habit of being flammable."

If we really want to protect ourselves from the end of the world, however, we might want to think about getting off it.

———————

As the richest man in the world, with a fortune worth around $130 billion as of early 2018, Amazon founder Jeff Bezos has a lot of money, and he could choose to spend it on anything on Earth. But he doesn't want to spend it on Earth. In 2000 Bezos founded Blue Origin, a private spaceflight company, and he has said he now liquidates about $1 billion in Amazon stock each year to fund its work.[35] Given that Bezos is Bezos, he likely has a business reason for spending billions on rocket ships—and if there's not one now, he'll find it soon enough. But the richest man in the world has a higher purpose. "We have to go to space to save Earth," Bezos said in 2017. "We have to hurry."[36]

Bezos has always been a space nerd. As an undergraduate at Princeton University, where he led the school's chapter of Students for Exploration and Development of Space, he was inspired by the work of the space colony theorist Gerald O'Neill, who was finishing his academic career at the college. The kicker of Bezos's high school valedictorian speech was this: "Space, the final frontier. Meet me there."

But Bezos isn't the only tech billionaire with a side hustle in space travel. When he isn't building self-driving cars or warning the world about the dangers of AI, Elon Musk also runs the private rocket company SpaceX, which has already begun flying missions for NASA. And like Bezos, Musk thinks that spreading to space is our destiny—though he has a characteristically dramatic way of putting it.

"If we were a multiplanetary species, that would reduce the possibility of some single event, man-made or natural, taking out civilization as we know it, as it did the dinosaurs," Musk told *Rolling Stone* in 2017.[37] "It makes the future far more inspiring if we are out there among the stars and you could move to another planet if you wanted to."

And Musk, it seems, wants to. He has said that SpaceX is working on an interplanetary rocket capable of making the trip to Mars, 34 million miles away, and plans to launch cargo vehicles to the Red Planet by 2022. It's all part of his master strategy to put one million people on Mars over the next several decades. "I've said I want to die on Mars," Musk told the audience at South by Southwest in 2013. "Just not on impact."[38]

Space colonization may be the next big thing in the tech world, but it's long been considered the ultimate solution to existential risk—and the ultimate destiny of human beings. Stephen Hawking, whose robotic voice was beamed into space even as his earthly ashes were interred between Charles Darwin and Isaac Newton at Westminster Abbey, once said that humanity had a thousand years to make it safely off the Earth. By 2017, shortly before his death, he had revised that deadline to just one hundred years. "I strongly believe we should start seeking alternative planets for possible habitation," he said at the Royal Society of London. "We are running out of space on earth and we need to break through technological limitations preventing us from living elsewhere in the universe."[39]

It's true that spreading off world would provide protection from a number of existential risks. Asteroids, supervolcanoes, even nuclear war—the Earth could be utterly destroyed and our space colonists would remain safe. Given the distances involved in interplanetary travel, even the most finely engineered disease would be unlikely to remain infectious long enough to kill off-worlders. Climate change, of course, is a problem for Earth and by Earth. "By definition, whatever causes us to go extinct will be something the likes of which we have not experienced so far," the Princeton astrophysicist J. Richard Gott wrote in 2007. "We simply may not be smart enough to know how best to spend our money on Earth to ensure the greatest chance of survival here. Spending money planting colonies in space simply gives us more chances."

That attention is again being paid to the possibility of space colonization is chiefly because two very rich men are willing to spend a lot of their money on the idea—even though NASA hasn't carried out

a manned spaceflight since retiring the shuttle in 2011. But Bezos—
who has predicted that one day there will be a trillion human beings
living throughout the solar system—is on to something. "If you take
current baseline energy usage, globally, and compound it at just a
few percent a year for just a few hundred years, you have to cover the
entire surface of the Earth in solar cells," Bezos said during a talk
at the Yale Club in early 2019. "Everybody on this planet is going to
want to be a first-world citizen using first-world amounts of energy,
and the people who are first-world citizens today using first-world
amounts of energy? We're going to want to use even more energy."[40]

We will need more raw materials, more energy, more space. Just
as we once grew by spreading across this planet, we will eventually
need to grow by leaving it. It might be a hundred years—it might be
far longer—but humankind's future as a technologically developing species means expansion, possibly endless expansion. To miss
out on this future—whether because of extinction or a catastrophic
setback—is to suffer what Nick Bostrom has called "astronomical
waste."[41] It's the loss of the cosmic inheritance—all that energy, all
that space—that could be ours.

This is the ultimate expression of existential hope: a human
civilization that endures for millions, even billions of years, growing powerful and energy-rich enough to support trillions of human
beings, all able to live lives of unimaginable material and intellectual
plenty. It's only going to happen, though, if we make it off this rock,
and tap the endless energy sources of the universe—and if no alien
civilizations, biological or artificial, have laid their claim first.

According to at least one theory, the universe has an expected
life span of 80 billion years, give or take a few billion,[42] which might
make it seem as if we have plenty of time to claim our inheritance.
Time lost is time we won't get back, however. Bostrom estimates that
for every century space colonization is delayed, we potentially lose

the lives of 10^{43} humans who could come into being only if we seize the galaxy.[44] This is our cosmic opportunity cost.

There's a difference, though, between fumbling the present so badly that we allow an extinction to occur on our watch, ending the human story prematurely, and delaying the launch of humanity on a cosmic voyage not all of us necessarily want. Space colonization may be our destiny, but it won't keep us safe—not for the foreseeable future. The energy and money that might be spent on nascent efforts to move off planet would be better used combatting existential threats that could end that future—and readying our survival, should the worst occur. That, as the existential risk expert Seth Baum wrote, "is doing good on a literally astronomic scale."[45]

Space is at best a distraction for now. Mars may be the other planet in the solar system most conducive to life, but it is still far more hostile than any place on Earth, save perhaps the bottom of the ocean. Just the round-trip to Mars and back would expose astronauts to up to two-thirds the radiation limits advised for space workers, putting them at unknown risk of cancer. Mars has been gradually losing its atmosphere for billions of years. It is very cold, and the air is unbreathable. There is a reason that as far as we know nothing currently lives on Mars. It's a bad neighborhood.

The major problem with Mars as an existential risk refuge—or anywhere else off planet—is that it is very difficult to imagine how it would be possible for the Earth to be screwed up so badly, by natural disaster or war or disease or climate change, that it would somehow become more hostile to humans than Mars. We could have runaway climate change and Earth would still be preferable to Mars. We could have a volcanic winter that lasts for years—and you could still grow more food on Earth than Mars. While proponents of colonization have said we could eventually terraform the Red Planet to make it more hospitable to humans—perhaps by thickening its atmosphere to retain heat—NASA scientists have confirmed that doing so is essentially impossible, at least with current technology.[46] Also, if we plan on terraforming a planet with a problematic climate to make it easier for us to live there, we might want to start with Earth. Even Bezos himself is deeply skeptical about what he has called the "Plan B" argument for space colonization. "Do me a favor, go live on top of Mount Everest for a year first, and see if you like it—because it's a garden paradise compared to Mars," Bezos said at the Yale Club talk.[47]

In March 2018, I visited my friend Ed Finn in Tempe, Arizona, where he was putting on a conference about human space settlement, which included a mockup of a city on the moon set almost 150 years in the future. Ed is the founding director of the Center for Science and the Imagination at Arizona State University. His job is to explore the intersection between art and science, and how they both might be used to predict and shape the future we want. He is someone who welcomes what's to come—but when I asked him about Musk's and Bezos's plans, even he was skeptical that space could serve as our species's escape pod.

"When we think about the climate crisis and other threats over the next hundred years, this space stuff is a luxury," Ed told me. "It is never going to solve our problems on Earth and it will never be a safety valve. Maybe in a few hundred years we'll be ready. But in the short term, space is an experiment, not a survival plan."

We still need survival plans.

10

THE END
Why We Fight

Some five hundred generations have passed since the first stirrings of what we now call human civilization.[1] Five hundred times a generation of fathers and mothers has ushered children into the world, who in time would succeed them, giving birth to children of their own who take their turn on the wheel. In that span of time—barely more than a blink against the life span of the cosmos—whole peoples have risen, fallen, and vanished from the face of the Earth. As mortality is to individuals, so collapse seems to be to human societies—simply part of our nature. Just as our children and our children's children continue our story, though, so those lost civilizations of the past leave something behind to be found by us, writing themselves into our time, as we hope to write ourselves into tomorrow.

But what happens when there's no one left to read the past, or write the future?

Bill Kitchen has one idea. Kitchen is an electrical engineer who made a fortune designing and selling amusement rides—you might know his iFly indoor skydiving attraction—but growing up as a young boy in Florida his passion was cosmology and astrology. He is a lifetime member of the SETI League, a nonprofit dedicated to privatizing the search for intelligent life, and like many of the other people we've met in this book, he is deeply worried about the fate of his species. "We humans are destroying the planet," he told me in his gentle southern accent. "It could be an asteroid or a nuclear war or rogue genetics or an AI that decides the Earth is better off without us. But humanity is in danger."

Kitchen wants to build a digital time capsule that would be a compendium of Earth's natural and anthropological history, a full suite of the best the human species has to offer, as well as a new Rosetta Stone that would explain our language and how to decode the information the time capsule contains. It would include the sequenced genomes of as many animals and plants on Earth as possible, a Noah's Ark for the ultimate flood, as well as the genomes of actual human beings. And all of this information will be beamed into space through a high-powered laser that Kitchen calls the Interstellar Beacon.[2] Kitchen plans for the beacon—which is still in the earliest of early stages—to be updated and broadcast continuously, to the moment of our eventual demise. "This will be our legacy," he said.

There have been other plans to create similar planetary archives. The Alliance to Rescue Civilization (ARC) advocated for the establishment of a database of humanity, complete with DNA samples of Earth life, to be stored on the moon and staffed by astronaut archivists. Lunar Mission One is a planned project to send a robotic probe to the moon's Shackleton Crater, where it will bore a hole and store public and private data about humanity—including the genetic data of people who will pay to fund the project. (A launch had been planned for 2024, but funding is fuzzy.) Memory of Mankind is an archival venture that will inscribe human knowledge on ceramic disks inspired by the Sumerian clay tablets that contain some of the oldest recorded information in human history. The tablets will be buried deep in a salt mine in Austria, there to be found by whoever or whatever might one day come across them. The Arch Mission Foundation is encoding human knowledge on 5-D data optical storage disks and plans to seed them around the solar system. The group's first disk is currently somewhere in space—it was launched on Elon Musk's SpaceX Falcon Heavy test flight in February 2018, tucked inside a racer-red Tesla Roadster. Fittingly, the disk contained digital copies of the sci-fi author Isaac Asimov's *Foundation* trilogy, which tells the story of a scientist who foresees the collapse of civilization and creates a compendium of all knowledge to give humans of the future a chance to rebuild.

These projects differ in their aims and methods. Some, like Memory of Mankind, are meant to be analog time capsules at a moment when information has become increasingly digitized, and therefore vulnerable to a technological collapse. Others, like Kitchen's Interstellar Beacon or Lunar Mission One, hold out the unlikely promise that individuals might actually be resurrected by some power in the future, human or otherwise. What they have in common is an awareness that our time may be coming to a close, and a hope that it will be possible to leave something of ourselves for a future that may yet be.

These beacons and depositories would differ from the pyramids and other monuments left by past civilizations. Like the "two vast and trunkless legs of stone" in Percy Shelley's "Ozymandias," those ancient shrines proclaimed the greatness of the kings and queens of their age, in full expectation that their people and their culture would live on indefinitely. But we, who have unearthed the colossal wrecks of the deep past and suspect the tenuousness of our shared future, should know better.

The field of existential risk dwells on the darkest of topics, but a strain of hope runs through it, hope that some of the same technologies that could doom us could also give us the power to cheat not just death but extinction itself, that we could be the generation that breaks the cycle of collapse altogether. No monuments would we need, for our ever-growing presence, stretching to every corner of space, would be all the proclamation that we require, until the universe itself goes dark. Perhaps. Myself, I'd settle for something humbler. Just a record of who we were. A permanent mark on the cosmos.

Back at the start of this book, I recounted the unveiling of the Doomsday Clock in 2018. The experts behind the clock—which has kept humanity's time since 1947—set the hands at 11:58. That was closer to midnight than any year since 1953, when both the United States and the Soviet Union tested their first thermonuclear warheads within nine months of each other. Announcing the new setting of the Doomsday Clock in the *Bulletin of the Atomic Scientists* in August 1953, Eugene Rabinowitch, the Russian-born American biophysicist and Manhattan Project alumnus who had cofounded the *Bulletin*, wrote that "only a few more swings of the pendulum, and, from Moscow to Chicago, atomic explosions will strike midnight for Western civilization."[3]

The *Bulletin* now updates the setting on the Doomsday Clock each year, and on January 24, 2019, the journal's Science and Security Board gathered again at the National Press Club in Washington, D.C., to reveal the new time. William Perry, the former U.S. secretary of defense whom we met in chapter 3, and ex–California governor Jerry Brown stood side by side on the stage as they unveiled the clock. The time was 11:58, two minutes to midnight—the same as the previous year. If the apocalypse wasn't any closer, it was no further away. The end of the world appeared to be in a holding pattern.

"Two minutes to midnight invokes memories of 1953," said Perry, who had lived and worked through every close call of the Cold War. "I know it because I was there." For his part, Brown, who had just finished his second stint as the governor of America's most populous state and had recently joined the *Bulletin* as its executive chair, called out the world leaders ignoring our growing existential peril. "The blindness and stupidity of the politicians and their consultants is truly shocking in the face of nuclear catastrophe and danger," Brown said. "It is two minutes to midnight. . . . It's hard to even feel or sense the peril and the danger we're in, but these scientists know what they're talking about. It's late and it's getting later, and we got to wake people up."[4]

If I had one objective in writing this book, it's that: wake people up. Wake them up to the reality of existential threats, whether from nature or the hand of man—and wake them up to the fact that we're not helpless in the face of those threats. The first part is easier. We may not fear thermonuclear war the way we once did, but the events of the past few years, the tensions between the United States and Russia, the way wild cards like North Korea have crashed the nuclear club, has at least snapped us out of the atomic amnesia of the post–Cold War years. With each passing month, climate change ever more indelibly imprints itself on the global consciousness, and with it, the deepening sense that we have irredeemably damaged our planet. You don't need to know the meaning of a climate tipping point to fear that we have passed it. And while we may not fully understand emerging technologies like artificial intelligence or synthetic biology, we know enough to worry about where they might take us. It's not an accident that our films and novels and TV shows mine the end of the world for material. We're afraid. But fear isn't sustainable—and fear isn't a strategy for survival.

In keeping the Doomsday Clock set at two minutes to midnight, the *Bulletin* made the only decision it could. The world hadn't gotten perceptibly worse over the course of 2018; there were even some improvements, like the beginnings of negotiation between the United States and North Korea on nuclear weapons. But on balance it hadn't gotten any better, either—arms control treaties between Washington and Moscow broke down, and the globe kept warming, even as carbon emissions reached a historic high. The *Bulletin* had a term for the state we now find ourselves in, circling the drain of Armageddon: a "new abnormal,"[5] as the physicist Robert Rosner put it the day of the clock's unveiling, "a disturbing reality in which things are not getting better and are not being effectively dealt with." It's not just that we find ourselves in a state of existential fear—we've had reason for such fear since the morning of July 16, 1945. But that fear has bred not passion but paralysis. Though we can imagine the end of the world all too easily, we can't imagine coming together to save it. And that creeping futility is what we must overcome.

———————

The Doomsday Clock is a brilliant symbol, but a symbol is all it is. There is no countdown to the end—at least not one we can hear. But if our current existential risks worsen with each passing year, and if we continue to add new ones, the odds of our long-term survival will be short.

The hopeful view is that what appears to be ever-increasing existential risk is actually a temporary bottleneck created by new technologies we can't yet control and by environmental challenges that are a function of our accelerating growth, like climate change. If we can make it through that bottleneck, we'll find safety on the other side. We just need a breakthrough. Maybe it will be artificial intelligence, ethical and controlled. Maybe it will be some mix of clean energy and carbon engineering that defuses climate change and gives us centuries more to grow on this planet. Maybe Elon Musk and Jeff Bezos are right, and the move to space will keep us safe. In this vision we survive—and thrive—not by slowing down, but by speeding up.

The age of existential risk has sharpened the stakes, but this has been the central human challenge for as long as there have been human beings. Faced with limits, we invent new technologies and new practices that allow us to grow, which then use up more resources and create new risks, forcing us to innovate again to keep one step ahead of our growing capacity for both success and destruction. This is how we put seven and a half billion people on this planet. This is how we reached a point where the life of the average human being is longer and healthier and richer and just plain better than it has ever been before, no matter how fed up and pessimistic we may feel on a day-to-day basis. But with each passing year the race becomes faster and harder to run. Sooner or later we may stumble, and be overtaken.

There is another option. We could deliberately choose to slow down, to select a more sustainable speed, to eschew both the potential risks and potential benefits of emerging technology. This, at its heart, is what environmentalists and conservationists have long called for us to do, and it applies not just to our energy use but to our mind-set, as individuals and as a species. What if we choose to live within our means?

It might work. But we would be surrendering much as well. In seeking to avoid existential risk, we would be giving up existential hope, a lottery ticket to a technological heaven where there are limits no longer. And we would be asking ourselves to alter what seems to be a basic drive of humanity: growth. Great religions claiming billions of adherents counsel humility and abstention, yet look around. The drive to grow and compete seems so hardwired into human beings that it can seem as if the only way we could change it would be to change our very DNA. Not by political activism, or moral suasion, but by rewriting our own source code.

The most startling conversation I had while researching this book wasn't with a nuclear warrior like William Perry or a gene-editing scientist like George Church. It was with a philosopher of ethics at the University of Oxford named Julian Savulescu. Savulescu believes that the new technologies I've highlighted and the thirst for growth have put humanity at risk of what he calls "Ultimate Harm"—meaning the end of the world. It might be the global threat of climate change, the nation-state threat of nuclear war, or the individual threat of bioengineered viruses. Savulescu's point is that as the power to inflict Ultimate Harm spreads, human ethics become the only emergency brake. If the world will blow up if just one of us pushes the self-destruct button, then we will survive only if human beings—all human beings—can be trusted not to push that button. The problem, Savulescu told me, is that we don't have the ethics to handle the dangerous world that we ourselves have created. "There's this growing mismatch between our cognitive and technological powers and our moral capacity to use them," he said.

Savulescu has a radical solution. He suggests that the same cutting-edge biotechnology that now poses an existential risk itself— the greatest looming existential risk, in my view—could one day be used to engineer more ethical and more moral human beings. As we learn to identify the genes associated with altruism and a sense of justice, we could upregulate them in the next generation, creating children who would innately possess the wisdom not to use that terrifying bioweapon, who would see the prudence in curbing their present-day consumption to ensure that future generations have a future at all. The options for self-destruction would still exist, but our morally bioenhanced offspring would be too good to choose them.

If that sounds like a desperate measure, well, so are the times. "I think we're at this point where we need to look at every avenue," Savulescu said. "And one of those avenues is not just looking to political reform—which we should be doing—but also to be looking at ourselves. We're the ones who cause the problems. We're the ones who make the choices. We're the ones who create these political systems. No one wants to acknowledge the elephant in the room, and that is that human beings may be the problem, not the political system."[6] It's not just ethical AI that we would need to create. It's ethical human beings.

This is a moral philosopher's thought experiment, not a concrete plan to begin editing genes for altruism into our babies—which, it should be noted, we're not close to knowing how to do, even if we wanted. But as we spoke—Savulescu in Oxford, me in Brooklyn—I realized he was trying to answer a question that had nagged me since I began reporting on environmental issues years ago, and one that followed me throughout the time I spent on this book: is it easier to change people or technology? If you hold out hope for people, then you believe that we can be persuaded to behave in a way that is more sustainable, even if it demands sacrifice. If you believe technology is more responsive, then you're in favor of running the race against risk faster, putting faith in our ability to innovate ahead of doom.

From what I have observed, most of us speak as if we believe it is people who can be changed, but behave as if technology will keep us ahead. We embrace a rhetoric of political change and personal responsibility, but the lives we actually live depend on technological and economic growth, whatever the consequences. Savulescu was trying to split the difference: use technology to change people.

That should tell you just how difficult it is to fundamentally change ourselves.

People have changed, of course. We've largely abandoned hideous practices like slavery, expanded the circle of human rights, and fought for the power to rule ourselves. But those changes mostly fed the engine of growth, and put more power in the hands of individuals, to be used for good or ill. Short of a fundamental political or even spiritual revolution, what I can't see changing is that primal human drive to expand.

Perhaps I'm suffering from a failure of imagination. The Marxist political theorist and literary critic Fredric Jameson, after all, once wrote that it is "easier to imagine the end of the world than to imagine the end of capitalism."[7] But everywhere I've traveled on this planet, I've seen people who want more. More for themselves, and more for their children. Who will tell them they can't have it, even if it may cost the world?

So we must run faster, as if we're running for our lives.

———————————

I began this book by describing a photograph of my wife, my father, my mother, myself, and my newborn son, taken just hours after his birth. It was the future coming into being, a single image that held three generations out of the hundreds that have walked this planet since our species launched, without ever realizing it, the ongoing project we call civilization.

There is another photograph of our son, attached to a corner of the refrigerator. It is an ultrasound, taken when he was around twelve weeks past conception. My wife had suffered three miscarriages before this pregnancy, three times hopes raised and denied. We had never made it this far before. The day that ultrasound was taken at the obstetrician's office, and every day after until Ronan was born six and a half months later, we worried. Neither of us had ever known we could want something so much. As I write these words, Ronan has passed his first birthday. He still has the button nose that was visible in that first image, and he's just as restless as he was in the womb, when we rejoiced with every kick. He is walking, and soon he will run his own race. He is the future made flesh, our own mark on the cosmos. One day my wife and I will be gone, and I pray that he will still be here. It would have seemed strange to hope for that before. Now nothing could seem more natural.

The present seems dark, the future seems darker. Yet I hope still. It may not line up with my reason or my research, but I do have hope. I have hope that we human beings can do what we have always done before and find a way to keep going, whether through innovation or through sacrifice or through some mix of both. I hope because the alternative is too grim to bear. I hope for him, and I hope for all of us.

ACKNOWLEDGMENTS

This book wouldn't have been possible without the love and encouragement of my wife, Siobhan O'Connor. She supported my decision to leave *Time* magazine to work on what became *End Times* and pushed me during those periods—which were not few—when I experienced doubts. She is the first person to hear my ideas—the good ones and the bad ones—and is the smartest writer and editor I know. She cares, deeply, about writing and books—which is why the one you're holding in your hands right now exists at all. Thank you for everything, love.

Our son Ronan was born a few months before I began working on *End Times* in earnest. Studying the end of the world doesn't make for easy thoughts, but knowing Ronan was here helped give me the strength to see this project through—even on those nights when he wasn't sleeping so easily. Now I get to watch him grow up, day by day, and every day I count myself lucky.

Todd Shuster, my agent at Aevitas Creative, took a chance on a thirty-nine-year-old would-be first-time author. He and his team—especially Justin Brouckaert—worked closely with me to refine my book proposal to the point where it was ready to be sold, and they've been with me every step of the way. And thanks as well to my *Time* colleague and Carroll Gardens neighbor Steve Koepp, who put me in touch with Todd in the first place.

Krishan Trotman, my editor at Hachette, saw the potential in *End Times* from the very beginning. She helped me refine my idea, guided my research and reporting, and provided invaluable insight all the way to publication day and beyond. Carrie Napolitano stepped in ably at the end to help push the manuscript through copy edits.

My publicity and marketing team at Hachette—Michelle Aielli, Sarah Falter, and Odette Fleming—brought *End Times* out into the world. You're reading this book in large part because of their efforts.

When I was looking for a fact checker to run through the finished manuscript, I had one person in mind: Barbara Maddux. Barbara worked closely with me as chief of reporters when I was the international editor of *Time* magazine, and without her help, I doubt we would have closed a single issue. I knew *End Times* would be in good hands when she brought out her red pen. My other former colleagues from more than fifteen years at *Time*, in New York and abroad—including Nancy Gibbs, Richard Stengel, Karl Taro Greenfeld, Jim Frederick, Jeffrey Kluger, and Radhika Jones—helped make me the writer and journalist I am today. A 2017 cover story I wrote in *Time* on pandemics helped form the foundation of my chapter on infectious disease, while another cover on synthetic biology the same year, commissioned by Matt McAllester of *Newsweek* magazine, did the same for the chapter on biotechnology.

There are too many people in the broader field of existential risk to thank individually, but I'm indebted to Phil Torres, Seth Baum, Olle Häggström, Milan Ćirković, and Nick Beckstead for taking the time to walk me through the basics of the subject in long conversations. Rachel Bronson of the *Bulletin of the Atomic Scientists* gave me invaluable insight into how the Doomsday Clock is set. Eric Christensen at the Catalina Sky Survey agreed to host me for a night of asteroid hunting, and Greg Leonard patiently answered my questions atop Mount Lemmon. NASA's Lindley Johnson took time away from defending the planet Earth to explain to me how we should be defending the planet Earth.

Jim Eckles at the White Sands Missile Range gave me an off-hours tour of the Trinity Site, and plenty of stories about the men and women of the Manhattan Project. Daniel Ellsberg and William J. Perry shared their experiences of the Cold War and nuclear near misses with me. Gernot Wagner and Martin Weitzman hosted me at Harvard University and helped shape my thinking about the extreme downside risks of climate change. Samuel Scheffler taught me about what we owe the people of the future, and Alexander Rose of the Long Now Foundation gave me a little bit of hope, as did Klaus Lackner and his carbon-sucking machines.

A trip to Cameroon for *Time* magazine in 2011 with the virologist Nathan Wolfe helped me understand the porous microbial border between animals and humans, while the work of my *Time* colleagues during the 2014 Ebola outbreak—led by Siobhan O'Connor—gave me incredible insight into the global threat of disease. Tom Inglesby and Marc Lipsitch explained the threat that some new tools of biotechnology could pose to the human race, while George Chuch took time to illustrate some of the unexpected pitfalls of these advances, as well as their benefits. Ginkgo Bioworks CEO Jason Kelly let me tour around his biological factory in South Boston, led by creative director Christina Agapakis.

Roman Yampolskiy broke down the fiendishly clever existential threat of advanced artificial intelligence, and the writings of Nick Bostom and Elizier Yudkowsky were key to the larger chapter. Seth Shostak at the SETI Institute welcomed me into his office to explain his lifelong search for intelligent life, Robin Hanson walked me through the Great Filter, and Adam Frank shared his thoughts about the possibility of interstellar climate change. David Denkenberger laid out the diet for a post-catastrophe world—heavy on the rats and the bacteria.

My friend Ed Finn—whom I first met working on the *Princeton Tiger* in the fall of 1998, before he moved on to bigger things—welcomed me to Arizona State University's Emerge festival, one of many initiatives he has spearheaded at the Center for Science and the Imagination. Mike Gardner provided me with a welcome ear as a fellow writer, and Paul Griffin was Paul Griffin. I wouldn't be doing this without my parents, and my brother Sean Walsh and sister-in-law Caroline Walsh.

NOTES

INTRODUCTION

1 "Global Health Observatory (GHO) data, World Health Organization, accessed March 29, 2019, https://www.who.int/gho/mortality_burden_disease/life_tables/situation_trends/en/.

2 Chris Micaud, "One in Seven Thinks End of World Is Coming: Poll," Reuters, May 1, 2012.

3 Melanie Randle and Richard Eckersley, "Public Perceptions of Future Threats to Humanity and Different Societal Responses: A Cross-National Study," Futures 72, (2015): 4–16, https://doi.org/10.1016/j.futures.2015.06.004.

4 Jacob Poushter, "Globally Is Life Better Today Than in Past?" Pew Research Center's Global Attitudes Project, September 17, 2018, accessed March 31, 2019, https://www.pewglobal.org/2017/12/05/worldwide-people-divided-on-whether-life-today-is-better-than-in-the-past/.

5 Jonathan Watts, "We Have 12 Years to Limit Climate Change Catastrophe, Warns UN," *Guardian,* October 8, 2018, accessed March 31, 2019, https://www.theguardian.com/environment/2018/oct/08/global-warming-must-not-exceed-15c-warns-landmark-un-report.

6 Homi Kharas and Kristofer Hamel, "A Global Tipping Point: Half the World Is Now Middle Class or Wealthier," Brookings, September 27, 2018, accessed March 31, 2019, https://www.brookings.edu/blog/future-development/2018/09/27/a-global-tipping-point-half-the-world-is-now-middle-class-or-wealthier/.

7 Elijah Wolfson, "How the World Got Better in 2018, in 15 Charts," Quartz, December 31, 2018, accessed March 31, 2019, https://qz.com/1506764/ways-the-world-improved-in-2018-in-charts/.

8 "The Visual History of Decreasing War and Violence," Our World In Data, accessed March 31, 2019, https://slides.ourworldindata.org/war-and-violence/#/6.

9 Max Roser, "Life Expectancy," Our World in Data, May 23, 2013, accessed March 31, 2019, https://ourworldindata.org/life-expectancy.

10 *Global Terrorism Index 2017: Measuring and Understanding the Impact of Terrorism,* Institute for Economics & Peace, 2017, accessed March 30, 2019.

11 Jonathan B. Wiener, "The Tragedy of the Uncommons: On the Politics of Apocalypse," *Global Policy,* June 6, 2016, accessed March 31, 2019, https://onlinelibrary.wiley.com/doi/full/10.1111/1758-5899.12319.

12 Max Roser and Esteban Ortiz-Ospina, "Global Extreme Poverty," Our World in Data, May 25, 2013, accessed March 31, 2019, https://ourworldindata.org/extreme-poverty.

13. John Leslie, *The End of the World: The Science and Ethics of Human Extinction* (London: Routledge, 2008).

14 Stephen Hawking, "Will Robots Outsmart Us? The Late Stephen Hawking Answers This and Other Big Questions Facing Humanity," *Sunday Times Magazine,* October 14, 2018, accessed March 31, 2019, https://www.thetimes.co.uk/article/stephen-hawking-ai-will-robots-outsmart-us-big-questions-facing-humanity-q95gdtq6w?_ga=2.72867923.1744012958.1539701503-136416986.1538527295.

15 Anders Sandberg and Nick Bostrom, *Global Catastrophic Risks Survey, Technical Report #2008-1,* Future of Humanity Institute, Oxford University, 2008, pp. 1–5, accessed March 31, 2019.

16 "How Likely Is an Existential Catastrophe?" *Bulletin of the Atomic Scientists,* December 11, 2018, accessed March 31, 2019, https://thebulletin.org/2016/09/how-likely-is-an-existential-catastrophe/.

17 Anders Sandberg, "Human Extinction from Natural Hazard Events," *Oxford Research Encyclopedia of Natural Hazard Science,* March 21, 2018, accessed March 31, 2019, http://oxfordre.com/naturalhazardscience/abstract/10.1093/acrefore/9780199389407.001.0001/acrefore-9780199389407-e-293?rskey=yYFZVm&result=2#acrefore-9780199389407-e-293-bibItem-0091.

18 Derek Parfit, *Reasons and Persons* (Oxford: Clarendon Press, 1987).

19 "Black Rhino," Extinction Is Forever, accessed March 31, 2019, https://www.extinctionisforever.org/.

20 Bostrom, Nick. "Existential Risk Prevention as Global Priority," *Global Policy* 4, no. 1 (2013): 15–31, doi:10.1111/1758-5899.12002.

CHAPTER 1: ASTEROID

1 Locke, Susannah. "Was the Dinosaurs' Long Reign on Earth a Fluke?" *Scientific American,* September 11, 2008.

2 Vikram Dodd, "Dinosaurs Were Killed by Isle of Wight–Sized Asteroid," *Guardian,* March 5, 2010, accessed March 31, 2019, https://www.theguardian.com/science/2010/mar/05/dinosaurs-asteroid-science-climate-change.

3 Nick Bostrom and Milan M. Ćirković, *Global Catastrophic Risks* (Oxford: Oxford University Press, 2018).

4 Luis W. Alvarez, Walter Alvarez, Frank Asaro, and Helen V. Michel, "Extraterrestrial Cause for the Cretaceous-Tertiary Extinction," *Science,* June 6, 1980, accessed March 31, 2019, http://science.sciencemag.org/content/208/4448/1095.

5 Leon Jaroff, "At Last, the Smoking Gun?" *Time*, June 24, 2001, accessed March 31, 2019, https://content.time.com/time/magazine/article/0,9171,157342,00. html.

6 Bigthinkeditor, "Neil DeGrasse Tyson (caught on Camera): The Universe Is Trying to Kill You," Big Think, October 6, 2018, accessed March 31, 2019, https:// bigthink.com/big-think-mentor/neil-degrasse-tyson-caught-on-camera-the-universe-is-trying-to-kill-you.

7 John R. Spencer and Jacqueline Mitton, *The Great Comet Crash: The Impact of Comet Shoemaker-Levy 9 on Jupiter* (New York: Cambridge University Press, 1995).

8 Susan W. Kieffer, *Biographical Memoirs: Eugene M. Shoemaker 1928–1997*, National Academy of Sciences, 2015, accessed March 31, 2019, http://www. nasonline.org/publications/biographical-memoirs/memoir-pdfs/shoemaker-eugene.pdf.

9 Spencer and Mitton, *The Great Comet Crash*.

10 "Astrogeology Science Center," Carolyn Shoemaker, USGS Astrogeology Science Center, accessed March 31, 2019, https://astrogeology.usgs.gov/people/carolyn-shoemaker/.

11 NASA, accessed March 31, 2019, https://nssdc.gsfc.nasa.gov/planetary/comet. html.

12 Christine Gorman, "The Comet That Battered Jupiter, and Shook Congress," *Scientific American*, February 1, 2016, accessed March 31, 2019, https://www. scientificamerican.com/article/s19-the-comet-that-battered-jupiter-and-shook-congress/.

13 Martin and Terry, "Shoemaker-Levy 9: Temperature, Diameter and Energy of Fireballs," SAO/NASA ADS: ADS Home Page, September 1, 1996, accessed March 31, 2019, http://adsabs.harvard.edu/abs/1996DPS....28.0814M.

14 Comet/Jupiter Collision FAQ—Post-Impact, accessed March 31, 2019, http:// www.physics.sfasu.edu/astro/sl9/cometfaq2.html#Q3.1

15 Mark R. Showalter, Matthew M. Hedman, and Joseph A. Burns, "The Impact of Comet Shoemaker-Levy 9 Sends Ripples Through the Rings of Jupiter," *Science*, May 6, 2011, accessed March 31, 2019, http://www.sciencemag.org/ content/332/6030/711.abstract.

16 Warren E. Leary, "Big Asteroid Passes Near Earth Unseen in a Rare Close Call," *New York Times*, April 20, 1989, accessed March 31, 2019, https://www.nytimes. com/1989/04/20/us/big-asteroid-passes-near-earth-unseen-in-a-rare-close-call. html.

17 Dan Quayle and the Asteroid Watch," *Chicago Tribune*, September 3, 2018, accessed March 31, 2019, http://articles.chicagotribune.com/1990-06-07/ news/9002160412_1.

18 Eric J. Lyman, "USA Today: 'Giggle Factor' Is No Laughing Matter to Scientists," accessed March 31, 2019. http://www.ericjlyman.com/usagigglefactor.html.

19 "Armageddon," rottentomatoes.com, accessed July 12, 2019, https://www.
   rottentomatoes.com/m/armageddon

20 "FAQs: Ten Frequently Asked Questions about NEO Impacts," accessed March
   31, 2019, https://www.boulder.swri.edu/~bottke/neo_faq.html.

21 Bryan Bender, "NASA's Asteroid Defense Program Aiming for More Impact,"
   Politico, September 21, 2018, accessed March 31, 2019, https://www.politico.
   com/story/2018/09/21/nasa-asteroid-defense-program-834651.

22 Phil Torres and Russell Blackford, *The End: What Science and Religion Tell Us
   about the Apocalypse,* Kindle ed. (Durham, NC: Pitchstone, 2016), p. 115, loc.
   2611.

23 R. G. Strom, "The Origin of Planetary Impactors in the Inner Solar System,"
   Science 309, no. 5742 (2005): 1847–50, doi:10.1126/science.1113544.

24 Craig Childs, *Apocalyptic Planet: Field Guide to the Future of the Earth,* Kindle
   ed. (New York: Vintage, 2013), p. 266, loc. 4358.

25 Carrie Nugent, *Asteroid Hunters,* Kindle ed. (London: TED Books, 2017), loc.
   109.

26 G. Krasinsky, "Hidden Mass in the Asteroid Belt," *Icarus* 158, no. 1 (2002):
   98–105, doi:10.1006/icar.2002.6837.

27 Tricia Talbert, "Planetary Defense Frequently Asked Questions," NASA,
   December 29, 2015, accessed March 31, 2019. https://www.nasa.gov/
   planetarydefense/faq.

28 Contributors, HowStuffWorks.com, "If You Were to Move All of the Matter
   in the Universe into One Corner, How Much Space Would It Take Up?"
   HowStuffWorks Science, March 8, 2018, accessed March 31, 2019, https://
   science.howstuffworks.com/dictionary/astronomy-terms/question221.htm.

29 Brian Dunbar, "Asteroid Fast Facts," NASA, March 24, 2015, accessed March 31,
   2019, https://www.nasa.gov/mission_pages/asteroids/overview/fastfacts.html.

30 H. Atkinson, C. Tickell, and D. Williams, eds., "Report of the Task Force on
   Potentially Hazardous Near Earth Objects," British National Space Centre,
   London, 2000, 54 pp.

31 Clark Chapman, "The Hazard of Near-Earth Asteroid Impacts on Earth," *Earth
   and Planetary Science Letters* 222 (2004): 1–15, doi:10.1016/j.epsl.2004.03.004.

32 Nugent, *Asteroid Hunters,* loc. 529.

33 Ibid., locs. 534–39.

34 Eric Christiensen, director, Catalina Sky Survey, in discussion with the author,
   March 2018.

35 "Charged Coupled Devices," McDonald Observatory, accessed March 31, 2019,
   https://mcdonaldobservatory.org/research/instruments/charged-coupled-
   devices.

36 Nugent, *Asteroid Hunters,* locs. 328–30.

37 Ibid., loc. 303.

38 Michael D. Lemonick, "Chicken Little Alert," *Time*, February 29, 2004, accessed March 31, 2019, https://content.time.com/time/magazine/article/0,9171,596132,00.html.

39 Contributors, HowStuffWorks.com. "Meteors Burn up When They Hit the Earth's Atmosphere. Why Doesn't the Space Shuttle?" HowStuffWorks Science, August 1, 2013, accessed March 31, 2019, https://science.howstuffworks.com/question308.htm.

40 Samantha Mathewson, "How Often Do Meteorites Hit the Earth?" Space.com, August 10, 2016, accessed March 31, 2019, https://www.space.com/33695-thousands-meteorites-litter-earth-unpredictable-collisions.html.

41 Interagency Working Group for Detecting and Mitigating the Impact of Earth-Bound Near-Earth Objects, *National Near-Earth Object Preparedness Strategy and Action Plan*, Washington, DC, 2018.

42 Michael Schirber, "Cities Cover More of Earth than Realized," LiveScience, March 11, 2005, accessed March 31, 2019, https://www.livescience.com/6893-cities-cover-earth-realized.html.

43 G. H. Stokes, D. K. Yeomans, et al., "Study to Determine the Feasibility of Extending the Search for Near-Earth Objects to Smaller Limiting Diameters," Report of the NASA NEO Science Definition Team, 2003.

44 "Congress Hears Options for Asteroid Defense: Pay Now or Pray Later," NBCNews.com, March 19, 2013, accessed March 31, 2019, http://cosmiclog.nbcnews.com/_news/2013/03/19/17373781-congress-hears-options-for-asteroid-defense-pay-now-or-pray-later?lite.

45 "Deep Impact," NASA, accessed March 31, 2019, https://www.jpl.nasa.gov/missions/deep-impact/.

46 Charles El Mir, Kt Ramesh, and Derek C. Richardson, "A New Hybrid Framework for Simulating Hypervelocity Asteroid Impacts and Gravitational Reaccumulation," *Icarus* 321 (2019): 1013–25, doi:10.1016/j.icarus.2018.12.032.

47 "It's Official: Try-hard Bruce Willis Could Not Save the World," *Astronomy & Geophysics* 53, no. 5 (2012), doi:10.1111/j.1468-4004.2012.53504_6.x.

48 Brian Dunbar, "An Innovative Solution to NASA's NEO Impact Threat Mitigation Grand Challenge and Flight Validation Mission Architecture Development," NASA, February 1, 2016, accessed March 31, 2019, https://www.nasa.gov/directorates/spacetech/niac/2012_phaseII_fellows_wie.html.

49 Government Accountability Office, *Nuclear Weapons: Actions Needed by NNSA to Clarify Dismantlement Performance Goal*, Washington, DC, 2014.

50 "Incoming!!!" Bad Astronomy, October 6, 2008, accessed March 31, 2019, http://blogs.discovermagazine.com/badastronomy/2008/10/06/incoming-2/#.W82ocBNKgfE.

51 NASA, accessed March 31, 2019, https://cneos.jpl.nasa.gov/news/news146.html.

52 NASA, accessed March 31, 2019, https://cneos.jpl.nasa.gov/sentry/details.html#?des=99942.

53 Harold Maass, "The Odds Are 11 Million to 1 That You'll Die in a Plane Crash," image, July 8, 2013, accessed March 31, 2019, https://theweek.com/articles/462449/odds-are-11-million-1-that-youll-die-plane-crash.

54 Phil Plait, "Will the Asteroid Apophis Hit Earth in 2036? No. Seriously, No," *Slate*, January 4, 2016, accessed March 31, 2019, https://slate.com/technology/2016/01/asteroid-apophis-will-not-hit-earth-in-2036.html.

55 Emma Luxton, "Which Countries Spend the Most on Space Exploration?" World Economic Forum, accessed March 31, 2019, https://www.weforum.org/agenda/2016/01/which-countries-spend-the-most-on-space-exploration/.

56 Robert Wickramatunga, "United Nations Office for Outer Space Affairs," COPUOS, accessed March 31, 2019, http://www.unoosa.org/oosa/en/ourwork/copuos/index.html.

57 Jakub Drmola and Miroslav Mareš, "Revisiting the Deflection Dilemma," *Astronomy & Geophysics* 56, no. 5 (2015), doi:10.1093/astrogeo/atv167.

58 Avery Thompson, "New Spacecraft Will Head to Binary Asteroid to Help Protect the Planet," *Popular Mechanics*, June 25, 2018, accessed March 31, 2019. https://www.popularmechanics.com/space/solar-system/a21937219/new-spacecraft-will-head-to-binary-asteroid-to-help-protect-the-planet/.

59 Paul K. Martin, NASA Office of Inspector General, *NASA's Efforts to Identify Near-Earth Objects and Mitigate Hazards*, Washington, DC, 2014.

60 Bryan Bender, Nancy Cook, and Jacqueline Klimas, "Trump versus the Killer Asteroids," *Politico*, September 27, 2018, accessed March 31, 2019, https://www.politico.com/story/2018/09/27/trump-killer-asteroids-space-nasa-813570.

61 "More Americans View Monitoring Climate or Asteroids as Top NASA Priorities than Do so for Sending Astronauts to the Moon or Mars," Pew Research Center: Internet, Science & Technology, June 1, 2018, accessed March 31, 2019, http://www.pewinternet.org/2018/06/06/majority-of-americans-believe-it-is-essential-that-the-u-s-remain-a-global-leader-in-space/ps_06-06-18_science-space-01/.

62 David Leonard, "Russian Fireball Explosion Shows Meteor Risk Greater Than Thought," Space.com, November 1, 2013, accessed March 31, 2019, https://www.space.com/23423-russian-fireball-meteor-airburst-risk.html.

63 Ellen Barry and Andrew E. Kramer, "Shock Wave of Fireball Meteor Rattles Siberia, Injuring 1,200," *New York Times*, February 15, 2013, accessed March 31, 2019, https://www.nytimes.com/2013/02/16/world/europe/meteorite-fragments-are-said-to-rain-down-on-siberia.html.

64 Large Synoptic Survey Telescope. "LSST General Public FAQs," LSST accessed March 31, 2019, https://www.lsst.org/content/lsst-general-public-faqs.

65 "Are We Prepared for an Asteroid Headed Straight to Earth?" Eos, June 29, 2018, accessed March 31, 2019, https://eos.org/articles/are-we-prepared-for-an-asteroid-headed-straight-to-earth.

66 Jackie Snow, "The Silicon Valley Summer Camp Trying to Save Us From Extinction," *Fast Company*, August 15, 2018, accessed March 31, 2019, https://

www.fastcompany.com/40498881/nasa-silicon-valley-ai-frontier-development-lab.

67 Jackie Snow, "Artificial Intelligence Just Discovered New Planets," *MIT Technology Review*, December 14, 2017, accessed March 31, 2019, https://www.technologyreview.com/the-download/609785/artificial-intelligence-just-discovered-new-planets/.

68 Jeff Foust, "White House Releases Near Earth Object Action Plan," SpaceNews.com, June 21, 2018, accessed March 31, 2019, http://spacenews.com/white-house-releases-near-earth-object-action-plan/.

69 David Kramer, "NASA Pushes for Asteroid Detection Satellite," *Physics Today*, January 12, 2019, accessed March 31, 2019, https://physicstoday.scitation.org/do/10.1063/PT.6.2.20190102a/full/.

70 "Danger of Death!" *Economist*, February 14, 2013, accessed March 31, 2019, https://www.economist.com/graphic-detail/2013/02/14/danger-of-death.

71 Doyle Rice, "Lightning Deaths at All-time Record Low in 2017," *USA Today*, January 2, 2018, accessed March 31, 2019, https://www.usatoday.com/story/weather/2018/01/02/lightning-deaths-all-time-record-low-2017/996949001/.

72 Eric Berger, "Just How Many Humans Have Space Rocks Killed, Anyway?" SciGuy, February 15, 2013, accessed March 31, 2019, https://blog.chron.com/sciguy/2013/02/just-how-many-humans-have-space-rocks-killed-anyway/.

73 Agam Bansal, Chandan Garg, Abhijith Pakhare, and Samiksha Gupta, "Selfies: A Boon or Bane?" *Journal of Family Medicine and Primary Care* 7, no. 4 (2018): 828, doi:10.4103/jfmpc.jfmpc10918.

74 *Defending Planet Earth Near-Earth Object Surveys and Hazard Mitigation Strategies*, Washington, DC: National Academies Press, 2010.

75 David Shepardson, "2017 Safest Year on Record for Commercial Passenger Air Travel: Groups," Reuters, January 1, 2018, accessed March 31, 2019, https://www.reuters.com/article/us-aviation-safety/2017-safest-year-on-record-for-commercial-passenger-air-travel-groups-idUSKBN1EQ17L.

76 Federal Aviation Administration, *Budget Estimates Fiscal Year 2018*, Washington, DC, 2017.

77 Mitchell Hartman, "Here's How Much Money There Is in the World—and Why You've Never Heard the Exact Number," *Business Insider*, November 17, 2017, accessed March 31, 2019, https://www.businessinsider.com/heres-how-much-money-there-is-in-the-world-2017-10.

78 Richard A. Posner, *Catastrophe: Risk and Response* (Oxford: Oxford University Press, 2006).

CHAPTER 2: VOLCANO

1 "Volcano Hazards Program YVO Yellowstone," USGS, accessed March 31, 2019, https://volcanoes.usgs.gov/volcanoes/yellowstone/yellowstone_sub_page_49.html.

2 Becky Oskin, "Magma 'Pancakes' May Have Fueled Toba Supervolcano," LiveScience, October 30, 2014, accessed March 31, 2019, https://www.livescience. com/48545-toba-supervolcano-layered-sills-reservoir.html.

3 Marie D. Jones and John M. Savino, *Supervolcano: The Catastrophic Event That Changed the Course of Human History,* rev. ed., Kindle ed. (Franklin Lakes, NJ: New Page Books, 2007), p. 8, loc. 93.

4 Ibid., pp. 116–17.

5 Greg Breining, *Super Volcano: The Ticking Time Bomb Beneath Yellowstone National Park,* Kindle ed. (St. Paul, MN: Voyageur, 2010), loc. 1595.

6 Jones and Savino, *Supervolcano,* p. 114, loc. 1477.

7 Cox, David. "Future—Would a Supervolcano Eruption Wipe Us Out?" BBC, July 24, 2017, accessed March 31, 2019, http://www.bbc.com/future/story/20170724-would-a-supervolcano-eruption-wipe-us-out.

8 Charles Mann, "State of the Species," *Orion* magazine, accessed March 31, 2019, https://orionmagazine.org/article/state-of-the-species/.

9 Jones and Savino, *Supervolcano,* pp. 94–95.

10 Bill McGuire, *Global Catastrophes: A Very Short Introduction,* Kindle ed. (Oxford: Oxford University Press, 2014), p. 68, loc. 1200.

11 Wynne Parry, "20 Years After Pinatubo: How Volcanoes Could Alter Climate," LiveScience, June 9, 2011, accessed March 31, 2019, https://www.livescience. com/14513-pinatubo-volcano-future-climate-change-eruption.html.

12 Michael R. Rampino and Stephen Self, "Volcanic Winter and Accelerated Glaciation following the Toba Super-eruption," *Nature* 359, no. 6390 (1992): 50-52, doi:10.1038/359050a0.

13 Alan Robock, Caspar M. Ammann, Luke Oman, Drew Shindell, Samuel Levis, and Georgiy Stenchikov, "Did the Toba Volcanic Eruption of ~74 Ka B.P. Produce Widespread Glaciation?" *Journal of Geophysical Research* 114, no. D10 (2009), doi:10.1029/2008jd011652.

14 Ibid.

15 Phil Torres and Russell Blackford, *The End: What Science and Religion Tell Us about the Apocalypse,* Kindle ed. (Durham, NC: Pitchstone, 2016), p. 113, loc. 2570.

16 Nick Bostrom, Milan M. Ćirković, and Martin J. Rees, *Global Catastrophic Risks,* Kindle ed. (Oxford: Oxford University Press, 2018), p. 13, loc. 706.

17 "The Uniformity Problem," accessed March 31, 2019, http://www.cs.unc. edu/~plaisted/ce/uniformity.html.

18 Ann Gibbons, "Pleistocene Population Explosions," *Science,* October 1, 1993, accessed March 31, 2019, http://science.sciencemag.org/content/262/5130/27.

19 "The Extinction Crisis," Center for Biological Diversity, accessed March 31, 2019, http://www.biologicaldiversity.org/programs/biodiversity/elements_of_biodiversity/extinction_crisis/.

20 Paul R. Renne, Courtney J. Sprain, Mark A. Richards, Stephen Self, Loÿc Vanderkluysen, and Kanchan Pande, "State Shift in Deccan Volcanism at the Cretaceous-Paleogene Boundary, Possibly Induced by Impact," *Science* 350, no. 6256 (2015): 76–78, doi:10.1126/science.aac7549.

21 Sarah Zielinski, "Massive Volcanic Eruptions Triggered Earth's 'Great Dying,'" Smithsonian.com, August 28, 2015, accessed March 31, 2019, https://www.smithsonianmag.com/science-nature/massive-volcanic-eruptions-triggered-earths-great-dying-180956431/.

22 Seth D. Burgess and Samuel A. Bowring, "High-precision Geochronology Confirms Voluminous Magmatism Before, During, and after Earth's Most Severe Extinction," *Science Advances* 1, no. 7 (2015), doi:10.1126/sciadv.1500470.

23 Michael Miller, "How Do You Stop the Next Mass Extinction? Look to the Past," Phys.org, November 10, 2017, accessed March 31, 2019, https://phys.org/news/2017-11-mass-extinction.html.

24 Peter Brannen, *The Ends of the World: Volcanic Apocalypses, Lethal Oceans and Our Quest to Understand Earth's Past Mass Extinctions* (London: Oneworld, 2018).

25 C. J. N.Wilson, S. Blake, B. L. A. Charlier, and A. N. Sutton. "The 26·5 ka Oruanui Eruption, Taupo Volcano, New Zealand: Development, Characteristics and Evacuation of a Large Rhyolitic Magma Body," *Journal of Petrology* 47, no. 1 (2005): 35–69, doi:10.1093/petrology/egi066.

26 James Morgan, "Supervolcano Eruption Mystery Solved," BBC News, January 6, 2014, accessed March 31, 2019, https://www.bbc.com/news/science-environment-25598050.

27 S. Sparks, S. Self, et al., "2005 Super-eruptions: Global effects and future threats. Report of a Geological Society of London Working Group, 2nd print ed.

28 Claudia Timmreck, Hans-F. Graf, Stephan J. Lorenz, Ulrike Niemeier, Davide Zanchettin, Daniela Matei, Johann H. Jungclaus, and Thomas J. Crowley, "Aerosol Size Confines Climate Response to Volcanic Super-eruptions," *Geophysical Research Letters* 37, no. 24 (2010), doi:10.1029/2010gl045464.

29 Eugene I. Smith, Zenobia Jacobs, Racheal Johnsen, Minghua Ren, Erich C. Fisher, Simen Oestmo, Jayne Wilkins, Jacob A. Harris, Panagiotis Karkanas, Shelby Fitch, Amber Ciravolo, Deborah Keenan, Naomi Cleghorn, Christine S. Lane, Thalassa Matthews, and Curtis W. Marean, "Humans Thrived in South Africa through the Toba Eruption about 74,000 Years Ago," *Nature* 555, no. 7697 (2018): 511–15, doi:10.1038/nature25967.

30 Gillen D'Arcy Wood, *Tambora: The Eruption That Changed the World,* Kindle ed. (Princeton, NJ: Princeton University Press, 2015), p. 16.

31 Robert Evans, "Blast from the Past," Smithsonian.com, July 1, 2002, accessed March 31, 2019, https://www.smithsonianmag.com/history/blast-from-the-past-65102374/.

32 Clive Oppenheimer, "Climatic, Environmental and Human Consequences of the Largest Known Historic Eruption: Tambora Volcano (Indonesia) 1815," *Progress in Physical Geography: Earth and Environment* 27, no. 2 (2003): 230–59, doi:10.1191/0309133303pp379ra.

33 R. B. Stothers, "The Great Tambora Eruption in 1815 and Its Aftermath," *Science* 224, no. 4654 (1984): 1191–98, doi:10.1126/science.224.4654.1191.

34 Gillen D'Arcy Wood, "The Volcano That Shrouded the Earth and Gave Birth to a Monster—Issue 31: Stress," *Nautilus*, December 31, 2015, accessed March 31, 2019, http://nautil.us/issue/31/stress/the-volcano-that-shrouded-the-earth-and-gave-birth-to-a-monster.

35 Gillen D'Arcy Wood, *Tambora: The Eruption That Changed the World*, Kindle ed. (Princeton, NJ: Princeton University Press, 2015).

36 Clive Oppenheimer, "Climatic, Environmental and Human Consequences of the Largest Known Historic Eruption: Tambora Volcano (Indonesia) 1815," *Progress in Physical Geography* 27, no. 2 (2003): 230–59.

37 Gillen D'Arcy Wood, "The Volcano That Shrouded the Earth and Gave Birth to a Monster—Issue 53: Monsters," *Nautilus*, October 5, 2017, accessed March 31, 2019, http://nautil.us/issue/53/monsters/the-volcano-that-shrouded-the-earth-and-gave-birth-to-a-monster-rp.

38 Jones and Savino, *Supervolcano*, p. 61, loc. 785.

39 William J. Broad, "A Volcanic Eruption That Reverberates 200 Years Later," *New York Times*, August 24, 2015, accessed March 31, 2019, https://www.nytimes.com/2015/08/25/science/mount-tambora-volcano-eruption-1815.html.

40 Lord Byron, "Darkness," Poetry Foundation, accessed March 31, 2019, https://www.poetryfoundation.org/poems/43825/darkness-56d222aeeee1b.

41 Torres and Blackford, *The End*, p. 112, loc. 2545.

42 "After Tambora," *Economist*, April 11, 2015, accessed March 31, 2019, https://www.economist.com/briefing/2015/04/11/after-tambora.

43 Evans, "Blast from the Past."

44 Gwynn Guilford, "Iceland's Last Giant Volcanic Eruption Cost the Global Economy $5bn. Is That About to Happen Again?" Quartz, August 19, 2014, accessed March 31, 2019, https://qz.com/251927/icelands-last-giant-volcanic-eruption-cost-the-global-economy-5bn-is-that-about-to-happen-again/.

45 C. Newhall, S. Self, and A. Robock, "Anticipating Future Volcanic Explosivity Index (VEI) 7 Eruptions and Their Chilling Impacts," *Geosphere* 14, no. 2 (2018).

46 Joseph G. Manning, Francis Ludlow, Alexander R. Stine, William R. Boos, Michael Sigl, and Jennifer R. Marlon, "Volcanic Suppression of Nile Summer Flooding Triggers Revolt and Constrains Interstate Conflict in Ancient Egypt," *Nature Communications* 8, no. 1 (2017), doi:10.1038/s41467-017-00957-y.

47 Ann Gibbons, "Why 536 Was 'the Worst Year to Be Alive,'" *Science*, November 16, 2018, accessed March 31, 2019, https://www.sciencemag.org/news/2018/11/why-536-was-worst-year-be-alive.

48 Chris Newhall, Stephen Self, and Alan Robock. 2018. "Anticipating Future Volcanic Explosivity Index (VEI) 7 Eruptions and Their Chilling Impacts." Geosphere14 (2): 572–603, https://pubs.geoscienceworld.org/gsa/geosphere/article/14/2/572/529016/anticipating-future-volcanic-explosivity-index-vei.

49 Morgan Warthin, "2017—Second Busiest Year on Record," National Parks Service, accessed March 31, 2019, https://www.nps.gov/yell/learn/news/18002.htm.

50 "Park Geology," Yellowstone Geology, accessed March 31, 2019, https://yellowstone.net/geology/.

51 Jones and Savino, *Supervolcano,* p. 191, loc. 2461.

52 Larry G. Mastin, Alexa R. Van Eaton, and Jacob B. Lowenstern, "Modeling Ash Fall Distribution from a Yellowstone Supereruption," *Geochemistry, Geophysics, Geosystems* 15, no. 8 (2014): 3459–75, doi:10.1002/2014gc005469.

53 Jones and Savino, *Supervolcano,* p. 29, loc. 367.

54 Mastin, Van Eaton, and Lowenstern. "Modeling Ash Fall Distribution from a Yellowstone Supereruption."

55 Jones and Savino, *Supervolcano,* p. 224, loc. 2899.

56 USGS, "Volcanic Ash Impacts & Mitigation—Transmission & Distribution," accessed March 31, 2019, https://volcanoes.usgs.gov/volcanic_ash/transmission_distribution.html.

57 Matthew L. Wald, "A Drill to Replace Crucial Transformers (Not the Hollywood Kind)," *New York Times,* March 15, 2012, accessed March 31, 2019, https://www.nytimes.com/2012/03/15/business/energy-environment/electric-industry-runs-transformer-replacement-test.html.

58 USGS, "Volcanic Ash Impacts & Mitigation—Agriculture—Plants & Animals," accessed March 31, 2019, https://volcanoes.usgs.gov/volcanic_ash/agriculture.html.

59 "Yellowstone Eruption: The Zones—Doc Zone—CBC-TV," CBCnews, accessed March 31, 2019, http://www.cbc.ca/doczone/features/the-zones.

60 "The 10 Most Expensive US Natural Disasters," *Christian Science Monitor,* June 27, 2013, accessed March 31, 2019, https://www.csmonitor.com/Environment/2013/0627/Billion-dollar-weather-The-10-most-expensive-US-natural-disasters/Hurricane-Katrina-August-2005-161.3-billion.

61 Robert B. Smith and Lee J. Siegel, *Windows into the Earth: The Geologic Story of Yellowstone and Grand Teton National Parks,* Kindle ed. (Oxford: Oxford University Press, 2000), loc. 2160.

62 C. Timmreck and H.-F. Graf, "The Initial Dispersal and Radiative Forcing of a Northern Hemisphere Mid-latitude Super Volcano: A Model Study," *Atmospheric Chemistry and Physics* 6, no. 1 (2006): 35–49, doi:10.5194/acp-6-35-2006.

63 John Vidal, "UN Warns of Looming Worldwide Food Crisis in 2013," *Guardian,* October 13, 2012, accessed March 31, 2019, https://www.theguardian.com/global-development/2012/oct/14/un-global-food-crisis-warning.

64 "World Hunger Statistics," Food Aid Foundation, accessed March 31, 2019, http://www.foodaidfoundation.org/world-hunger-statistics.html.

65 Hans-PeterPlag et al. *Extreme Geohazards: Reducing Disaster Risk and Increasing Resilience,* European Science Foundation, 2015.

66 Sebastian Farquhar et al., *Existential Risk: Diplomacy and Governance*, Global Priorities Project, 2017.

67 USGS, "Volcano Hazards Program YVO Yellowstone," accessed March 31, 2019, https://volcanoes.usgs.gov/volcanoes/yellowstone/faqs_supervolcanoes.html.

68 S. Sparks, S. Self, et al., "2005: Super-eruptions: global effects and future threats," Report of Geological Society of London Working Group, 2nd print ed.

69 Department of the Interior, *USGS Funding*, Washington, DC, 2018.

70 Plag et al. *Extreme Geohazards*.

71 Brian Wilcox et al., *Defending Human Civilization from Supervolcanic Eruptions*, Jet Propulsion Laboratory, California Institute of Technology, 2015.

72 Eric Klemetti, "No, NASA Isn't Going to Drill to Stop Yellowstone from Erupting," Rocky Planet, October 11, 2017, accessed March 31, 2019, http://blogs.discovermagazine.com/rockyplanet/2017/08/31/2250904/#.WzD6UhJKi8o.

73 Sparks, Self, et al., "2005: Super-eruptions: global effects and future threats."

74 Jonathan Rougier, R. Stephen, J. Sparks, Katharine V. Cashman, and Sarah K. Brown, "The Global Magnitude–frequency Relationship for Large Explosive Volcanic Eruptions," *Earth and Planetary Science Letters* 482 (2018): 621–29, doi:10.1016/j.epsl.2017.11.015.

75 Ibid.

76 Milan M. Ćirković, Anders Sandberg, and Nick Bostrom, "Anthropic Shadow: Observation Selection Effects and Human Extinction Risks," *Risk Analysis* 30, no. 10 (2010): 1495–1506, doi:10.1111/j.1539-6924.2010.01460.x.

CHAPTER 3: NUCLEAR

1 Jim Eckles, discussion with author, April 2018.

2 "Manhattan Project: The Manhattan Project and the Second World War, 1939–1945," accessed March 31, 2019, https://www.osti.gov/opennet/manhattan-project-history/Events/1945/retrospect.htm.

3 Alex Wellerstein, "The Light of Trinity, the World's First Nuclear Bomb," *New Yorker,* July 12, 2018, accessed March 31, 2019, https://www.newyorker.com/tech/elements/the-first-light-of-the-trinity-atomic-test.

4 Ferenc Morton Szasz, *The Day the Sun Rose Twice: The Story of the Trinity Site Nuclear Explosion, July 16, 1945*, Kindle ed. (Albuquerque: University of New Mexico Press, 1995), p. 147, loc. 2127.

5 Ibid., p. 65, loc. 1020.

6 Wellerstein, "The Light of Trinity."

7 Peter Goodchild, "Meet the Real Dr Strangelove," *Guardian,* April 1, 2004, accessed March 31, 2019, https://www.theguardian.com/science/2004/apr/01/science.research1.

8 Kabir Chibber, "Meet the People Out to Stop Humanity from Destroying Itself," Quartz, May 13, 2015, accessed March 31, 2019, https://qz.com/386637/meet-the-people-trying-to-prevent-humanity-from-destroying-itself.

9   Szasz, *The Day the Sun Rose Twice,* p. 57.

10 "Manhattan Project. LA-602: Ignition of the Atmosphere with Nuclear Bombs, 1946." Available at http://www.fas.org/sgp/othergov/doe/lanl/docs1/00329010. pdf, accessed March 24, 2011.

11 Szasz, *The Day the Sun Rose Twice,* p. 58.

12 Richard Rhodes, *The Making of the Atomic Bomb,* Kindle ed. (New York: Simon & Schuster, 2012), locs. 13882–85.

13 Ibid., locs. 13888–89.

14 Manhattan Project: The Trinity Test, July 16, 1945, accessed March 31, 2019, https://www.osti.gov/opennet/manhattan-project-history/Events/1945/trinity.htm.

15 "On the 50th Anniversary of the Atomic Bomb," Radiation at Trinity Site, History of the Atomic Age, accessed March 31, 2019, http://www.atomicarchive.com/History/trinity/radiation.shtml.

16 Szasz, *The Day the Sun Rose Twice,* p. 117, loc. 1592.

17 Trinity Atomic website, accessed March 31, 2019, https://www.abomb1.org/trinity/trinity1.html.

18 Anna E. Holmes, "A Crimson Fracture in the Sky," Topic, March 18, 2019, accessed March 31, 2019, https://www.topic.com/a-crimson-fracture-in-the-sky.

19 Szasz, *The Day the Sun Rose Twice,* p. 83.

20 Rhodes, *The Making of the Atomic Bomb,* locs. 14135–37.

21 Ibid., loc. 14441.

22 Ibid., loc. 14811.

23 Ibid., loc. 14950.

24 "The Nobel Peace Prize 2017," NobelPrize.org, accessed March 31, 2019, https://www.nobelprize.org/prizes/peace/2017/ican/lecture/.

25 Ken Ringle, "A Fallout over Numbers," *Washington Post,* August 5, 1995, accessed March 31, 2019, https://www.washingtonpost.com/archive/politics/1995/08/05/a-fallout-over-numbers/d9c5fb21-880b-4c6c-85f1-b80e16a0a0ee/?utm_term=.ea5c24c60040.

26 Eric Schlosser, "The Growing Dangers of the New Nuclear-Arms Race," *New Yorker,* May 24, 2018, accessed March 31, 2019, https://www.newyorker.com/news/news-desk/the-growing-dangers-of-the-new-nuclear-arms-race.

27 Rhodes, *The Making of the Atomic Bomb,* loc. 107.

28 "Fact Sheets & Briefs," Nuclear Testing Tally, Arms Control Association, accessed March 31, 2019, https://www.armscontrol.org/factsheets/nucleartesttally.

29 Schlosser, "The Growing Dangers of the New Nuclear-Arms Race."

30 Fred Kaplan, "Daniel Ellsberg's Memoir About Life as a Nuclear War Planner Would Be Terrifying Even If Trump Weren't President," *Slate,* December 4, 2017, accessed March 31, 2019, http://www.slate.com/articles/arts/books/2017/12/the_doomsday_machine_daniel_ellsberg_s_sobering_new_memoir_about_life_as.html.

31 *Life*, September 15, 1961.

32 Raymond J. Mauer, *Duck and Cover*. Directed by Anthony Rizzo. Archer
    Productions, 1952.

33 "Neither Snow, Nor Sleet, Nor Heat, Nor Fallout…." *Bulletin of the Atomic
    Scientists*, September 1998.

34 Zachary Keck, "America's Insane Plan to Survive a Russian Nuclear Attack,"
    National Interest, July 6, 2017, accessed March 31, 2019, http://nationalinterest.
    org/blog/the-buzz/americas-insane-plan-survive-soviet-nuclear-attack-21399.

35 Lee Clarke's Mission Improbable: Using Fantasy Documents to Tame Disaster,
    accessed March 31, 2019, http://leeclarke.com/mipages/guillen.html.

36 Philip Boffey, "Social Scientists Believe Leaders Lack a Sense of War's Reality,"
    New York Times, September 7, 1982, accessed March 31, 2019, https://www.
    nytimes.com/1982/09/07/science/social-scientists-believe-leaders-lack-a-sense-
    of-war-s-reality.html.

37 Scott Slovic and Paul Slovic, "The Arithmetic of Compassion," *New York
    Times*, December 4, 2015, accessed March 31, 2019, https://www.nytimes.
    com/2015/12/06/opinion/the-arithmetic-of-compassion.html.

38 Shankar Vedantam, "Mass Suffering and Why We Look the Other Way,"
    Washington Post, January 5, 2009, accessed March 31, 2019, http://www.
    washingtonpost.com/wp-dyn/content/article/2009/01/04/AR2009010401307.
    html.

39 Robert Jay Lifton, *The Climate Swerve: Reflections on Mind, Hope, and Survival*,
    Kindle ed. (New York: New Press, 2017), loc. 458.

40 Herman Kahn, *On Thermonuclear War* (Princeton, NJ: Princeton University
    Press, 1960).

41 Kaplan, "Daniel Ellsberg's Memoir."

42 Daniel Ellsberg, *The Doomsday Machine: Confessions of a Nuclear War Planner*
    (New York: Bloomsbury, 2017).

43 Louis Menand, "Fat Man," *New Yorker*, June 18, 2017, accessed March 31, 2019,
    https://www.newyorker.com/magazine/2005/06/27/fat-man.

44 Hans M. Kristensen and Robert S. Norris, "Global Nuclear Weapons
    Inventories, 1945–2013," *Bulletin of the Atomic Scientists* 69, no. 5 (2013): 75–81,
    doi:10.1177/0096340213501363.

45 Yasmeen Serhan, "When the World Lucked Out of a Nuclear War," *Atlantic*,
    October 31, 2017, accessed March 31, 2019, https://www.theatlantic.com/
    international/archive/2017/10/when-the-world-lucked-out-of-nuclear-
    war/544198/.

46 Raffi Khatchadourian, "The Doomsday Invention," *New Yorker*, September 7,
    2017, accessed March 31, 2019, http://www.newyorker.com/magazine/2015/11/23/
    doomsday-invention-artificial-intelligence-nick-bostrom.

47 Tucker Davey, "55 Years After Preventing Nuclear Attack, Arkhipov Honored
    with Inaugural Future of Life Award," Future of Life Institute, June 5, 2018,

accessed March 31, 2019, https://futureoflife.org/2017/10/27/55-years-preventing-nuclear-attack-arkhipov-honored-inaugural-future-life-award/.

48 Baker Spring, "What Americans Need to Know About Missile Defense: We're Not There Yet," Heritage Foundation, accessed March 31, 2019, https://www.heritage.org/missile-defense/report/what-americans-need-know-about-missile-defense-were-not-there-yet.

49 Pavel Aksenov, "Stanislav Petrov: The Man Who May Have Saved the World," BBC News, September 26, 2013, accessed March 31, 2019, https://www.bbc.com/news/world-europe-24280831.

50 Khatchadourian, "The Doomsday Invention."

51 "JFK on Nuclear Weapons and Non-Proliferation," Carnegie Endowment for International Peace, accessed March 31, 2019, https://carnegieendowment.org/2003/11/17/jfk-on-nuclear-weapons-and-non-proliferation-pub-14652.

52 "New Evidence on the Origins of Overkill," Nuclear Vault, accessed March 31, 2019, https://nsarchive2.gwu.edu/nukevault/ebb236/.

53 *Worldwide Effects of Nuclear War*, Washington, DC. U.S. Arms Control and Disarmament Agency, 1975.

54 R. P. Turco, O. B. Toon, T. P. Ackerman, J. B. Pollack, and C. Sagan. "Nuclear Winter: Global Consequences of Multple Nuclear Explosions," *Science* 222, no. 4630 (1983): 1283–92, doi:10.1126/science.222.4630.1283.

55 Jill Lepore, "The Atomic Origins of Climate Science," *New Yorker*, October 12, 2018, accessed March 31, 2019, https://www.newyorker.com/magazine/2017/01/30/the-atomic-origins-of-climate-science.

56 Carl Sagan, *Pale Blue Dot: A Vision of the Human Future in Space* (New York: Random House, 1994).

57 Turco et al., "Nuclear Winter."

58 Alan Robock, Luke Oman, and Georgiy L. Stenchikov, "Nuclear Winter Revisited with a Modern Climate Model and Current Nuclear Arsenals: Still Catastrophic Consequences," *Journal of Geophysical Research: Atmospheres* 112, no. D13 (2007), doi:10.1029/2006jd008235.

59 "Richard P. Turco," RSS, accessed March 31, 2019, https://www.macfound.org/fellows/288/.

60 Kristensen and Norris. "Global Nuclear Weapons Inventories, 1945–2013."

61 John F. Harris and Bryan Bender, "Bill Perry Is Terrified. Why Aren't You?" *Politico*, January 6, 2017, accessed March 31, 2019, https://www.politico.com/magazine/story/2017/01/william-perry-nuclear-weapons-proliferation-214604.

62 "A New Era," *Bulletin of the Atomic Scientists,* December 1991.

63 Adam Nagourney, David E. Sanger, and Johana Barr, "Hawaii Panics After Alert About Incoming Missile Is Sent in Error," *New York Times,* January 13, 2018, accessed March 31, 2019, https://www.nytimes.com/2018/01/13/us/hawaii-missile.html.

64 Tulsi Gabbard, "HAWAII - THIS IS A FALSE ALARM. THERE IS NO
INCOMING MISSILE TO HAWAII. I HAVE CONFIRMED WITH OFFICIALS
THERE IS NO INCOMING MISSILE. Pic.twitter.com/DxfTXIDOQs," Twitter,
January 13, 2018, accessed March 31, 2019, https://twitter.com/TulsiGabbard/
status/952243723525677056.

65 Ankit Panda, "False Alarms of the Apocalypse," *Atlantic,* January 14,
2018, accessed March 31, 2019, https://www.theatlantic.com/international/
archive/2018/01/what-the-hell-happened-in-hawaii/550514/.

66 Madison Park, "Here's What Went Wrong with the Hawaii False Alarm," CNN,
January 31, 2018, accessed March 31, 2019, https://www.cnn.com/2018/01/31/
us/hawaii-false-alarm-investigation-findings/index.html.

67 "Notes on the Hawaii False Alarm, One Year Later," Restricted Data: The
Nuclear Secrecy Blog, accessed March 31, 2019, http://blog.nuclearsecrecy.
com/2019/01/13/notes-on-the-hawaii-false-alarm-one-year-later/.

68 "North Korea Nuclear Timeline Fast Facts," CNN, March 22, 2019, accessed
March 31, 2019, https://www.cnn.com/2013/10/29/world/asia/north-korea-
nuclear-timeline—fast-facts/index.html.

69 Jaweed Kaleem, Raoul Rañoa, and Angelica Quintero, "As North Korean Threat
Grows, Hawaii Prepares for Nuclear Attack," *Los Angeles Times*, November 10,
2017, accessed March 31, 2019, http://www.latimes.com/nation/la-na-hawaii-
nuke-2017-story.html.

70 "Amid an Escalating War of Words, Let's Check In on Nuclear Stockpiles,"
NBCNews.com, accessed March 31, 2019, https://www.nbcnews.com/news/
us-news/how-many-nuclear-weapons-exist-united-nations-calls-total-
elimination-n804721.

71 Chronological Listing of Above Ground Nuclear Detonations: Explanation and
Summary, accessed March 31, 2019, http://www.johnstonsarchive.net/nuclear/
atest00.html.

72 "City to Remove Outdated Fallout Shelter Signs," WNYC, accessed March 31,
2019, https://www.wnyc.org/story/city-remove-outdated-fallout-shelter-signs/.

73 Jeet Heer, "Is This the Dawn of a New Era of Nuclear Proliferation?" *New
Republic,* May 15, 2018, accessed March 31, 2019, https://newrepublic.com/
article/148413/dawn-new-era-nuclear-proliferation.

74 "JFK on Nuclear Weapons and Non-Proliferation," Carnegie Endowment for
International Peace, accessed March 31, 2019, https://carnegieendowment.
org/2003/11/17/jfk-on-nuclear-weapons-and-non-proliferation-pub-14652.

75 Thomas C. Frohlich and Alexander Kent,"Countries Spending the Most on
the Military," *USA Today,* July 12, 2014, accessed March 31, 2019, https://www.
usatoday.com/story/money/business/2014/07/12/countries-spending-most-on-
military/12491639/.

76 *Nuclear Posture Review*, Department of Defense, Washington, DC, 2018.

77 Ellis Mishulovich, "Why Does Russia Build So Many Doomsday Weapons?"
National Review, April 19, 2018, accessed March 31, 2019, https://www.
nationalreview.com/2018/04/russian-doomsday-weapons-cold-war-history/.

78 Vladimir Putin, "Presidential Address to the Federal Assembly," President of Russia, March 1, 2018, accessed March 31, 2019, http://en.kremlin.ru/events/president/news/56957.

79 Eliot Marshall, "More Precise U.S. Nukes Could Raise Tensions with Russia," *Science*, December 8, 2017, accessed March 31, 2019, https://www.sciencemag.org/news/2017/03/more-precise-us-nukes-could-raise-tensions-russia?utm_campaign=news_daily_2017-03-22&et_rid=289403812&et_cid=1231346.

80 David E. Sanger and William J. Broad. "To Counter Russia, U.S. Signals Nuclear Arms Are Back in a Big Way," *New York Times*, February 5, 2018, accessed March 31, 2019, https://www.nytimes.com/2018/02/04/us/politics/trump-nuclear-russia.html.

81 Kate Hudson, "Trump's Nuclear Posture Review Makes War More Likely," *Huffington Post UK,* April 2, 2018, accessed March 31, 2019, https://www.huffingtonpost.co.uk/entry/trumps-nuclear-posture-review-makes-war-more-likely_uk_5a749795e4b0fc3e14a4e856?guccounter=1.

82 *Defense Primer: Command and Control of Nuclear Forces*, Congressional Research Service, Washington, DC, 2018.

83 "U.S. Nuclear General Says Would Resist 'illegal' Trump Strike Order," Reuters, November 19, 2017, accessed March 31, 2019, https://www.reuters.com/article/us-usa-nuclear-commander/u-s-nuclear-general-says-would-resist-illegal-trump-strike-order-idUSKBN1DI0QV.

84 Kat Eschner, "How the Presidency Took Control of America's Nuclear Arsenal," Smithsonian.com, January 5, 2018, accessed March 31, 2019, https://www.smithsonianmag.com/history/how-the-presidency-took-control-americas-nuclear-arsenal-180967747/.

85 "Texts of Accounts by Lucas and Considine on Interviews with MacArthur in 1954," *New York Times,* April 9, 1964, accessed March 31, 2019, https://www.nytimes.com/1964/04/09/archives/texts-of-accounts-by-lucas-and-considine-on-interviews-with.html.

86 David E. Sanger, "U.S. General Considered Nuclear Response in Vietnam War, Cables Show," *New York Times,* October 6, 2018, accessed March 31, 2019, https://www.nytimes.com/2018/10/06/world/asia/vietnam-war-nuclear-weapons.html?action=click&module=Top Stories&pgtype=Homepage.

87 Richard Rhodes, "Absolute Power," *New York Times*, March 21, 2014, accessed March 31, 2019, https://www.nytimes.com/2014/03/23/books/review/thermonuclear-monarchy-by-elaine-scarry.html.

88 Annie Karni, Michael Crowley, and Louis Nelson, "Washington's Growing Obsession: The 25th Amendment," *Politico*, January 4, 2018, accessed March 31, 2019, https://www.politico.com/story/2018/01/03/trump-25th-amendment-mental-health-322625.

89 Uri Friedman, "'Then What Happens?': Congress Questions the President's Authority to Wage Nuclear War," *Atlantic*, November 15, 2017, accessed March 31, 2019, https://www.theatlantic.com/international/archive/2017/11/trump-nuclear-weapons-senate/545846/.

90 Christine Blackman, "Chance of Nuclear War Is Greater than You Think: Stanford Engineer Makes Risk Analysis," Stanford University, July 17, 2009, accessed March 31, 2019, https://news.stanford.edu/news/2009/july22/hellman-nuclear-analysis-071709.html.

91 Martin J. Rees, *Our Final Hour: A Scientist's Warning: How Terror, Error, and Environmental Disaster Threaten Humankind's Future in This Century on Earth and Beyond* (New York: Basic Books, 2003).

CHAPTER 4: CLIMATE CHANGE

1 NASA, accessed April 1, 2019, https://science.nasa.gov/earth-science/oceanography/living-ocean.

2 I. Joughin and B. E. Smith, "Further Summer Speedup of Jakobshavn Isbræ," *Cryosphere Discussions* 7, no. 6 (2013): 5461–73, doi:10.5194/tcd-7-5461-2013.

3 Ala Khazendar, Ian G. Fenty, Dustin Carroll, Alex Gardner, Craig M. Lee, Ichiro Fukumori, Ou Wang, Hong Zhang, Hélène Seroussi, Delwyn Moller, Brice P. Y. Noël, Michiel R. Van Den Broeke, Steven Dinardo, and Josh Willis, "Interruption of Two Decades of Jakobshavn Isbrae Acceleration and Thinning as Regional Ocean Cools," *Nature Geoscience* 12, no. 4 (2019): 277–83, doi:10.1038/s41561-019-0329-3.

4 "Greenland Ice Sheet," *Encyclopædia Britannica*, April 16, 2018, accessed March 31, 2019, https://www.britannica.com/place/Greenland-Ice-Sheet.

5 Bryan Walsh, "Unfrozen Tundra," *Time*, September 25, 2008, accessed March 31, 2019, https://content.time.com/time/magazine/article/0,9171,1844549,00.html.

6 "Climate Change Evidence: How Do We Know?" NASA, March 26, 2019, accessed March 31, 2019, https://climate.nasa.gov/evidence/.

7 Ibid.

8 Jason Samenow, "North Pole Surges above Freezing in the Dead of Winter, Stunning Scientists," *Washington Post,* February 26, 2018, accessed March 31, 2019, https://www.washingtonpost.com/news/capital-weather-gang/wp/2018/02/26/north-pole-surges-above-freezing-in-the-dead-of-winter-stunning-scientists/.

9 Nick Watts et al, "The 2018 Report of the *Lancet* Countdown on Health and Climate Change: Shaping the Health of Nations for Centuries to Come," *Lancet* 392, no. 10163 (2018): 2479–2514, doi.org/10.1016/S0140-6736(18)32594-7.

10 Doyle Rice, "Climate Change's Impact on Human Health Is Already Here —and Is 'Potentially Irreversible,' Report Says," *USA Today,* October 31, 2017, accessed March 31, 2019, https://www.usatoday.com/story/news/health/2017/10/30/climate-changes-impact-human-health-already-here-and-potentially-irreversible-report-says/815123001/.

11 Umair Irfan, "California's Wildfires Are Hardly "Natural"—Humans Made Them Worse at Every Step," Vox, November 19, 2018, accessed March 31, 2019, https://www.vox.com/2018/8/7/17661096/california-wildfires-2018-camp-woolsey-climate-change.

12 Global Warming of 1.5 °C, accessed March 31, 2019, https://www.ipcc.ch/sr15/.

13 Jeff Goodell, "Welcome to the Age of Climate Migration," *Rolling Stone,* June 25, 2018, accessed March 31, 2019, https://www.rollingstone.com/politics/politics-news/welcome-to-the-age-of-climate-migration-202221/.

14 Jonathan Watts, "We Have 12 Years to Limit Climate Change Catastrophe, Warns UN," *Guardian,* October 8, 2018, accessed March 31, 2019, https://www.theguardian.com/environment/2018/oct/08/global-warming-must-not-exceed-15c-warns-landmark-un-report.

15 John Schwartz, "Greenland's Melting Ice Nears a 'Tipping Point,' Scientists Say," *New York Times,* January 21, 2019, accessed March 31, 2019, https://www.nytimes.com/2019/01/21/climate/greenland-ice.html.

16 Mirko Orlić and Zoran Pasarić, "Semi-Empirical versus Process-Based Sea-Level Projections for the Twenty-First Century," *Nature Climate Change* 3, no. 8 (2013), 735–38, doi:10.1038/nclimate1877.

17 Elizabeth Kolbert, *The Sixth Extinction: An Unnatural History* (New York: Henry Holt, 2014).

18 Gernot Wagner and Martin Weitzman, *Climate Shock: The Economic Consequences of a Hotter Planet,* Kindle ed. (Princeton, NJ: Princeton University Press, 2016), loc. 1032.

19 "Climate Change Conference Opens in Bali," *Sydney Morning Herald,* December 3, 2007, http://www.smh.com.au/environment/climate-change-conference-opens-in-bali-20071204-gdrqkg.html.

20 Nigel Morris, "All Roads Lead to Copenhagen as 100 World Leaders Head to Summit," *Independent,* October 22, 2011, http://www.independent.co.uk/environment/climate-change/all-roads-lead-to-copenhagen-as-100-world-leaders-head-to-summit-1834631.html?action=Popup.

21 "CO2 Emissions (Metric Tons per Capita)," *Data,* data.worldbank.org/indicator/EN.ATM.CO2E.PC.

22 "World Development Indicators," DataBank, databank.worldbank.org/data/reports.aspx?source=2&series=EN.ATM.CO2E.KT&country=.

23 "Developing Countries Are Responsible for 63 Percent of Current Carbon Emissions," Center for Global Development, http://www.cgdev.org/media/developing-countries-are-responsible-63-percent-current-carbon-emissions.

24 Lisa Lerer, "Obama's Dramatic Climate Meet," *Politico,* December 19, 2009, http://www.politico.com/story/2009/12/obamas-dramatic-climate-meet-030801.

25 "Information Provided by Parties to the Convention Relating to the Copenhagen Accord," UNFCCC, unfccc.int/process/conferences/pastconferences/copenhagen-climate-change-conference-december-2009/statements-and-resources/information-provided-by-parties-to-the-convention-relating-to-the-copenhagen-accord.

26 Robert Rapier, "China Emits More Carbon Dioxide than the U.S. and EU Combined," *Forbes,* July 1, 2018, http://www.forbes.com/sites/

rrapier/2018/07/01/china-emits-more-carbon-dioxide-than-the-u-s-and-eu-combined/#7b401329628c.

27 "The Paris Agreement," UNFCCC, unfccc.int/process-and-meetings/the-paris-agreement/the-paris-agreement.

28 "Statement by the President on the Paris Climate Agreement," National Archives and Records Administration, obamawhitehouse.archives.gov/the-press-office/2015/12/12/statement-president-paris-climate-agreement.

29 Anne C. Mulkern, "COPENHAGEN: D.C. Blizzard Makes for a Rotten Time in Denmark," December 21, 2009, http://www.eenews.net/stories/85935.

30 Bryan Walsh, "COP15: Climate-Change Conference," *Time*, December 21, 2009, content.time.com/time/specials/packages/article/0,28804,1929071_1929070_1949054,00.html.

31 "Statement by President Trump on the Paris Climate Accord," White House, www.whitehouse.gov/briefings-statements/statement-president-trump-paris-climate-accord/.

32 "Report: World Must Cut 25% from Predicted 2030 Emissions," United Nations, http://www.un.org/sustainabledevelopment/blog/2016/11/report-world-must-cut-further-25-from-predicted-2030-emissions/.

33 "How a 2 C Temperature Increase Could Change the Planet," *CBCnews*, CBC/Radio Canada, http://www.cbc.ca/news2/interactives/2degrees/?utm_source=hootsuite.

34 Brad Plumer and Nadja Popovich, "Here's How Far the World Is from Meeting Its Climate Goals," *New York Times*, November 6, 2017, http://www.nytimes.com/interactive/2017/11/06/climate/world-emissions-goals-far-off-course.html.

35 Brady Dennis and Chris Mooney, "'We Are in Trouble.' Global Carbon Emissions Reached a Record High in 2018," *Washington Post*, December 5, 2018, http://www.washingtonpost.com/energy-environment/2018/12/05/we-are-trouble-global-carbon-emissions-reached-new-record-high/?utm_term=.392aa91e4039.

36 "Global Energy & CO2 Status Report," GECO 2019, http://www.iea.org/geco/?utm_content=buffer9da4a&utm_medium=social&utm_source=twitter.com&utm_campaign=buffer.

37 "Fossil Fuel Energy Consumption (% of Total)," *Data*, data.worldbank.org/indicator/eg.use.comm.fo.zs.

38 K. Caldeira, "Climate Sensitivity Uncertainty and the Need for Energy Without CO2 Emission," *Science* 299, no. 5615 (2003): 2052–54, doi:10.1126/science.1078938.

39 James Temple, "At This Rate, It's Going to Take Nearly 400 Years to Transform the Energy System," *MIT Technology Review*, February 20, 2019, http://www.technologyreview.com/s/610457/at-this-rate-its-going-to-take-nearly-400-years-to-transform-the-energy-system/.

40 Global Warming of 1.5° C, accessed March 31, 2019, https://www.ipcc.ch/sr15/.

41 Concordia University, "Carbon Emissions Linked to Global Warming in

Simple Linear Relationship," *ScienceDaily*, http://www.sciencedaily.com/releases/2009/06/090610154453.htm (accessed March 31, 2019).

42 "Renewable Energy," Energy Economics, BP Global, http://www.bp.com/en/global/corporate/energy-economics/statistical-review-of-world-energy/renewable-energy.html.

43 "The Montreal Protocol on Substances That Deplete the Ozone Layer," U.S. Department of State, http://www.state.gov/e/oes/eqt/chemicalpollution/83007.htm.

44 Michael Weisskopf, "CFCs Rise and Fall of Chemical 'Miracle," *Washington Post*, April 10, 1988, http://www.washingtonpost.com/archive/politics/1988/04/10/cfcs-rise-and-fall-of-chemical-miracle/9dc7f67b-8ba9-4e11-b247-a36337d5a87b/?utm_term=.c29eda325187.

45 "Du Pont Studies CFC Substitutes," *Christian Science Monitor*, August 9, 1989, http://www.csmonitor.com/1989/0809/apont.html.

46 Cass R. Sunstein, "Of Montreal and Kyoto: A Tale of Two Protocols," *ELR News & Analysis* August, 2018.

47 Paul Voosen, "Meet Vaclav Smil, the Man Who Has Quietly Shaped How the World Thinks about Energy," *Science*, March 21, 2018, http://www.sciencemag.org/news/2018/03/meet-vaclav-smil-man-who-has-quietly-shaped-how-world-thinks-about-energy?linkId=50146334.

48 "Al Gore's Speech on Renewable Energy," NPR, July 17, 2008, http://www.npr.org/templates/story/story.php?storyId=92638501.

49 Hannah Ritchie and Max Roser, "Energy Production & Changing Energy Sources," Our World in Data, March 28, 2014, ourworldindata.org/energy-production-and-changing-energy-sources.

50 *BP Statistical Review of World Energy*, London, June 2018.

51 Ibid.

52 "World Energy Outlook 2018," http://www.iea.org/weo2018/.

53 "Global Greenhouse Gas Emissions Data," Environmental Protection Agency, April 13, 2017, http://www.epa.gov/ghgemissions/global-greenhouse-gas-emissions-data.

54 Voosen, "Meet Vaclav Smil."

55 "7 Million Premature Deaths Annually Linked to Air Pollution," World Health Organization, March 25, 2014, http://www.who.int/mediacentre/news/releases/2014/air-pollution/en/.

56 Emma Foehringer Merchant, "2017 Was Another Record-Busting Year for Renewable Energy, but Emissions Still Increased," Greentech Media, June 4, 2018, http://www.greentechmedia.com/articles/read/2017-another-record-busting-year-for-global-renewable-energy-capacity#gs.BBznDOc.

57 David L. Chandler and MIT News Office, "Explaining the Plummeting Cost of Solar Power," *MIT News*, November 20, 2018, news.mit.edu/2018/explaining-dropping-solar-cost-1120.

58 Vaclav Smil, *Global Catastrophes and Trends: The Next Fifty Years,* Kindle ed. (Cambridge, MA: MIT Press, 2013), p. 77, loc. 1339.

59 Dan Reicher, Jeff Brown, and David Fedor. *Derisking Carbonization: Making Green Energy Investments Blue Chip*, Stanford Steyer-Taylor Center for Energy Policy and Finance, October 27, 2017.

60 "Global Military Spending Remains High at $1.7 Trillion," SIPRI, http://www. sipri.org/media/press-release/2018/global-military-spending-remains-high-17-trillion.

61 Bill McKibben, "A World at War," *New Republic*, August 15 2016, newrepublic. com/article/135684/declare-war-climate-change-mobilize-wwii.

62 "Life on Earth Is Under Assault—But There's Still Hope," *National Geographic*, March 23, 2018, news.nationalgeographic.com/2018/03/ipbes-biodiversity-report-conservation-climate-change-spd/.

63 "Living Planet Report 2014," World Wildlife Fund, http://www.worldwildlife. org/pages/living-planet-report-2014.

64 Elizabeth Boakes and David Redding, "Earth's Species Are Now Going Extinct 1,000 Times Faster than the Natural Rate," *ScienceAlert*, http://www. sciencealert.com/biodiversity-loss-extinction-natural-rate-1000-times-faster-habitat-loss.

65 Malcolm L. Mccallum, "Amphibian Decline or Extinction? Current Declines Dwarf Background Extinction Rate," *Journal of Herpetology* 41, no. 3 (2007): 483–91, doi:10.1670/0022-1511(2007)41[483:adoecd]2.0.co;2.

66 L. Lebreton et al., "Evidence That the Great Pacific Garbage Patch Is Rapidly Accumulating Plastic," *Scientific Reports* 8, no. 1 (2018), doi:10.1038/s41598-018-22939-w.

67 Chris Mooney, "Humans Are Putting 8 Million Metric Tons of Plastic in the Oceans—Annually," *Washington Post*, February 12, 2015, http://www. washingtonpost.com/news/energy-environment/wp/2015/02/12/humans-are-putting-8-million-metric-tons-of-plastic-in-the-oceans-annually/.

68 Angus Maddison, *Contours of the World Economy, 1-2030 AD: Essays in Macro-Economic History* (Oxford: Oxford University Press, 2013).

69 "Overview," World Bank, http://www.worldbank.org/en/topic/poverty/ overview.

70 Max Roser, "Human Development Index (HDI)," Our World in Data, July 25, 2014, ourworldindata.org/human-development-index.

71 Renae Merle, "A Guide to the Financial Crisis—10 Years Later," *Washington Post*, September 10, 2018, http://www.washingtonpost.com/business/economy/a-guide-to-the-financial-crisis--10-years-later/2018/09/10/114b76ba-af10-11e8-a20b-5f4f84429666_story.html?utm_term=.65de73d9c67b.

72 Ibid.

73 Mahiben Maruthappu et al., "Economic Downturns, Universal Health Coverage, and Cancer Mortality in High-Income and Middle-Income Countries,

1990–2010: A Longitudinal Analysis," *Lancet* 388, no. 10045 (2016): 684–95, doi:10.1016/s0140-6736(16)00577-8.

74 "The Human Costs of Financial Crises: Linking Market Sentiment to Human Capital Loss," Phys.org, November 7, 2017, phys.org/news/2017-11-human-financial-crises-linking-sentiment.html.

75 William D. Nordhaus, "An Analysis of the Dismal Theorem," *Cowles Foundation Discussion Papers*, Cowles Foundation for Research in Economics, Yale University, February 2, 2009, ideas.repec.org/p/cwl/cwldpp/1686.html.

76 Robbie Gonzalez, "The Potential Pitfalls of Sucking Carbon From the Atmosphere," *Wired*, June 13, 2018, http://www.wired.com/story/the-potential-pitfalls-of-sucking-carbon-from-the-atmosphere/.

77 "Topic 2: Future Changes, Risks and Impacts," *IPCC 5th Assessment Synthesis Report*, ar5-syr.ipcc.ch/topic_futurechanges.php.

78 "Working Group I: The Physical Science Basis," IPCC, http://www.ipcc.ch/working-group/wg1/.

79 "Climate Science Special Report: Climate Models, Scenarios, and Projections," science2017.globalchange.gov/chapter/4/.

80 Michael E. Mann, "The 'Fat Tail' of Climate Change Risk," *Huffington Post*, December 7, 2017, http://www.huffingtonpost.com/michael-e-mann/the-fat-tail-of-climate-change-risk_b_8116264.html.

81 Philip Lynch, "Hot Summer in Australia, Where Fruit Is Cooking on the Trees," *Irish Times*, January 21, 2019, http://www.irishtimes.com/life-and-style/abroad/hot-summer-in-australia-where-fruit-is-cooking-on-the-trees-1.3762965.

82 Will Steffen et al., "Trajectories of the Earth System in the Anthropocene," *PNAS*, August 14, 2018, http://www.pnas.org/content/115/33/8252.

83 Ute Kehse, "Global Warming Doesn't Stop When the Emissions Stop," Phys.org, October 3, 2017, phys.org/news/2017-10-global-doesnt-emissions.html.

84 Lacie Glover, "Why Does an MRI Cost So Darn Much?," *Time*, July 16, 2014, time.com/money/2995166/why-does-mri-cost-so-much.

85 "How Is Our Self Related to Midline Regions and the Default-Mode Network?" *NeuroImage*, Academic Press, May 15, 2011, http://www.sciencedirect.com/science/article/pii/S1053811911005167.

86 Jane McGonigal. "Our Puny Human Brains Are Terrible at Thinking About the Future," *Slate*, April 13, 2017, http://www.slate.com/articles/technology/future_tense/2017/04/why_people_are_so_bad_at_thinking_about_the_future.html.

87 Neil Irwin, "Climate Change's Giant Impact on the Economy: 4 Key Issues," *New York Times*, January 17, 2019, http://www.nytimes.com/2019/01/17/upshot/how-to-think-about-the-costs-of-climate-change.html.

88 "GDP Growth (Annual %)," *Data*, data.worldbank.org/indicator/NY.GDP.MKTP.KD.ZG.

89 Derek Parfit, *Reasons and Persons* (Oxford: Clarendon Press, 1984), p. 357.

90 Juliette Jowit and Patrick Wintour, "Cost of Tackling Global Climate Change Has Doubled, Warns Stern," *Guardian*, June 25, 2008, http://www.theguardian.com/environment/2008/jun/26/climatechange.scienceofclimatechange.

91 Gernot Wagner and Martin Weitzman, *Climate Shock: The Economic Consequences of a Hotter Planet*, Kindle ed. (Princeton, NJ: Princeton University Press, 2016), pp. 62–63.

92 "FAE: Trends," FAE: Dashboard, explorer.usaid.gov/aid-trends.html.

93 Wendell Bell, *Foundations of Futures Studies: Human Science for a New Era* (New York: Transaction, 1997).

94 Richard Fisher, "Future—The Perils of Short-Termism: Civilisation's Greatest Threat," *BBC*, January 10, 2019, http://www.bbc.com/future/story/20190109-the-perils-of-short-termism-civilisations-greatest-threat.

95 P. D. James, *The Children of Men*, Kindle ed. (New York: Knopf, 2018), p. 11.

96 "Home," Revive & Restore, February 1, 2019, reviverestore.org/.

97 David Montgomery, "How Human Land Use Has Transformed the Earth—Animated," CityLab, January 11, 2019, http://www.citylab.com/environment/2019/01/farming-climate-change-land-use-urbanization-over-time-map/579724/.

98 Stewart Brand, *Whole Earth Discipline: An Ecopragmatist Manifesto* (Boston: Atlantic Books, 2010).

99 Elizabeth Kolbert, "Can Carbon Dioxide Removal Save the World?" *New Yorker*, May 31, 2018, http://www.newyorker.com/magazine/2017/11/20/can-carbon-dioxide-removal-save-the-world.

100 Direct Air Capture of CO_2 with Chemicals: A Technology Assessment for the APS Panel on Public Affairs, APS Physics. Princeton, NJ, June 2011.

101 "Global Carbon Project (GCP)," Carbon Budget, http://www.globalcarbonproject.org/carbonbudget/.

102 David W. Keith et al., "A Process for Capturing CO2 from the Atmosphere," *Joule* 2, no. 8 (2018): 1635, doi:10.1016/j.joule.2018.06.010.

103 "Gas Prices Canada—Understanding Gas Prices," Petro, retail.petro-canada.ca/en/fuelsavings/gas-taxes-canada.aspx.

104 Alden Woods, "Klaus Lackner Didn't Set out to Save the World, but He Thinks His Machine Could Help," *Azcentral*, March 8, 2018, http://www.azcentral.com/story/news/local/arizona-best-reads/2018/03/08/klaus-lackner-did-not-set-out-save-world-but-he-thinks-his-machine-could-help-solve-global-warming/397954002/.

105 "Mt. Pinatubos Cloud Shades Global Climate," Free Library, http://www.thefreelibrary.com/Mt. Pinatubos cloud shades global climate.-a012467057.

106 David Rotman, "Meet the Man with a Cheap and Easy Plan to Stop Global Warming," *MIT Technology Review*, March 28, 2016, http://www.technologyreview.com/s/511016/a-cheap-and-easy-plan-to-stop-global-warming/.

CHAPTER 5: DISEASE

1 Trail, Antiquities and Monuments Office, accessed April 1, 2019, https://www.amo.gov.hk/en/trails_sheungwan1.php?tid=24.

2 E. G. Pryor, "The Great Plague of Hong Kong," *Journal of the Hong Kong Branch of the Royal Asiatic Society* 15 (1975): 69.

3 HONG KONG: Population Growth of the Whole Country, accessed April 1, 2019, http://www.populstat.info/Asia/hongkonc.htm.

4 Elisabeth Rosenthal, "Chickens Killed in Hong Kong to Combat Flu," *New York Times*, December 29, 1997, accessed April 1, 2019, https://www.nytimes.com/1997/12/29/world/chickens-killed-in-hong-kong-to-combat-flu.html.

5 "Update 95—SARS: Chronology of a Serial Killer," World Health Organization, July 24, 2015, accessed April 1, 2019, http://www.who.int/csr/don/2003_07_04/en/.

6 Bryan Walsh, "Outbreak in Asia," *Time*, March 17, 2003, accessed April 1, 2019, https://content.time.com/time/magazine/article/0,9171,433331,00.html.

7 Wayne Kondro, "SARS Virus Claims Its Third Victim in Canada," *Lancet* 361, no. 9363 (2003): 1106, doi:10.1016/s0140-6736(03)12913-3.

8 Outbreak of Severe Acute Respiratory Syndrome (SARS) at Amoy Gardens, Kowloon Bay, Hong Kong Main Findings of the Investigation, Government of Hong Kong, 2003.

9 "When SARS First Spread, Guangzhou Residents Sought Out Vinegar," *South China Morning Post,* February 21, 2013, accessed April 1, 2019, https://www.scmp.com/news/china/article/1155640/when-sars-first-spread-guangzhou-residents-sought-out-vinegar.

10 Rob Stein, "SARS Travel Advisory Is Expanded," *Washington Post*, April 24, 2003, accessed April 1, 2019, https://www.washingtonpost.com/archive/politics/2003/04/24/sars-travel-advisory-is-expanded/0c6f10d3-e2d2-457d-b59e-5113a0121b8e/?utm_term=.f53986913a2a.

11 "Facts and Figures, HKIA at a Glance," Hong Kong International Airport, accessed April 1, 2019, https://www.hongkongairport.com/en/the-airport/hkia-at-a-glance/fact-figures.page.

12 Peter Shadbolt, "SARS 10 Years On: How Dogged Detective Work Defeated an Epidemic," CNN, February 21, 2013, accessed April 1, 2019, https://www.cnn.com/2013/02/21/world/asia/sars-amoy-gardens/index.html.

13 Bernd Sebastian Kamps, SARS Reference, Virology, accessed April 1, 2019, http://sarsreference.com/sarsref/virol.htm.

14 Richard D. Smith, "Responding to Global Infectious Disease Outbreaks: Lessons from SARS on the Role of Risk Perception, Communication and Management," *Social Science & Medicine* 63, no. 12 (2006): 3113–23, doi:10.1016/j.socscimed.2006.08.004.

15 Shadbolt, "SARS 10 Years On."

16 David Cyranoski, "Bat Cave Solves Mystery of Deadly SARS Virus—and Suggests New Outbreak Could Occur," *Nature News,* December 1, 2017, accessed April 1, 2019, https://www.nature.com/articles/d41586-017-07766-9.

17 Ian Sample and John Gittings, "In China the Civet Cat Is a Delicacy—and May Have Caused SARS," *Guardian*, May 24, 2003, accessed April 1, 2019, https://www.theguardian.com/world/2003/may/24/china.sars.

18 Gary Wong, Wenjun Liu, Yingxia Liu, Boping Zhou, Yuhai Bi, and George F. Gao, "MERS, SARS, and Ebola: The Role of Super-Spreaders in Infectious Disease," *Cell Host & Microbe* 18, no. 4 (2015): 398–401, doi:10.1016/j.chom.2015.09.013.

19 Hannah Beech, "Unmasking a Crisis," *Time*, April 13, 2003, accessed April 1, 2019, https://content.time.com/time/magazine/article/0,9171,443205,00.html.

20 John Pomfret, "Beijing Told Doctors to Hide SARS Victims," *Washington Post*, April 20, 2003, accessed April 1, 2019, https://www.washingtonpost.com/archive/politics/2003/04/20/beijing-told-doctors-to-hide-sars-victims/3eb7d1aa-d2ff-477b-bc15-d0164377b123/?utm_term=.0b56deed6c46.

21 "Airline Industry—Passenger Traffic Worldwide 2019: Statistics," Statista, accessed April 1, 2019, https://www.statista.com/statistics/564717/airline-industry-passenger-traffic-globally/.

22 "Update 95—SARS: Chronology of a Serial Killer," World Health Organization, July 24, 2015, accessed April 1, 2019, http://www.who.int/csr/don/2003_07_04/en/.

23 "Origin of HIV & AIDS," Avert, October 19, 2018, accessed April 1, 2019, https://www.avert.org/professionals/history-hiv-aids/origin.

24 "SARS: Frequently Asked Questions," Centers for Disease Control and Prevention, accessed April 1, 2019, https://www.cdc.gov/sars/about/faq.html.

25 "Consensus Document on the Epidemiology of Severe Acute Respiratory Syndrome (SARS)," World Health Organization, January 1, 1970, accessed April 1, 2019, http://apps.who.int/iris/handle/10665/70863.

26 Stacey Knobler, *Learning from SARS: Preparing for the Next Disease Outbreak: Workshop Summary* (Washington, DC: National Academies Press, 2004).

27 John Whitfield, "Portrait of a Serial Killer," *Nature News*, October 3, 2002, accessed April 1, 2019, https://www.nature.com/news/2002/021003/full/news021001-6.html.

28 "Number of Malaria Deaths," World Health Organization, February 26, 2019, accessed April 1, 2019, http://www.who.int/gho/malaria/epidemic/deaths/en/.

29 "Two of History's Deadliest Plagues Were Linked, with Implications for Another Outbreak," *National Geographic*, January 12, 2018, accessed April 1, 2019, https://news.nationalgeographic.com/news/2014/01/140129-justinian-plague-black-death-bacteria-bubonic-pandemic/.

30 *Bugs, Drugs and Smoke: Stories from Public Health* (Geneva: World Health Organization, 2011).

31 C. Thèves, P. Biagini, and E. Crubézy, "The Rediscovery of Smallpox," *Clinical Microbiology and Infection* 20, no. 3 (2014): 210–18, doi:10.1111/1469-0691.12536.

32 Peter Spreeuwenberg, Madelon Kroneman, and John Paget, "Reassessing the Global Mortality Burden of the 1918 Influenza Pandemic," *American Journal of Epidemiology*, 2019, doi:10.1093/aje/kwz041.

33 "Spanish Flu," History.com, October 12, 2010, accessed April 1, 2019, https://www.history.com/topics/1918-flu-pandemic.

34 "Global HIV and AIDS Statistics," Avert, October 8, 2018, accessed April 1, 2019, https://www.avert.org/global-hiv-and-aids-statistics.

35 Alfred Crosby, *America's Forgotten Pandemic: The Influenza of 1918* (Cambridge: Cambridge University Press, 2003).

36 Bryan Walsh, "Scientists Discover Why the 1918 Flu Pandemic Was So Deadly," *Time,* April 29, 2014, accessed April 1, 2019, http://time.com/79209/solving-the-mystery-flu-that-killed-50-million-people/.

37 Vincent J. Cirillo, "Two Faces of Death: Fatalities from Disease and Combat in America's Principal Wars, 1775 to Present," *Perspectives in Biology and Medicine* 51, no. 1 (2007): 121–33, doi:10.1353/pbm.2008.0005.

38 Sebastian Farquhar et al., *Existential Risk: Diplomacy and Governance*, Global Priorities Project, 2017.

39 "Smallpox," World Health Organization, May 2, 2018, accessed April 1, 2019, http://www.who.int/csr/disease/smallpox/en/.

40 Tania Browne, "Destroying the Last Samples of Smallpox Virus Could Prove Short-sighted," *Guardian*, July 10, 2014, accessed April 1, 2019, https://www.theguardian.com/science/blog/2014/jul/10/smallpox-virus-vials-variola-research.

41 D. M. Morens et al., "Predominant role of bacterial pneumonia as a cause of death in pandemic influenza: Implications for pandemic influenza preparedness," *Journal of Infectious Diseases* 198, no. 7 (2008): 962–70, doi: 10.1086/591708.

42 Fatimah S. Dawood et al., "Estimated global mortality associated with the first 12 months of 2009 pandemic influenza A H1N1 virus circulation: A modelling study," *Lancet Infectious Diseases* 12, no. 9 (2012): 687–95.

43 "Influenza (Seasonal)," World Health Organization, accessed April 1, 2019, http://www.who.int/news-room/fact-sheets/detail/influenza-(seasonal).

44 "The 1918 Influenza Pandemic," accessed April 1, 2019, https://virus.stanford.edu/uda/.

45 Sonia Shah, "The Fight against Infectious Diseases Is Still an Uphill Battle," *Science News,* December 14, 2016, accessed April 1, 2019, https://www.sciencenews.org/article/infectious-diseases-sonia-shah.

46 Kate E. Jones et al., "Global Trends in Emerging Infectious Disease," *Nature* 451 (2008): 990–94.

47 "Global Rise in Human Infectious Disease Outbreaks," *Journal of the Royal Society*, accessed April 1, 2019, https://royalsocietypublishing.org/doi/full/10.1098/rsif.2014.0950.

48 Max Roser and Esteban Ortiz-Ospina, "World Population Growth," Our World in Data, May 9, 2013, accessed April 1, 2019, https://ourworldindata.org/world-population-growth.

49 Mia Armstrong, "How Ready Are We for the Next Pandemic?" *Slate*, October 23, 2018, accessed April 1, 2019, https://slate.com/technology/2018/10/sonia-shah-interview-pandemic-epidemic-spanish-flu.html.

50 Carly Cassella, "A Lot More People Are Already Dying from Superbugs Than You Realize," ScienceAlert, accessed April 1, 2019, https://www.sciencealert.com/more-people-dying-superbugs-antibiotic-resistance-than-you-might-think.

51 Robin McKie, "'Antibiotic Apocalypse': Doctors Sound Alarm over Drug Resistance," *Guardian*, October 8, 2017, accessed April 1, 2019, https://www.theguardian.com/society/2017/oct/08/world-faces-antibiotic-apocalypse-says-chief-medical-officer.

52 Bryan Walsh, "The World Is Not Ready for the Next Pandemic," *Time*, May 4, 2017, accessed April 1, 2019, http://time.com/magazine/us/4766607/may-15th-2017-vol-189-no-18-u-s/.

53 "World Bank Group Launches Groundbreaking Financing Facility to Protect Poorest Countries Against Pandemics," World Bank, accessed April 1, 2019, http://www.worldbank.org/en/news/press-release/2016/05/21/world-bank-group-launches-groundbreaking-financing-facility-to-protect-poorest-countries-against-pandemics.

54 Rachel Nuwer, "Future—What If a Deadly Influenza Pandemic Broke Out Today?" BBC, November 22, 2018, accessed April 1, 2019, http://www.bbc.com/future/story/20181120-what-if-a-deadly-influenza-pandemic-broke-out-today.

55 Kate Wheeling, "Are Anti-Vaxxers a Major Health Threat? The World Health Organization Says Yes," Pacific Standard, January 16, 2019, accessed April 1, 2019, https://psmag.com/news/are-anti-vaxxers-a-major-health-threat-the-world-health-organization-says-yes.

56 Andy Coghlan, "Africa's Road-Building Frenzy Will Transform Continent," *New Scientist*, accessed April 1, 2019, https://www.newscientist.com/article/mg22129512-800-africas-road-building-frenzy-will-transform-continent/.

57 "Going Viral," Global Risks 2019, accessed April 1, 2019, http://reports.weforum.org/global-risks-2019/going-viral/#view/fn-4.

58 Bryan Walsh, "Virus Hunter," *Time*, November 7, 2011, accessed April 1, 2019, http://content.time.com/time/subscriber/article/0,33009,2097962-1,00.html.

59 Margaret Chan, "Ebola Virus Disease in West Africa—No Early End to the Outbreak," *New England Journal of Medicine* 371, no. 13 (2014): 1183–85, doi:10.1056/nejmp1409859.

60 Press release, Centers for Disease Control and Prevention, June 8, 2009, http://www.cdc.gov/media/pressrel/2009/r090608.htm.

61 Julia Belluz, "This Is What Keeps Tom Frieden, the CDC Director, Up at Night," *Vox*, January 17, 2017, http://www.vox.com/science-and-health/2017/1/16/14042500/tom-frieden-vaccines-ebola-zika-trump-cdc.

62 "WHO Declares End of Ebola Outbreak in Nigeria," World Health Organization, October 21, 2014, http://www.who.int/mediacentre/news/statements/2014/nigeria-ends-ebola/en/.

63 Diane Dewar, "The US Health Economy Is Big, but Is It Better?" *Conversation*, March 4, 2019, theconversation.com/the-us-health-economy-is-big-but-is-it-better-80593.

64 Ed Yong, "The Next Plague Is Coming. Is America Ready?" *Atlantic*, July 25, 2018, http://www.theatlantic.com/magazine/archive/2018/07/when-the-next-plague-hits/561734/.

65 Lena H. Sun, "CDC to Cut by 80 Percent Efforts to Prevent Global Disease Outbreak," *Washington Post*, February 1, 2018, http://www.washingtonpost.com/news/to-your-health/wp/2018/02/01/cdc-to-cut-by-80-percent-efforts-to-prevent-global-disease-outbreak.

66 Lauren Weber, "Sudden Departure of White House Global Health Security Head Has Experts Worried," *Huffington Post*, May 10, 2018, http://www.huffingtonpost.com/entry/tim-ziemer-global-health-security-leaves_us_5af37dfbe4b0859d11d02290.

67 George W. Bush, "George W. Bush: PEPFAR Saves Millions of Lives in Africa. Keep It Fully Funded," *Washington Post*, April 7, 2017, http://www.washingtonpost.com/opinions/george-w-bush-pepfar-saves-millions-of-lives-in-africa-keep-it-fully-funded/2017/04/07/2089fa46-1ba7-11e7-9887-1a5314b56a08_story.html?utm_term=.ad37276c8c20.

68 Lenny Bernstein, "Trump Wanted to Keep Americans Critically Ill with Ebola out of the U.S.," *Washington Post,* August 24, 2016, accessed April 1, 2019, https://www.washingtonpost.com/news/to-your-health/wp/2016/08/24/trump-wanted-to-deny-u-s-care-to-americans-critically-ill-with-ebola/?utm_term=.f013d8905046.

69 Donald J. Trump, "Ebola Is Much Easier to Transmit than the CDC and Government Representatives Are Admitting. Spreading All over Africa-and Fast. Stop Flights," Twitter, October 2, 2014, accessed April 1, 2019, https://twitter.com/realdonaldtrump/status/517613167359574016?lang=en.

70 Sheila Kaplan, "Dr. Brenda Fitzgerald, C.D.C. Director, Resigns Over Tobacco and Other Investments," *New York Times*, January 31, 2018, accessed April 1, 2019, https://www.nytimes.com/2018/01/31/health/cdc-brenda-fitzgerald-resigns.html.

71 Dan Diamond, "Price Resigns from HHS after Facing Fire for Travel," *Politico*, September 30, 2017, accessed April 1, 2019, https://www.politico.com/story/2017/09/29/price-has-resigned-as-health-and-human-services-secretary-243315.

72 Laignee Barron, "The EPA's Website After a Year of Climate Change Censorship," *Time,* March 1, 2018, accessed April 1, 2019, http://time.com/5075265/epa-website-climate-change-censorship/.

73 Michael Osterholm, *Deadliest Enemy* (New York: Little, Brown, 2017).

74 Ibid.

75 Dimitrios Gouglas et al., "Estimating the Cost of Vaccine Development against Epidemic Infectious Diseases: A Cost Minimisation Study," *Lancet Global Health* 6, no. 12 (2018), doi:10.1016/s2214-109x(18)30346-2.

76 "Ebola (Ebola Virus Disease)," Centers for Disease Control and Prevention, March 8, 2019, accessed April 1, 2019, https://www.cdc.gov/vhf/ebola/history/2014-2016-outbreak/case-counts.html.

77 Centers for Disease Control and Prevention, accessed April 1, 2019, https://www. cdc.gov/flu/about/disease/2014-15.htm.

78 "Morbidity and Mortality Weekly Report (MMWR)," Centers for Disease Control and Prevention, February 15, 2018, accessed April 1, 2019, https:// www.cdc.gov/mmwr/volumes/67/wr/mm6706a2.htm.

79 Mackenzie Bean, "CDC: 2017–18 Flu Season Had 'Record-Breaking' Hospitalization Rates," *Becker's Hospital Review,* accessed April 1, 2019, https:// www.beckershospitalreview.com/quality/cdc-2017-18-flu-season-had-record-breaking-hospitalization-rates.html.

80 Centers for Disease Control and Prevention, September 27, 2017, accessed April 1, 2019, https://www.cdc.gov/flu/about/viruses/change.htm.

81 *Report to the President on Reengineering the Influenza Vaccine Production Enterprise to Meet the Challenges of Pandemic Influenza,* President's Council of Advisors on Science and Technology, Washington, DC, August 2010.

82 Bryan Walsh, "The World Is Not Ready for the Next Pandemic," *Time,* May 4, 2017, accessed April 1, 2019, http://time.com/magazine/us/4766607/may-15th-2017-vol-189-no-18-u-s/.

83 Jon Cohen, "Universal Flu Vaccine Remains 'an Alchemist's Dream,'" *Science,* December 3, 2018, accessed April 1, 2019, https://www.sciencemag.org/ news/2018/11/universal-flu-vaccine-remains-alchemist-s-dream.

84 Bryan Walsh, "Universal Flu Vaccine: A Miracle Within Trump's Reach," Bloomberg.com, March 10, 2017, accessed April 1, 2019, https://www.bloomberg. com/view/articles/2017-03-10/a-miracle-within-trump-s-reach-universal-flu-vaccine.

85 Lisa Schnirring, "WHO Declares Ebola a Public Health Emergency," CIDRAP, August 8, 2014, accessed April 1, 2019, http://www.cidrap.umn.edu/news-perspective/2014/08/who-declares-ebola-public-health-emergency.

86 Ebola Virus Disease Democratic Republic of Congo: External Situation Report 34, World Health Organization, Geneva, March 26, 2019.

87 Osterholm, *Deadliest Enemy.*

88 Shruti Ravindran, "Why the Rest of the World Doesn't Suffer from Leprosy like India Does," Quartz India, February 18, 2015, accessed April 1, 2019, https:// qz.com/328951/why-indias-poor-continue-to-be-ravaged-by-the-worlds-oldest-disease/.

89 Bill Gates, "Transcript of 'The Next Outbreak? We're Not Ready,'" TED, accessed April 1, 2019, https://www.ted.com/talks/bill_gates_the_next_disaster_we_re_not_ready/transcript?language=en.

90 Marc Lipsitch (Harvard Medical School), personal communication with author, May 2018.

CHAPTER 6: BIOTECHNOLOGY

1 Clade X details: Julia Cizek, "Clade X, a Tabletop Exercise Hosted by the Center for Health Security," Johns Hopkins Center for Health Security, January 7,

2019, accessed April 1, 2019, http://www.centerforhealthsecurity.org/our-work/events/2018_clade_x_exercise/.

2 Kevin Loria, "Bill and Melinda Gates Think a Weaponized Disease May Be the Biggest Threat to Humanity—Here's How Worried You Should Be," *Business Insider*, March 12, 2018, accessed April 1, 2019, https://www.businessinsider.com/pandemic-risk-to-humanity-2017-9.

3 *Biodefense in the Age of Synthetic Biology* (Washington, DC: National Academies Press, 2018).

4 Stefan Riedel, "Biological Warfare and Bioterrorism: A Historical Review," *Proceedings* (Baylor University Medical Center), October 2004, accessed April 1, 2019, https://www.ncbi.nlm.nih.gov/pmc/articles/PMC1200679/.

5 Nicholas Kristof, "Unmasking Horror—A Special Report; Japan Confronting Gruesome War Atrocity," *New York Times*, March 17, 1995, accessed April 1, 2019, https://www.nytimes.com/1995/03/17/world/unmasking-horror-a-special-report-japan-confronting-gruesome-war-atrocity.html.

6 Helen Thompson, "In 1950, the U.S. Released a Bioweapon in San Francisco," Smithsonian.com, July 6, 2015, accessed April 1, 2019, https://www.smithsonianmag.com/smart-news/1950-us-released-bioweapon-san-francisco-180955819/.

7 Milton Leitenberg, Jens H. Kuhn, and Raymond A. Zilinskas, *The Soviet Biological Weapons Program: A History* (Cambridge, MA: Harvard University Press, 2012).

8 Michelle Rozo and Gigi Kwik Gronvall, "The Reemergent 1977 H1N1 Strain and the Gain-of-Function Debate," *Mbio* 6, no. 4 (2015), doi:10.1128/mbio.01013-15.

9 "Statement on Chemical and Biological Defense Policies and Programs," American Presidency Project, November 25, 1969, accessed April 1, 2019, https://www.presidency.ucsb.edu/documents/statement-chemical-and-biological-defense-policies-and-programs.

10 Piers Millett and Andrew Snyder-Beattie, "Existential Risk and Cost-Effective Biosecurity," *Health Security* 15, no. 4 (2017): 373–83, doi:10.1089/hs.2017.0028.

11 "Summary," U.S. Bureau of Labor Statistics, April 13, 2018, accessed April 1, 2019, https://www.bls.gov/ooh/life-physical-and-social-science/microbiologists.htm, and "17-2161 Nuclear Engineers," U.S. Bureau of Labor Statistics, March 29, 2019, accessed April 1, 2019, https://www.bls.gov/oes/current/oes172161.htm.

12 National Research Council, *Biotechnology Research in an Age of Terrorism* (Washington, DC: National Academies Press, 2004), https://doi.org/10.17226/10827.

13 "Leo Szilard's Fight to Stop the Bomb," Atomic Heritage Foundation, July 15, 2016, accessed April 1, 2019, https://www.atomicheritage.org/history/leo-szilards-fight-stop-bomb.

14 "Smallpox," National Institute of Allergy and Infectious Diseases, March 27, 2019, accessed April 1, 2019, https://www.niaid.nih.gov/diseases-conditions/smallpox.

15 Center for Biosecurity, "Dark Winter," Johns Hopkins Center for Health Security, June 22, 2018, accessed April 1, 2019, http://www.centerforhealthsecurity.org/our-work/events/2001_dark-winter/.

16 Antonio Regalado, "See What's Behind One Man's Proposal to Recode Life as We Know It," *MIT Technology Review,* June 2, 2016, accessed April 1, 2019, https://www.technologyreview.com/s/601610/plan-to-fabricate-a-genome-raises-questions-on-designer-humans/.

17 Reference Links for Key Numbers in Biology, accessed April 1, 2019, https://bionumbers.hms.harvard.edu/KeyNumbers.aspx.

18 "Blue Whale," *National Geographic,* September 21, 2018, accessed April 1, 2019, http://www.nationalgeographic.com/animals/mammals/b/blue-whale/.

19 James M. Heather and Benjamin Chain, "The Sequence of Sequencers: The History of Sequencing DNA," *Genomics* 107, no. 1 (2016): 1–8, doi:10.1016/j.ygeno.2015.11.003.

20 "Human Genome Project Completion: Frequently Asked Questions," National Human Genome Research Institute (NHGRI), accessed April 1, 2019, https://www.genome.gov/11006943/human-genome-project-completion-frequently-asked-questions/.

21 Andrew Pollack, "Scientists Announce HGP-Write, Project to Synthesize the Human Genome," *New York Times*, June 2, 2016, accessed April 1, 2019, https://www.nytimes.com/2016/06/03/science/human-genome-project-write-synthetic-dna.html?_r=0.

22 Kai Kupferschmidt, "How Canadian Researchers Reconstituted an Extinct Poxvirus for $100,000 Using Mail-order DNA," *Science*, December 8, 2017, accessed April 1, 2019, https://www.sciencemag.org/news/2017/07/how-canadian-researchers-reconstituted-extinct-poxvirus-100000-using-mail-order-dna.

23 Ryan S. Noyce, Seth Lederman, and David H. Evans, "Construction of an Infectious Horsepox Virus Vaccine from Chemically Synthesized DNA Fragments," *Plos One* 13, no. 1 (2018), doi:10.1371/journal.pone.0188453.

24 Nell Greenfieldboyce, "Did Pox Virus Research Put Potential Profits Ahead of Public Safety?" NPR, February 17, 2018, accessed April 1, 2019, https://www.npr.org/sections/health-shots/2018/02/17/585385308/did-pox-virus-research-put-potential-profits-ahead-of-public-safety.

25 Kupferschmidt, "How Canadian Researchers Reconstituted an Extinct Poxvirus for $100,000 Using Mail-order DNA."

26 Nick Bostrom, "Information Hazards: A Typology of Potential Harms from Knowledge," *Review of Contemporary Philosophy* 10, no. 44 (2012).

27 Joshua Gans, "'Information Wants to Be Free': The History of That Quote," Digitopoly, October 25, 2015, accessed April 1, 2019, https://digitopoly.org/2015/10/25/information-wants-to-be-free-the-history-of-that-quote/.

28 N. Bostrom, T. Douglas, and A. Sandberg, "The Unilateralist's Curse and the Case for a Principle of Conformity," *Social Epistemology* 30, no. 4 (2016): 350–71, doi:10.1080/02691728.2015.1108373.

29 Gregory Lewis, "Horsepox Synthesis: A Case of the Unilateralist's Curse?" *Bulletin of the Atomic Scientists,* June 28, 2018, accessed April 1, 2019, https://thebulletin.org/2018/02/horsepox-synthesis-a-case-of-the-unilateralists-curse/.

30 "The Scientific Tragedy of the Atomic Bomb," Ashbrook, accessed April 1, 2019, https://ashbrook.org/publications/respub-v8n1-cook/.

31 Owen Dyer, "Scientists Call for Moratorium on Editing Heritable Genes," *Bmj,* 2019, L1256, doi:10.1136/bmj.l1256.

32 Alison Young, "Hundreds of Bioterror Lab Mishaps Cloaked in Secrecy," *USA Today,* August 17, 2014, accessed April 1, 2019, https://www.usatoday.com/story/news/nation/2014/08/17/reports-of-incidents-at-bioterror-select-agent-labs/14140483/.

33 Sabrina Tavernise and Donald G. McNeil, "C.D.C. Details Anthrax Scare for Scientists at Facilities," *New York Times,* June 19, 2014, accessed April 1, 2019, https://www.nytimes.com/2014/06/20/health/up-to-75-cdc-scientists-may-have-been-exposed-to-anthrax.html.

34 Ai Ee Ling, "Editorial on Laboratory-Acquired Incidents in Taipei, Taiwan and Singapore Following the Outbreak of SARS Coronavirus," *Applied Biosafety* 12, no. 1 (2007): 17, doi:10.1177/153567600701200103.

35 Richard D. Henkel, Thomas Miller, and Robbin S. Weyant, "Monitoring Select Agent Theft, Loss and Release Reports in the United States—2004–2010," *Applied Biosafety* 17, no. 4 (2012): 171–80, doi:10.1177/153567601201700402.

36 "Cumulative Number of Confirmed Human Cases of Avian Influenza A(H5N1) Reported to WHO," World Health Organization, February 25, 2019, accessed April 1, 2019, https://www.who.int/influenza/human_animal_interface/H5N1_cumulative_table_archives/en/.

37 Marc Lipsitch and Thomas V. Inglesby, "Moratorium on Research Intended to Create Novel Potential Pandemic Pathogens," *MBio*5, no. 6 (2014), doi:10.1128/mbio.02366-14.

38 Ibid.

39 "Heart Disease and Stroke Statistics: 2017 Update," American College of Cardiology, accessed April 1, 2019, https://www.acc.org/latest-in-cardiology/ten-points-to-remember/2017/02/09/14/58/heart-disease-and-stroke-statistics-2017.

40 Lenny Bernstein, "U.S. Lifts Research Moratorium on Enhancing Germs' Danger," *Washington Post,* December 19, 2017, accessed April 1, 2019, https://www.washingtonpost.com/national/health-science/us-announces-new-policy-for-pathogens-that-could-cause-a-pandemic/2017/12/19/bb715f7c-e4b5-11e7-833f-155031558ff4_story.html?utm_term=.fdb555758c6f.

41 Jocelyn Kaiser, "NIH Lifts 3-Year Ban on Funding Risky Virus Studies," *Science,* December 19, 2017, accessed April 1, 2019, https://www.sciencemag.org/news/2017/12/nih-lifts-3-year-ban-funding-risky-virus-studies.

42 Jocelyn Kaiser, "Exclusive: Controversial Experiments That Could Make Bird Flu More Risky Poised to Resume," *Science,* February 11, 2019, accessed April

1, 2019, https://www.sciencemag.org/news/2019/02/exclusive-controversial-experiments-make-bird-flu-more-risky-poised-resume.

43 Marc Lipsitch and Tom Inglesby, "The U.S. Is Funding Dangerous Experiments It Doesn't Want You to Know About," *Washington Post,* February 27, 2019, accessed April 1, 2019, https://www.washingtonpost.com/opinions/the-us-is-funding-dangerous-experiments-it-doesnt-want-you-to-know-about/2019/02/27/5f60e934-38ae-11e9-a2cd-307b06d0257b_story.html?utm_term=.28d02f787dbc.

44 National Science Foundation, National Center for Science and Engineering Statistics, Survey of Earned Doctorates.

45 Zhu Liu and Yong Geng, "Is China Producing Too Many PhDs?" *Nature* 474, no. 7352 (2011): 450, doi:10.1038/474450b.

46 Megan Molteni, "With Embryo Base Editing, China Gets Another Crispr First," *Wired,* August 27, 2018, accessed April 1, 2019, https://www.wired.com/story/crispr-base-editing-first-china/.

47 "Pig Organs for Human Patients: A Challenge Fit for CRISPR," Wyss Institute, May 30, 2018, accessed April 1, 2019, https://wyss.harvard.edu/pig-organs-for-human-patients-a-challenge-fit-for-crispr/.

48 "Gene-Therapy Trials Must Proceed with Caution," *Nature* 534, no. 7609 (2016): 590, doi:10.1038/534590a.

49 Rob Stein, "Facing Backlash, Chinese Scientist Defends Gene-Editing Research on Babies," NPR, November 28, 2018, accessed April 1, 2019, https://www.npr.org/sections/health-shots/2018/11/28/671375070/facing-backlash-chinese-scientist-defends-gene-editing-research-on-babies.

50 "2012 : What Is Your Favorite Deep, Elegant, or Beautiful Explanation?" Edge.org, January 1, 1970, accessed April 1, 2019, https://www.edge.org/response-detail/10898.

51 Monique Brouillette, "Would You Feel Sexy Wearing Eau de Extinction?" *MIT Technology Review,* December 6, 2016, accessed April 1, 2019, https://www.technologyreview.com/s/602899/would-you-feel-sexy-wearing-eau-de-extinction/.

52 Bayer AG Communications, "Bayer and Ginkgo Bioworks Unveil Joint Venture, Joyn Bio, and Establish Operations in Boston and West Sacramento," Bayer News, accessed April 1, 2019, https://media.bayer.com/baynews/baynews.nsf/id/Bayer-Ginkgo-Bioworks-unveil-joint-venture-Joyn-Bio-establish-operations-Boston-West-Sacramento.

53 Rebecca Spalding, "The DNA Cops Who Make Sure the World's Deadliest Viruses Aren't Rebuilt," Bloomberg.com, June 27, 2018, accessed April 1, 2019, https://www.bloomberg.com/news/features/2018-06-27/these-dna-cops-make-sure-deadly-viruses-don-t-get-rebuilt.

54 Ibid.

55 Freeman Dyson, "Our Biotech Future," *New York Review of Books,* July 19, 2007, accessed April 1, 2019, https://www.nybooks.com/articles/2007/07/19/our-biotech-future/.

56 Nick Bostrom, Milan M. Ćirković, and Martin J. Rees, *Global Catastrophic Risks,* Kindle ed. (Oxford: Oxford University Press, 2018), p. 338, loc. 7282.

57 "Global Cost of Cybercrime Exceeded $600 Billion in 2017, Report Estimates," Security Intelligence, accessed April 1, 2019, https://securityintelligence.com/news/global-cost-of-cybercrime-exceeded-600-billion-in-2017-report-estimates/.

58 John Lott, "This Marks the End of Gun Control," Fox News, July 20, 2018, accessed April 1, 2019, http://www.foxnews.com/opinion/2018/07/20/making-guns-on-3d-printers-is-blow-against-gun-control.html.

59 Tiffany Hsu, "3-D Printed Gun Advocate Cody Wilson Quits Company He Founded," *New York Times*, September 25, 2018, accessed April 1, 2019, https://www.nytimes.com/2018/09/25/business/printed-gun-cody-wilson-defense-distributed.html.

CHAPTER 7: ARTIFICIAL INTELLIGENCE

1 Chris Smith, "IPhone 7 vs. IPhone 7 Plus: Does 3GB of RAM Really Make a Difference?" *BGR*, September 20, 2016, bgr.com/2016/09/20/iphone-7-3gb-ram-iphone-7-plus/ and "Apple IIe—All Computer Information and Tech Specs," IGotOffer, December 11, 2017, igotoffer.com/apple/apple-iie.

2 Naomi Eide, "After 50 Years of Intel, Moore's Law Is Alive and Well—Sort of," CIO Dive, May 18, 2018, accessed April 1, 2019, http://www.ciodive.com/news/after-50-years-of-intel-moores-law-is-alive-and-well-sort-of/523856/.

3 Marc Andreessen, "Why Software Is Eating the World," Andreessen Horowitz, December 21, 2018, accessed April 1, 2019, https://a16z.com/2011/08/20/why-software-is-eating-the-world/.

4 Cade Metz, "Inside the Poker AI That Out-Bluffed the Best Humans," *Wired*, February 2, 2018, accessed April 1, 2019, http://www.wired.com/2017/02/libratus/; Joel Lehman et al., "The Surprising Creativity of Digital Evolution," 2018 Conference on Artificial Life, 2018, doi:10.1162/isal_a_00016.

5 Ian Tattersall, "Homo Sapiens," *Encyclopædia Britannica*, January 11, 2019, accessed April 1, 2019, http://www.britannica.com/topic/Homo-sapiens.

6 "Elon Musk: Artificial Intelligence Is the 'Greatest Risk We Face as a Civilization,'" *Fortune,* accessed April 1, 2019, http://fortune.com/2017/07/15/elon-musk-artificial-intelligence-2/.

7 Matt McFarland, "Elon Musk: 'With Artificial Intelligence We Are Summoning the Demon,'" *Washington Post,* October 24, 2014, accessed April 1, 2019, http://www.washingtonpost.com/news/innovations/wp/2014/10/24/elon-musk-with-artificial-intelligence-we-are-summoning-the-demon/?utm_term=.bde9a3b77d88.

8 Rory Cellan-Jones, "Stephen Hawking Warns Artificial Intelligence Could End Mankind," BBC News, December 2, 2014, accessed April 1, 2019, http://www.bbc.com/news/technology-30290540.

9 Huw Price, "Cambridge, Cabs and Copenhagen: My Route to Existential Risk," *New York Times*, January 27, 2013, accessed April 1, 2019, https://opinionator.

blogs.nytimes.com/2013/01/27/cambridge-cabs-and-copenhagen-my-route-to-existential-risk/.

10 Cotton-Barratt and Toby Ord, *Existential Risk and Existential Hope: Definitions*, Future of Humanity Institute—Technical Report #2015-1.

11 Ibid.

12 Catherine Clifford, "Google CEO: A.I. Is More Important than Fire or Electricity," CNBC, February 1, 2018, accessed April 1, 2019, http://www.cnbc.com/2018/02/01/google-ceo-sundar-pichai-ai-is-more-important-than-fire-electricity.html.

13 Cade Metz, "Tech Giants Are Paying Huge Salaries for Scarce A.I. Talent," *New York Times*, August 7, 2018, accessed April 1, 2019, http://www.nytimes.com/2017/10/22/technology/artificial-intelligence-experts-salaries.html.

14 "Introducing Google AI," Google AI blog, May 7, 2018, accessed April 1, 2019, https://ai.googleblog.com/2018/05/introducing-google-ai.html.

15 Max Haldevang, "Stephen Hawking Left Us Bold Predictions on AI, Superhumans, and Aliens," Quartz, October 15, 2018, accessed April 1, 2019, https://qz.com/1423685/stephen-hawking-says-superhumans-will-take-over-ai-is-a-threat-and-humans-will-conquer-space.

16 John McCarthy, Marvin L. Minsky, Nathaniel Rochester, and Claude E. Shannon, "A Proposal for the Dartmouth Summer Research Project on Artificial Intelligence, August 31, 1955," *AI Magazine*, December 15, 2006, accessed April 1, 2019, https://aaai.org/ojs/index.php/aimagazine/article/view/1904.

17 Yongdong Wang, "Your Next New Best Friend Might Be a Robot—Issue 52: The Hive," *Nautilus*, September 14, 2017, accessed April 1, 2019, http://nautil.us/issue/52/the-hive/your-next-new-best-friend-might-be-a-robot-rp.

18 A. M. Turing, "Intelligent Machinery, A Heretical Theory," *Philosophia Mathematica* 4, no. 3 (1996): 256–60, doi:10.1093/philmat/4.3.256.

19 Roland Pease, "Alan Turing: Inquest's Suicide Verdict 'Not Supportable'," BBC News, June 26, 2012, accessed April 1, 2019, https://www.bbc.com/news/science-environment-18561092.

20 Irving John Good, "Speculations Concerning the First Ultraintelligent Machine," *Advances in Computers* 6 (1966): 31–88, doi:10.1016/s0065-2458(08)60418-0.

21 Glenn Elert, "Volume of a Human Brain," Physics Factbook, accessed April 1, 2019, https://hypertextbook.com/facts/2001/ViktoriyaShchupak.shtml.

22 James Barrat, *Our Final Invention: Artificial Intelligence and the End of the Human Era* (New York: Thomas Dunne Books, 2015).

23 Kasey Panetta, "5 Trends Emerge in the Gartner Hype Cycle for Emerging Technologies, 2018," Smarter with Gartner, August 16, 2018, accessed April 1, 2019, https://www.gartner.com/smarterwithgartner/5-trends-emerge-in-gartner-hype-cycle-for-emerging-technologies-2018/.

24 Kevin Kelly, "The Three Breakthroughs That Have Finally Unleashed AI on the World," *Wired*, June 22, 2018, accessed April 1, 2019, https://www.wired.com/2014/10/future-of-artificial-intelligence/.

25 Daniel Oberhaus, "How Garry Kasparov Learned to Stop Worrying and Love AI," Motherboard, August 2, 2017, accessed April 1, 2019, https://motherboard.vice.com/en_us/article/vbe4e9/how-garry-kasparov-learned-to-stop-worrying-and-love-ai.

26 Jennifer Kahn, "It's Alive!" *Wired*, March 2002, accessed April 1, 2019, https://www.wired.com/2002/03/everywhere/.

27 Hunter Schwarz, "23 Oddly Specific Netflix Categories That Only Have One Show You Can Watch," *BuzzFeed*, January 11, 2014, accessed April 1, 2019, https://www.buzzfeed.com/hunterschwarz/23-oddly-specific-netflix-categories-that-only-have-one-show.

28 Michelle Castillo, "Netflix Takes Up 8 Percent of the Time Spent Watching Video, but the Company Wants to Change That," CNBC, July 17, 2018, accessed April 1, 2019, https://www.cnbc.com/2018/07/17/netflix-small-portion-of-overall-watch-time-and-competition-is-stiff.html.

29 CD-ROMS had 682 megabytes, of 682 million bytes. Figure comes from dividing into 2.5 quintillion bytes: "CD-ROM / DVD," accessed April 1, 2019, https://www.interdatarecovery.com/supported-media/cd-rom-dvd.

30 Richard Harris, "More Data Will Be Created in 2017 than the Previous 5,000 Years of Humanity," *App Developer Magazine,* December 23, 2016, accessed April 1, 2019, https://appdevelopermagazine.com/more-data-will-be-created-in-2017-than-the-previous-5,000-years-of-humanity-/.

31 Alexis C. Madrigal, "How Netflix Reverse-Engineered Hollywood," *Atlantic*, January 2, 2014, accessed April 1, 2019, https://www.theatlantic.com/technology/archive/2014/01/how-netflix-reverse-engineered-hollywood/282679/.

32 Max Tegmark, *Life 3.0: Being Human in the Age of Artificial Intelligence,* Kindle ed. (London: Penguin, 2018), loc. 1541.

33 David Silver et al., "Mastering the Game of Go without Human Knowledge," *Nature* 550, no. 7676 (2017): 354–59, doi:10.1038/nature24270.

34 "AlphaZero: Shedding New Light on the Grand Games of Chess, Shogi and Go," DeepMind, accessed April 1, 2019, https://deepmind.com/blog/alphazero-shedding-new-light-grand-games-chess-shogi-and-go/.

35 Satinder Singh, Andy Okun, and Andrew Jackson, "Learning to Play Go from Scratch," *Nature* 550, no. 7676 (2017): 336–37, doi:10.1038/550336a.

36 Michael Shinzaki, "A New Poker A.I. Eviscerated Its Human Competition, and It's a Beautiful Thing," *Slate*, February 1, 2017, accessed April 1, 2019, http://www.slate.com/blogs/future_tense/2017/02/01/libratus_a_poker_artificial_intelligence_eviscerated_its_human_competition.html.

37 James Maynard, "CMU AI Claudico Is Good at Poker But Not Good Enough for World's Best Human Players," *Tech Times*, May 10, 2015, accessed April 1, 2019, https://www.techtimes.com/articles/51946/20150510/cmu-ai-claudico-good-poker-enough-worlds-best-human-players.htm.

38 James Vincent, "DeepMind's AI Agents Conquer Human Pros at StarCraft II," *Verge,* January 24, 2019, accessed April 1, 2019, https://www.theverge. com/2019/1/24/18196135/google-deepmind-ai-starcraft-2-victory.

39 Timothy Revell, "AI Will Be Able to Beat Us at Everything by 2060, Say Experts," *New Scientist,* May 31, 2017, accessed April 1, 2019, https://www.newscientist. com/article/2133188-ai-will-be-able-to-beat-us-at-everything-by-2060-say-experts/.

40 "100 Years Ago, Some People Were REALLY Hostile to the Introduction of the Automobile," DangerousMinds, May 11, 2015, accessed April 1, 2019, https:// dangerousminds.net/comments/100_years_ago_some_people_were_really_ hostile_to_the_introduction.

41 Adrienne LaFrance, "When the Telephone Was Dangerous," *Atlantic,* September 6, 2015, accessed April 1, 2019, https://www.theatlantic.com/notes/2015/09/ when-the-telephone-was-dangerous/403609/.

42 Rachel Layne, "AI May Be One Reason Some Wages Are Sinking," CBS 42, January 8, 2018, accessed April 1, 2019, https://www.cbsnews.com/news/ai-in-workplace-sinking-wages/.

43 "Jobs Lost, Jobs Gained: What the Future of Work Will Mean for Jobs, Skills, and Wages," McKinsey & Company, accessed April 1, 2019, https://www.mckinsey. com/featured-insights/future-of-work/jobs-lost-jobs-gained-what-the-future-of-work-will-mean-for-jobs-skills-and-wages.

44 Alan Berube, "City and Metropolitan Income Inequality Data Reveal Ups and Downs Through 2016," Brookings, March 28, 2018, accessed April 1, 2019, https://www.brookings.edu/research/city-and-metropolitan-income-inequality-data-reveal-ups-and-downs-through-2016/.

45 Lisa Fu and Stacy Jones, "Tech Giants Bring In More Revenue but Fewer Workers," *Fortune,* June 22, 2017, accessed April 1, 2019, http://fortune. com/2017/06/22/tech-automation-jobs/.

46 Doug Menuez, "Enthusiasts and Skeptics Debate Artificial Intelligence," Hive, January 30, 2015, accessed April 1, 2019, https://www.vanityfair.com/news/ tech/2014/11/artificial-intelligence-singularity-theory.

47 Mihai Diaconeasa, "Proceedings of the First International Colloquium on Catastrophic and Existential Risk," B. John Garrick Institute for the Risk Sciences, December 31, 2017, accessed April 1, 2019, https://www.risksciences. ucla.edu/news-events/2018/1/2/proceedings-of-the-first-international-colloquium-on-catastrophic-and-existential-risk.

48 "Keep Killer Robots Science Fiction," Ban Lethal Autonomous Weapons, March 26, 2019, accessed April 1, 2019, http://autonomousweapons.org/slaughterbots/.

49 Michaela Ross and Zachary Sherwood, "News & Reports," Bloomberg GOV, accessed April 1, 2019, https://about.bgov.com/blog/pentagon-bridge-artificial-intelligences/.

50 Open Letter Signed by Google Employees, "We Work for Google. It Shouldn't Be in the Business of War," *Guardian,* April 5, 2018, accessed April 1, 2019, https://

www.theguardian.com/commentisfree/2018/apr/04/google-ceo-drones-ai-war-surveillance.

51 Sundar Pichai, "AI at Google: Our Principles," Google, June 7, 2018, accessed April 1, 2019, https://www.blog.google/technology/ai/ai-principles/.

52 "An Open Letter to the United Nations Convention on Certain Conventional Weapons," Future of Life Institute, accessed April 1, 2019, https://futureoflife.org/autonomous-weapons-open-letter-2017/.

53 Aaron Gregg, "Microsoft, Amazon Pledge to Work with Pentagon following Anonymous Online Rebukes," *Washington Post,* October 26, 2018, accessed April 1, 2019, https://www.washingtonpost.com/business/2018/10/26/microsoft-amazon-pledge-work-with-pentagon-following-anonymous-online-rebukes/?utm_term=.1bc082d286f1.

54 Oriana Pawlyk, "China Leaving US Behind on Artificial Intelligence: Air Force General," Military.com, July 30, 2018, accessed April 1, 2019, https://www.military.com/defensetech/2018/07/30/china-leaving-us-behind-artificial-intelligence-air-force-general.html.

55 "Facial and Emotional Recognition; How One Man Is Advancing Artificial Intelligence," CBS News, January 13, 2019, accessed April 1, 2019, https://www.cbsnews.com/news/60-minutes-ai-facial-and-emotional-recognition-how-one-man-is-advancing-artificial-intelligence/.

56 "The State of Artificial Intelligence in China," Nanalyze, February 6, 2019, accessed April 1, 2019, https://www.nanalyze.com/2019/01/artificial-intelligence-china/.

57 Peter J. Brown, "Can Blockchain Power a Cashless World?," *Asia Times,* February 15, 2019, accessed April 1, 2019, https://www.asiatimes.com/2019/02/article/can-blockchain-power-a-cashless-world/.

58 Nicholas Thompson and Ian Bremmer, "The AI Cold War That Threatens Us All," *Wired,* November 19, 2018, accessed April 1, 2019, https://www.wired.com/story/ai-cold-war-china-could-doom-us-all/?mbid=social_twitter.

59 Mara Hvistendahl, "In China, a Three-Digit Score Could Dictate Your Place in Society," *Wired,* December 5, 2018, accessed April 1, 2019, https://www.wired.com/story/age-of-social-credit/.

60 Tom Simonite, "George Soros Attacks China's AI Push as 'Mortal Danger,'" *Wired,* January 25, 2019, accessed April 1, 2019, https://www.wired.com/story/mortal-danger-chinas-push-into-ai/.

61 *The Malicious Use of Artificial Intelligence: Forecasting, Prevention, Mitigation,* Future of Humanity Institute et al., February 2018.

62 Pete Pachal, "Google Photos Identified Two Black People as 'Gorillas,'" Mashable, July 1, 2015, accessed April 1, 2019, http://mashable.com/2015/07/01/google-photos-black-people-gorillas/#M.3Be4dBfuqB.

63 "Whoops, Alexa Plays Porn Instead of a Kids Song!" *Entrepreneur,* January 3, 2017, accessed April 1, 2019, https://www.entrepreneur.com/video/287281.

64 James Vincent, "Twitter Taught Microsoft's Friendly AI Chatbot to Be a Racist Asshole in Less than a Day," *Verge,* March 24, 2016, accessed April 1, 2019, https://www.theverge.com/2016/3/24/11297050/tay-microsoft-chatbot-racist.

65 "Why Uber's Self-Driving Car Killed a Pedestrian," *Economist,* May 29, 2018, accessed April 1, 2019, https://www.economist.com/the-economist-explains/2018/05/29/why-ubers-self-driving-car-killed-a-pedestrian.

66 Johana Bhuiyan, "Elon Musk Says Tesla Crashes Shouldn't Be Front-Page News Because There Are More Human-Driven Fatalities. That's Not an Accurate Comparison," Recode, May 17, 2018, accessed April 1, 2019, https://www.recode.net/2018/5/17/17362308/elon-musk-tesla-self-driving-autopilot-fatalities.

67 "U.S. DOT Announces 2017 Roadway Fatalities Down," NHTSA, October 4, 2018, accessed April 1, 2019, https://www.nhtsa.gov/press-releases/us-dot-announces-2017-roadway-fatalities-down.

68 Casey Ross and Ike Swetlitz, "IBM's Watson Recommended 'Unsafe and Incorrect' Cancer Treatments," *STAT,* July 30, 2018, accessed April 1, 2019, https://www.statnews.com/2018/07/25/ibm-watson-recommended-unsafe-incorrect-treatments/.

69 John Naughton, "Magical Thinking about Machine Learning Won't Bring the Reality of AI Any Closer," *Guardian,* August 5, 2018, accessed April 1, 2019, https://www.theguardian.com/commentisfree/2018/aug/05/magical-thinking-about-machine-learning-will-not-bring-artificial-intelligence-any-closer.

70 Mattha Busby, "'Killer Robots' Ban Blocked by US and Russia at UN Meeting," *Independent,* September 3, 2018, accessed April 1, 2019, https://www.independent.co.uk/life-style/gadgets-and-tech/news/killer-robots-un-meeting-autonomous-weapons-systems-campaigners-dismayed-a8519511.html.

71 "DeepMind Ethics & Society," DeepMind, accessed April 1, 2019, https://deepmind.com/applied/deepmind-ethics-society/.

72 "A Principled AI Discussion in Asilomar," Future of Life Institute, June 5, 2018, accessed April 1, 2019, https://futureoflife.org/2017/01/17/principled-ai-discussion-asilomar/.

73 Jeremy Kahn, "Facebook Endows AI Ethics Institute at German University TUM," Bloomberg.com, January 20, 2019, accessed April 1, 2019, https://www.bloomberg.com/news/articles/2019-01-20/facebook-endows-ai-ethics-institute-at-german-university-tum.

74 "Changes in Funding in the AI Safety Field," Centre for Effective Altruism, March 9, 2017, accessed April 1, 2019, https://www.centreforeffectivealtruism.org/blog/changes-in-funding-in-the-ai-safety-field/.

75 "AlphaZero: Shedding New Light on the Grand Games of Chess, Shogi and Go," DeepMind, accessed April 1, 2019, https://deepmind.com/blog/alphazero-shedding-new-light-grand-games-chess-shogi-and-go/.

76 Nick Bostrom, *Superintelligence: Paths, Dangers, Strategies,* Kindle ed. (New York: Oxford University Press, 2014).

77 Nate Soares, "Ensuring Smarter-than-Human Intelligence Has a Positive Outcome," Future of Life Institute, April 18, 2017, accessed April 1, 2019, https://futureoflife.org/2017/04/18/ensuring-smarter-than-human-intelligence/.

78 Mark Harris, "Inside the First Church of Artificial Intelligence," *Wired*, February 2, 2018, accessed April 1, 2019, https://www.wired.com/story/anthony-levandowski-artificial-intelligence-religion/.

79 Christianna Reedy, "Kurzweil Claims That the Singularity Will Happen by 2045," Futurism, October 16, 2017, accessed April 1, 2019, https://futurism.com/kurzweil-claims-that-the-singularity-will-happen-by-2045.

80 Dom Galeon, "Softbank CEO: The Singularity Will Happen by 2047," Futurism, March 2, 2017, accessed April 1, 2019, https://futurism.com/softbank-ceo-the-singularity-will-happen-by-2047/.

81 Caleb Garling, "Andrew Ng: Why 'Deep Learning' Is a Mandate for Humans, Not Just Machines," *Wired*, August 25, 2015, accessed April 1, 2019, https://www.wired.com/brandlab/2015/05/andrew-ng-deep-learning-mandate-humans-not-just-machines/.

82 Bostrom, *Superintelligence,* loc. 1530.

83 J. G. Jenkin, "Atomic Energy Is 'Moonshine': What Did Rutherford Mean?" *Physics in Perspective* (2011): 128, https://doi.org/10.1007/s00016-010-0038-1.

84 "The Artificial Intelligence Revolution: Part 2," Wait But Why, December 20, 2017, accessed April 1, 2019, https://waitbutwhy.com/2015/01/artificial-intelligence-revolution-2.html.

85 Stephen Hawking, "Stephen Hawking: 'Are We Taking Artificial Intelligence Seriously,'" *Independent*, October 23, 2017, accessed April 1, 2019, https://www.independent.co.uk/news/science/stephen-hawking-transcendence-looks-at-the-implications-of-artificial-intelligence-but-are-we-taking-9313474.html.

86 Nick Bostrom, "Transcript of 'What Happens When Our Computers Get Smarter than We Are?'" TED, accessed April 1, 2019, https://www.ted.com/talks/nick_bostrom_what_happens_when_our_computers_get_smarter_than_we_are/transcript?language=en.

87 Bostrom, Superintelligence, loc. 2834.

88 Decisionproblem.com, accessed April 1, 2019, http://www.decisionproblem.com/paperclips/index2.html.

89 "Specification Gaming Examples in AI," Victoria Krakovna, June 5, 2018, accessed April 1, 2019, https://vkrakovna.wordpress.com/2018/04/02/specification-gaming-examples-in-ai/.

90 Julian Benson, "Elite's AI Created Super Weapons and Started Hunting Players. Skynet Is Here," Kotaku UK, June 3, 2016, accessed April 1, 2019, http://www.kotaku.co.uk/2016/06/03/elites-ai-created-super-weapons-and-started-hunting-players-skynet-is-here.

91 Tegmark, *Life 3.0.*

92 Tom Simonite, "AI Is the Future—But Where Are the Women?" *Wired,* August 17, 2018, accessed April 1, 2019, https://www.wired.com/story/artificial-intelligence-researchers-gender-imbalance/.

93 "Coherent Extrapolated Volition," Lesswrongwiki, accessed April 1, 2019, https://wiki.lesswrong.com/wiki/Coherent_Extrapolated_Volition.

94 Eliezer Yudkowsky, *Coherent Extrapolated Volition,* Singularity Institute, San Francisco, 2004.

95 Eric Johnson, "Full Q&A: Y Combinator's Sam Altman and Recode's Kara Swisher Discuss Tech Ethics, Addiction and Facebook," Recode, December 10, 2018, accessed April 1, 2019, https://www.recode.net/2018/12/10/18134926/sam-altman-kara-swisher-recode-decode-live-mannys-podcast-transcript-facebook-zuckerberg-ethics.

CHAPTER 8: ALIENS

1 Jonathan B. Wiener, "The Tragedy of the Uncommons: On the Politics of Apocalypse," *Global Policy7* (2016): 67–80, doi:10.1111/1758-5899.12319.

2 Joel Achenbach, "NASA Estimates 1 Billion 'Earths' in Our Galaxy Alone," *Washington Post,* July 24, 2015, accessed April 1, 2019, https://www.washingtonpost.com/news/speaking-of-science/wp/2015/07/24/nasa-estimates-1-billion-earths-in-our-galaxy-alone/.

3 "Is There Anybody out There?" Planetary Society Blog, accessed April 1, 2019, http://www.planetary.org/blogs/jason-davis/2017/20171025-seti-anybody-out-there.html.

4 "Jill Tarter Searches for Extraterrestrial Life," TED, accessed April 1, 2019, https://www.ted.com/participate/ted-prize/prize-winning-wishes/setilive.

5 Giuseppe Cocconi and Philip Morrison, "7. Searching for Interstellar Communications," *A Source Book in Astronomy and Astrophysics, 1900–1975,* doi:10.4159/harvard.9780674366688.c9.

6 Elizabeth Howell, "Drake Equation: Estimating the Odds of Finding E.T.," Space.com, April 6, 2018, accessed April 1, 2019, https://www.space.com/25219-drake-equation.html.

7 Marina Koren, "Congress Is Quietly Nudging NASA to Look for Aliens," *Atlantic,* May 9, 2018, accessed April 1, 2019, https://www.theatlantic.com/science/archive/2018/05/seti-technosignatures-nasa-jill-tarter/558512/.

8 Paul Bignell, "42: The Answer to Life, the Universe and Everything," *Independent,* March 26, 2018, accessed April 1, 2019, https://www.independent.co.uk/lifestyle/history/42-the-answer-to-life-the-universe-and-everything-2205734.html.

9 Michele Johnson, "How Many Exoplanets Has Kepler Discovered?," NASA, April 9, 2015, accessed April 1, 2019, https://www.nasa.gov/kepler/discoveries.

10 "Hunt for Other Worlds: 3,500 Exoplanets and Counting," *Christian Science Monitor,* August 8, 2017, accessed April 1, 2019, https://www.csmonitor.com/Science/Spacebound/2017/0808/Hunt-for-other-worlds-3-500-exoplanets-and-counting.

11 Eric Berger, "Number of Potentially Habitable Planets in Our Galaxy: Tens of Billions," Ars Technica, May 11, 2016, accessed April 1, 2019, https://arstechnica.com/science/2016/05/number-of-potentially-habitable-planets-in-our-galaxy-tens-of-billions/.

12 Mike Wall, "Breakthrough Listen to Search 1 Million Stars for ET Signals Using South African Scopes," Space.com, October 2, 2018, accessed April 1, 2019, https://www.space.com/41998-breakthrough-listen-alien-life-search-meerkat.html.

13 "Is There Anybody out There?" Planetary Society Blog, accessed April 1, 2019, http://www.planetary.org/blogs/jason-davis/2017/20171025-seti-anybody-out-there.html.

14 Andrew Siemion, director, Breakthrough Listen, communication with author, April 2018.

15 Nola Taylor Redd, "How Old Is the Universe?" Space.com, June 8, 2017, accessed April 1, 2019, https://www.space.com/24054-how-old-is-the-universe.html.

16 "METI: Should We Be Shouting at the Cosmos?" Science 2.0, August 27, 2014, accessed April 1, 2019, http://www.science20.com/brinstorming/meti_should_we_be_shouting_cosmos-114283.

17 Genevieve Valentine, "An Awkward History of Our Space Transmissions," Gizmodo, June 18, 2013, accessed April 1, 2019, https://gizmodo.com/an-awkward-history-of-our-space-transmissions-5780084.

18 Ibid.

19 Steven Johnson, "Greetings, E.T. (Please Don't Murder Us)," *New York Times*, June 28, 2017, accessed April 1, 2019, https://www.nytimes.com/2017/06/28/magazine/greetings-et-please-dont-murder-us.html.

20 "Ambassador for Earth," *Nature* 443, no. 7112 (2006): 606, doi:10.1038/443606a.

21 Falk, Dan. "It Could Be a Really Bad Idea for Us to Contact Aliens," *Slate,* March 29, 2015, accessed April 1, 2019, https://slate.com/technology/2015/03/active-seti-should-we-reach-out-to-extraterrestrial-life-or-are-aliens-dangerous.html.

22 Michio Kaku, "Is There Intelligent Life in the Universe?" Big Think, October 6, 2018, accessed April 1, 2019, https://bigthink.com/dr-kakus-universe/is-there-intelligent-life-in-the-universe.

23 Seth Shostak, *Confessions of an Alien Hunter: A Scientist's Search for Extraterrestrial Intelligence,* Kindle ed. (Washington, DC: National Geographic Society, 2009), locs. 210–13.

24 Johnson, "Greetings, E.T. (Please Don't Murder Us)."

25 Robert A. Freitas Jr., *Xenology: An Introduction to the Scientific Study of Extraterrestrial Life, Intelligence, and Civilization*, 1st ed. (Sacramento, CA: Xenology Research Institute, 1979).

26 Alexander Koch, Chris Brierley, Mark M. Maslin, and Simon L. Lewis, "Earth System Impacts of the European Arrival and Great Dying in the Americas

after 1492," *Quaternary Science Reviews* 207 (2019): 13–36, doi:10.1016/j. quascirev.2018.12.004.

27 Ibid.

28 "Regarding Messaging to Extraterrestrial Intelligence (METI) / Active Searches for Extraterrestrial Intelligence (Active SETI)," statement regarding METI/ Active SETI, accessed April 1, 2019, https://setiathome.berkeley.edu/meti_ statement_0.html.

29 Jared M. Diamond, *The Third Chimpanzee: The Evolution and Future of the Human Animal* (New York: HarperCollins, 1992).

30 "Ronald Reagan Speech About Alien Invasion," YouTube, April 24, 2010, accessed April 1, 2019, https://www.youtube.com/watch?v=chVKSY2gT00.

31 Lee Speigel, "Could Earth Defend Itself from an ET Invasion?" *Huffington Post,* December 7, 2017, accessed April 1, 2019, https://www.huffingtonpost. com/2013/10/07/alien-invasion-earth-is-defenseless_n_4046659.html.

32 "Project Blue Book—Unidentified Flying Objects," National Archives and Records Administration, accessed April 1, 2019, https://www.archives.gov/ research/military/air-force/ufos.html.

33 Helene Cooper, Ralph Blumenthal, and Leslie Kean, "Glowing Auras and 'Black Money': The Pentagon's Mysterious U.F.O. Program," *New York Times,* December 16, 2017, accessed April 1, 2019, https://www.nytimes.com/2017/12/16/ us/politics/pentagon-program-ufo-harry-reid.html?module=inline.

34 Eric Benson, "Harry Reid on What the Government Knows About UFOs," Intelligencer, March 21, 2018, accessed April 1, 2019, https://nymag.com/ intelligencer/2018/03/harry-reid-on-what-the-government-knows-about-ufos. html.

35 "Area 51," *Encyclopædia Britannica,* January 7, 2019, accessed April 1, 2019, https://www.britannica.com/place/Area-51.

36 H. G. Wells and M. A. Danahay, *The War of the Worlds* (Peterborough, Ontario: Broadview Press, 2003).

37 Nathan Nunn and Nancy Qian, "The Colombian Exchange: A History of Disease, Food and Ideas," *Journal of Economic Perspectives* 24, no. 2 (2010): 163–88.

38 arXiv:1802.02180 [astro-ph.IM].

39 G. D. Brin, "The 'Great Silence': The Controversy Concerning Extraterrestrial Intelligence Life," *Quarterly Journal of the Royal Astronomical Society* 24, no. 3 (1983): 283–309.

40 E. M. Fri Jones, "'Where Is Everybody': An Account of Fermi's Question," doi:10.2172/5746675, https://www.osti.gov/servlets/purl/5746675.

41 Robert H. Gray, "The Fermi Paradox Is Not Fermi's, and It Is Not a Paradox," *Scientific American* blog, January 29, 2016, accessed April 1, 2019, https://blogs. scientificamerican.com/guest-blog/the-fermi-paradox-is-not-fermi-s-and-it-is-not-a-paradox.

42 Robert Roy Britt, "Milky Way's Age Narrowed Down," Space.com, August 17, 2004, accessed April 1, 2019, https://www.space.com/263-milky-age-narrowed. html.

43 Ibid.

44 "Astronomers Have Developed a Computer Model That Simulates How Difficult It Would Be for Aliens to Ignore Humanity," *MIT Technology Review,* September 6, 2016, accessed April 1, 2019, https://www.technologyreview.com/s/602302/ galactic-model-simulates-how-et-civilizations-could-be-deliberately-avoiding-earth/.

45 "U.S. Energy Information Administration—EIA—Independent Statistics and Analysis," History of Energy Consumption in the United States, 1775–2009, U.S. Energy Information Administration (EIA), accessed April 1, 2019, https:// www.eia.gov/todayinenergy/detail.php?id=10.

46 Anders Sandberg, Stuart Armstrong, and Milan Cirkovic, "That is not dead which can eternal lie: The aestivation hypothesis for resolving Fermi's paradox," *JBIS: Journal of the British Interplanetary Society* 69 (2017).

47 Seth Shostak, SETI senior astronomer, personal communication with author, April 2018.

48 A. Wolszczan and D. A. Frail, "A Planetary System around the Millisecond Pulsar PSR1257 12," *Nature* 355, no. 6356 (1992): 145–47, doi:10.1038/355145a0.

49 Phil Torres and Martin Rees, *Morality, Foresight, and Human Flourishing: An Introduction to Existential Risks,* Kindle ed. (Durham, NC: Pitchstone, 2017), loc. 688.

50 Stephanie Pappas, "What Was the First Life on Earth?" LiveScience, March 1, 2017, accessed April 1, 2019, https://www.livescience.com/57942-what-was-first-life-on-earth.html.

51 Mike Wall, "Curiosity Rover Finds Ancient 'Building Blocks for Life' on Mars," Space.com, June 7, 2018, accessed April 1, 2019, https://www.space.com/40819-mars-methane-organics-curiosity-rover.html.

52 Dirk Schulze-Makuch, "Where to Look for Life on Titan," *Air & Space,* June 19, 2018, accessed April 1, 2019, https://www.airspacemag.com/daily-planet/ where-look-life-titan-180969409/.

53 Ben K. D. Pearce, Ralph E. Pudritz, Dmitry A. Semenov, and Thomas K. Henning, "Origin of the RNA World: The Fate of Nucleobases in Warm Little Ponds," *Proceedings of the National Academy of Sciences* 114, no. 43 (2017): 11327–32, doi:10.1073/pnas.1710339114.

54 Helen Macdonald, "Nathalie Cabrol Searches the Earth for the Secrets of Life on Mars," *New York Times,* March 22, 2018, accessed April 1, 2019, https://www.nytimes.com/interactive/2018/03/22/magazine/voyages-nathalie-cabrol-searching-mars-life-on-earth.html.

55 *Astrobiology Strategy,* NASA, Washington, DC, 2015.

56 Shostak, *Confessions of an Alien Hunter,* loc. 138.

57 Nick Bostrom, "Where Are They?" MIT Technology Review, July 11, 2015, accessed April 1, 2019, https://www.technologyreview.com/s/409936/where-are-they/.

58 Adam Stevens, Duncan Forgan, and Jack O'Malley James, "Observational Signatures of Self-Destructive Civilizations," *International Journal of Astrobiology* 15, no. 4 (2015): 333–44, doi:10.1017/s1473550415000397.

59 A. Frank, Jonathan Carroll-Nellenback, M. Alberti, and A. Kleidon. "The Anthropocene Generalized: Evolution of Exo-Civilizations and Their Planetary Feedback," *Astrobiology* 18, no. 5 (2018): 503–18, doi:10.1089/ast.2017.1671.

60 Adam Frank, "How Do Aliens Solve Climate Change?" *Atlantic*, May 30, 2018, accessed April 1, 2019, https://www.theatlantic.com/science/archive/2018/05/how-do-aliens-solve-climate-change/561479/.

61 Joshua Cooper, "Bioterrorism and the Fermi Paradox," *International Journal of Astrobiology* 12, no. 02 (2013): 144–48, doi:10.1017/s1473550412000511.

62 Stevens, Forgan, and James, "Observational Signatures of Self-Destructive Civilizations."

CHAPTER 9: SURVIVAL

1 "Practicing Your Crisis Response: Handling What Comes Right of Boom," Security Intelligence, February 8, 2019, accessed April 1, 2019, https://securityintelligence.com/practicing-your-crisis-response-how-well-can-you-handle-right-of-boom/.

2 Evan Osnos, "Survival of the Richest," *New Yorker*, November 16, 2018, accessed April 1, 2019, https://www.newyorker.com/magazine/2017/01/30/doomsday-prep-for-the-super-rich.

3 Olivia Carville, "The Super Rich of Silicon Valley Have a Doomsday Escape Plan," Bloomberg.com, September 5, 2018, accessed April 1, 2019, https://www.bloomberg.com/features/2018-rich-new-zealand-doomsday-preppers/?utm_source=nextdraft&utm_medium=email.

4 "Why Are Huge Numbers of Americans Preparing for Doomsday?" MSN, January 2, 2018, accessed April 1, 2019, https://www.msn.com/en-us/news/us/why-are-huge-numbers-of-americans-preparing-for-doomsday/ar-BBHNhW1.

5 Jakob Schiller, "What Doomsday Preppers Would Eat After the Apocalypse," *Wired,* June 3, 2017, accessed April 1, 2019, https://www.wired.com/2015/03/doomsday-preppers-eat-apocalypse/.

6 "20 Prepper Gift Ideas for Father's Day," *Prepper Journal,* January 15, 2015, accessed April 1, 2019, http://www.theprepperjournal.com/2013/06/03/20-prepper-gift-ideas-for-fathers-day/.

7 National Geographic Channel, "If It's the End of the World As We Know It . . . New Survey Reveals Americans Are Not Prepared!," PR Newswire, February 7, 2012, accessed April 1, 2019, https://www.prnewswire.com/news-releases/if-its-the-end-of-the-world-as-we-know-it-new-survey-reveals-americans-are-not-prepared-national-geographic-channel-releases-survey-timed-to-new-series-doomsday-preppers-138865199.html.

8 David Denkenberger and Joshua M. Pearce, *Feeding Everyone No Matter What: Managing Food Security After Global Catastrophe*, Kindle ed. (San Diego: Elsevier Science, 2014), loc. 995.

9 "Atlas Survival Shelters," Atlas Survival Shelters, accessed April 1, 2019, https://www.atlassurvivalshelters.com/concrete-dome.

10 "The Prepster, Two-Person, 3-Day Emergency Kit Bag," Preppi, accessed April 1, 2019, https://www.preppi.co/products/the-prepster?variant=4869981700125.

11 Garrett M. Graff, "The Doomsday Diet," Eater, December 12, 2017, accessed April 1, 2019, https://www.eater.com/2017/12/12/16757660/doomsday-biscuit-all-purpose-survival-cracker.

12 Douglas Rushkoff, "Survival of the Richest," Medium, August 25, 2018, accessed April 1, 2019, https://medium.com/s/futurehuman/survival-of-the-richest-9ef6cddd0cc1.

13 Denkenberger and Pearce, *Feeding Everyone No Matter What*, loc. 947.

14 Denkenberger. "We Have to Act Now If We Want to Feed Everyone When There Is a Global Catastrophe," *Huffington Post UK*, January 31, 2018, accessed April 1, 2019, https://www.huffingtonpost.co.uk/entry/we-have-to-act-now-if-we-want-to-feed-everyone-when-the-world-ends_uk_5a6f5e60e4b06e25326a63c3.

15 "GDP (current US$)," Data, accessed April 1, 2019, https://data.worldbank.org/indicator/NY.GDP.MKTP.CD.

16 Denkenberger and Pearce, *Feeding Everyone No Matter What*.

17 Ibid., loc. 1969.

18 Ibid., loc. 1973.

19 "UN Urges People to Eat Insects to Fight World Hunger," BBC News, May 13, 2013, accessed April 1, 2019, http://www.bbc.com/news/world-22508439.

20 Bryan Walsh, "Eating Bugs," *Time*, May 29, 2008, accessed April 1, 2019, http://content.time.com/time/magazine/article/0,9171,1810336,00.html.

21 "The Weight of All Those Creepy Crawlies," *Globe and Mail*, February 21, 2018, accessed April 1, 2019, https://www.theglobeandmail.com/opinion/the-weight-of-all-those-creepy-crawlies/article4461850/.

22 Walsh, "Eating Bugs."

23 Denkenberger and Pearce, *Feeding Everyone No Matter What*, loc. 2582.

24 Ibid., loc. 1360.

25 Holly Graham et al., "It's a Man Eat Man World," *Journal of Physics Special Topics*, November 19, 2017.

26 Sarah Scoles, "A Rare Journey into the Cheyenne Mountain Complex, a Super-Bunker That Can Survive Anything," *Wired*, June 3, 2017, accessed April 1, 2019, https://www.wired.com/2017/05/rare-journey-cheyenne-mountain-complex-super-bunker-can-survive-anything/.

27 "Norway: 'Doomsday' Vault Where World's Seeds Are Kept Safe," *Time*, accessed April 1, 2019, http://time.com/doomsday-vault/.

28 Damian Carrington, "Arctic Stronghold of World's Seeds Flooded after Permafrost Melts," *Guardian*, May 19, 2017, accessed April 1, 2019, https://www.theguardian.com/environment/2017/may/19/arctic-stronghold-of-worlds-seeds-flooded-after-permafrost-melts.

29 R. P. Murray-McIntosh, B. J. Scrimshaw, P. J. Hatfield, and D. Penny, "Testing Migration Patterns and Estimating Founding Population Size in Polynesia by Using Human MtDNA Sequences," *Proceedings of the National Academy of Sciences* 95, no. 15 (1998): 9047–52.

30 Rachel Feltman, "We Could Move to Another Planet with a Spaceship like This," *Popular Science,* May 4, 2018, accessed April 1, 2019, https://www.popsci.com/realistic-generational-spaceship.

31 TEDx Talks, "Humanity on the Edge of Extinction: Anders Sandberg: TEDxVienna," YouTube, December 14, 2017, accessed April 1, 2019, https://www.youtube.com/watch?v=O-WXOaAnipM.

32 Katherine J. Curtis White, "Declining Fertility among North American Hutterites: The Use of Birth Control within a Dariusleut Colony," *Biodemography and Social Biology* 49, nos. 1–2 (2002): 58–73, doi:10.1080/19485565.2002.99890 49.

33 Zaria Gorvett, "Future—Could Just Two People Repopulate Earth?" BBC, January 13, 2016, accessed April 1, 2019, http://www.bbc.com/future/story/20160113-could-just-two-people-repopulate-earth.

34 Seth D. Baum, "Long-Term Trajectories of Human Civilization," *Foresight* 21, no. 1 (2019): 53–83, doi 10.1108/FS-04-2018-0037, March 11, 2019.

35 Steven Levy, "Jeff Bezos Wants Us All to Leave Earth—for Good," *Wired,* November 20, 2018, accessed April 1, 2019, https://www.wired.com/story/jeff-bezos-blue-origin/.

36 Dom Galeon, "Jeff Bezos: 'We Have to Go to Space to Save Earth,'" Futurism, November 6, 2017, accessed April 1, 2019, https://futurism.com/jeff-bezos-space-save-earth.

37 Neil Strauss, "Elon Musk: The Architect of Tomorrow," *Rolling Stone,* June 25, 2018, accessed April 1, 2019, https://www.rollingstone.com/culture/culture-features/elon-musk-the-architect-of-tomorrow-120850/.

38 Elien Blue Becque, "Elon Musk Wants to Die on Mars," *Vanity Fair,* January 30, 2015, accessed April 1, 2019, https://www.vanityfair.com/news/tech/2013/03/elon-musk-die-mars.

39 Rahul Kalvapalle, "Humanity Has 100 Years to Colonize Other Planets or Die Out: Stephen Hawking," *Global News,* May 22, 2017, accessed April 1, 2019, https://globalnews.ca/news/3468885/stephen-hawking-humanity-should-leave-earth-in-100-years/.

40 Dave Mosher, "Jeff Bezos Just Gave a Private Talk in New York. From Utopian Space Colonies to Dissing Elon Musk's Martian Dream, Here Are the Most Notable Things He Said," *Business Insider,* February 23, 2019, accessed April 1, 2019, https://www.businessinsider.com/jeff-bezos-blue-origin-wings-club-presentation-transcript-2019-2#he-added-that-his-vision-is-to-move-heavy-

industry-and-energy-generation-off-earth-that-way-our-planet-starts-pristine-and-we-dont-have-to-suffer-through-shortages-and-rationing-as-populations-grow-and-people-increasingly-live-first-world-lifestyles-15.

41 Nick Bostrom, "Astronomical Waste: The Opportunity Cost of Delayed Technological Development," *Utilitas* 15, no. 03 (2003): 308–14, doi:10.1017/s0953820800004076.

42 Dave Goldberg, "Will the Universe End in a 'Big Rip'?" Io9, December 16, 2015, accessed April 1, 2019, https://io9.gizmodo.com/will-the-universe-end-in-a-big-rip-476467549.

43 Bostrom, "Astronomical Waste."

44 Seth Baum, "Why Colonizing the Galaxy Is the Highest Good—Facts So Romantic," *Nautilus*, April 14, 2017, accessed April 1, 2019, http://nautil.us/blog/why-colonizing-the-galaxy-is-the-highest-good.

45 Alok Jha, "Mars Mission Astronauts Face Radiation Exposure Risk," *Guardian*, May 31, 2013, accessed April 1, 2019, https://www.theguardian.com/science/2013/may/30/radiation-dose-mars-mission-safe-lifetime-limit.

46 Adam Rogers, "Sorry, Nerds: Terraforming Might Not Work on Mars," *Wired*, July 30, 2018, accessed April 1, 2019, https://www.wired.com/story/co2-terraforming-mars/.

47 Mosher, "Jeff Bezos Just Gave a Private Talk in New York."

CHAPTER 10: THE END

1 James Hayes-Bohanan, "More Food For Thought," Food for Thought, accessed April 1, 2019, http://webhost.bridgew.edu/jhayesboh/Morefood.htm.

2 "Backup Humanity," Interstellarbeacon, accessed April 1, 2019, https://www.interstellarbeacon.org/.

3 Rabinowitch, Eugene. "The Narrowing Way," *Bulletin of the Atomic Scientists*, August 1953.

4 "Doomsday Clock," *Bulletin of the Atomic Scientists*, accessed April 1, 2019, https://thebulletin.org/doomsday-clock/.

5 "Current Time," *Bulletin of the Atomic Scientists*, accessed April 1, 2019, https://thebulletin.org/doomsday-clock/current-time/.

6 Julian Savulescu and Ingmar Persson, "Moral Enhancement," *Philosophy Now: A Magazine of Ideas*, 2012, accessed April 1, 2019, https://philosophynow.org/issues/91/Moral_Enhancement.

7 Fredric Jameson, "Fredric Jameson, Future City, NLR 21, May–June 2003," *New Left Review*, May/June 2003, accessed April 1, 2019, https://newleftreview.org/II/21/fredric-jameson-future-city.